Theory of Automatic Robot Assembly and Programming

Theory of Automatic Robot Assembly and Programming

Bartholomew O. Nnaji

Professor and Director
Automation and Robotics Laboratory
Department of Industrial Engineering
University of Massachusetts
Amherst
USA

CHAPMAN & HALL
London · Glasgow · New York · Tokyo · Melbourne · Madras

Published by Chapman & Hall, 2–6 Boundary Row, London SE1 8HN

Chapman & Hall 2–6 Boundary Row, London SE1 8HN, UK

Blackie Academic & Professional, Wester Cleddens Road, Bishopbriggs, Glasgow G64 2NZ, UK

Chapman & Hall Inc., 29 West 35th Street, New York NY10001

Chapman & Hall Japan, Thomson Publishing Japan, Hirakawacho Nemoto Building, 6F, 1-7-11 Hirakawa-cho, Chiyoda-ku, Tokyo 102, Japan

Chapman & Hall Australia, Thomas Nelson Australia, 102 Dodds Street, South Melbourne, Victoria 3205, Australia

Chapman & Hall India, R. Seshadri, 32 Second Main Road, CIT East, Madras 600 035, India

First edition 1993

© 1993 Chapman & Hall

Typeset in 10/12pt Palatino by Thomson Press (India) Ltd, New Delhi
Printed in Great Britain by the University Press, Cambridge

ISBN 0 412 39310 7   0 442 31663 1   (USA)

Apart from any fair dealing for the purposes of research or private study, or criticism or review, as permitted under the UK Copyright Designs and Patents Act, 1988, this publication may not be reproduced, stored, or transmitted, in any form or by any means, without the prior permission in writing of the publishers, or in the case of reprographic reproduction only in accordance with the terms of the licences issued by the Copyright Licensing Agency in the UK, or in accordance with the terms of licences issued by the appropriate Reproduction Rights Organization outside the UK. Enquiries concerning reproduction outside the terms stated here should be sent to the publishers at the London address printed on this page.
  The publisher makes no representation, express or implied, with regard to the accuracy of the information contained in this book and cannot accept any legal responsibility for any errors or omissions that may be made.

A catalogue record for this book is available from the British Library

Library of Congress Cataloging-in-Publication data available

Nnaji, Bartholomew O., 1956–
    Theory of automatic robot assembly and programming/Bartholomew O. Nnaji. – 1st ed.
        p.   cm.
    Includes bibliographical references and index.
    ISBN 0–442–31663–1
    1. Robots, Industrial. 2. Robots – Programming. I. Title.
  TS191.8.N59   1992
  670.42'72 – dc20   92–30586
                                                              CIP

∞ Printed on permanent acid-free text paper, manufactured in accordance with the proposed ANSI/NISO Z 39.48-199X and ANSI Z 39.48-1984

To
Professor Richard J. Giglio
for his courage and vision

Contents

List of tables	xiii
Acknowledgements	xv
Preface	xvii

1 Machine programming — 1
  1.1 Introduction — 1
  1.2 Problems of machine reasoning — 2
  1.3 Robot programming — 2
    1.3.1 Explicit robot programming — 3
  1.4 Machine task-level programming — 7
  1.5 Organization of this book — 9

2 CAD in automatic machine programming — 10
  2.1 Introduction — 10
  2.2 Desired CAD data — 11
  2.3 Feature, feature classification and representation — 12
    2.3.1 Feature concept — 12
    2.3.2 The Euler formula for features — 14
    2.3.3 Feature classification — 14
    2.3.4 Feature representation — 17
    2.3.5 Feature representation for mechanical components — 21
  2.4 Feature reasoning for mechanical components — 22
    2.4.1 Notations and definitions in feature reasoning — 22
    2.4.2 Envelope determination — 25
    2.4.3 Coordinate frame mapping — 28
    2.4.4 Extraction of form features — 30
    2.4.5 Feature recognition — 31
    2.4.6 Criteria used for feature recognition — 32
    2.4.7 Generic feature primitives — 34
    2.4.8 Pattern matching — 36
    2.4.9 Feature decomposition — 37

3 Spatial relationships — 41
  3.1 Introduction — 41
  3.2 Background — 43
    3.2.1 Importance of a solid modeler in spatial reasoning — 43
    3.2.2 Spatial relationships developments — 44

		3.2.2.1	Mathematical representation	45
		3.2.2.2	Expressions for positions of bodies in terms of relations between features	46
	3.2.3	Features and spatial relationships		48
3.3	Spatial relationships			48
	3.3.1	Assembly locations		49
	3.3.2	Types of spatial relationship		50
	3.3.3	Degrees of freedom		54
	3.3.4	Intersection of degrees of freedom		56
	3.3.5	Inferring mating frames		60
3.4	Product specification attributes			63
3.5	Applications			65
	3.5.1	Bill of materials and precedence constraints		65
	3.5.2	Feasible approach directions		65
	3.5.3	World modeling		66

4 Structure of an automatic robot programmer — 67

4.1	Introduction	67
4.2	An overview of RALPH	71
4.3	World knowledge database	72
4.4	RALPH commands	72
	4.4.1 Task-level commands	73
	4.4.2 Mid-level internal commands	76
	4.4.3 General robot-level commands	76
4.5	Mathematical consideration	79
4.6	Task planner	82
	4.6.1 The task-level planner	82
	4.6.2 The mid-level planner	85
4.7	An example of assembly task	87
4.8	Programming issues	91
4.9	Discussion	92

5 Sensors and representation — 93

5.1	Background	93
	5.1.1 Tactile sensors	94
	5.1.2 Force sensors	95
	5.1.3 Proximity sensors	95
	5.1.4 Ultrasonic ranging	95
	5.1.5 Infrared	95
	5.1.6 Vision	95
5.2	Internal and external sensors	97
	5.2.1 Internal sensors	97
	5.2.2 External sensors	97
5.3	Sensor fusion	99
5.4	Sensor architecture	100

		5.4.1	General sensor level	102
		5.4.2	Generic sensor level	103
			5.4.2.1 Non-contact sensors	104
			5.4.2.2 Contact sensors	105
			5.4.2.3 Internal sensors	106
		5.4.3	Specific sensor level	106
	5.5	Representation		106
		5.5.1	Planning	107
			5.5.1.1 General sensor planner	107
			5.5.1.2 Generic sensor planner	110
	5.6	Probability of sensor usage		110
		5.6.1	Generic plan	111
			5.6.1.1 Specific sensor planner	112
		5.6.2	Sensor range	113
		5.6.3	Set considerations	116
		5.6.4	Individual sensor properties	118
		5.6.5	Gripper considerations	118
	5.7	Processing		119
		5.7.1	Specific level processor	119
		5.7.2	Generic level processor	121
		5.7.3	General level processor	125
		5.7.4	Summary	128
6	**World modeling and task specification**			**131**
	6.1	World modeling		131
		6.1.1	Geometric description	131
		6.1.2	Parametric world modeler	133
		6.1.3	Physical description	134
		6.1.4	Kinematics of linkages	134
		6.1.5	Description of the robot characteristics	135
		6.1.6	Complexity of the world model	135
	6.2	Task specification		135
	6.3	Assembly stability model		136
		6.3.1	Quantitative approach to analyzing stability	139
		6.3.2	Description of the two-block system	139
		6.3.3	Equations of motion of the two-block system	140
		6.3.4	Simulation of the two-block assembly	143
		6.3.5	Equivalent parameters for transformation of a multi-body system to a single body	146
	6.4	Designing for stability		148
		6.4.1	Singularity and stability	148
		6.4.2	Inertial effects and non-linearities	149
		6.4.3	Generalized centroid	149
	6.5	Relative stability		150
	6.6	Summary		150

Contents

7 Gross motion planning and collision avoidance — 151
- 7.1 Introduction — 151
- 7.2 Gross motion in RALPH — 151
- 7.3 Robot motion planning problems — 152
  - 7.3.1 Findspace and findpath problems — 152
    - 7.3.1.1 The findspace problem — 152
    - 7.3.1.2 The findpath problem — 152
    - 7.3.1.3 Compliant motion with uncertainty — 154
  - 7.3.2 Configuration space — 154
    - 7.3.2.1 Definition — 154
  - 7.3.3 Computation of the configuration space obstacles — 155
    - 7.3.3.1 Advantages and disadvantages of C-space — 157
  - 7.3.4 Path planning algorithms — 157
    - 7.3.4.1 Visibility graph — 157
    - 7.3.4.2 Hypothesize and test — 157
    - 7.3.4.3 Voronoi diagram — 158
    - 7.3.4.4 Cell decomposition — 159
    - 7.3.4.5 Potential field — 161
- 7.4 The path planning algorithm — 162
  - 7.4.1 Outline of collision repelling algorithm — 162
  - 7.4.2 Creating the configuration space — 163
    - 7.4.2.1 Finding the position of the axis — 164
  - 7.4.3 Intersecting link j with the obstacles — 164
    - 7.4.3.1 Updating the bitmap — 168
  - 7.4.4 Finding the path for the arm — 169
    - 7.4.4.1 Creating the retracted free C-space — 169
    - 7.4.4.2 Creating the numerical potential field in C^R_{free} — 172
    - 7.4.4.3 Effects of varying ϱ — 173
  - 7.4.5 Moving the end-effector — 173
    - 7.4.5.1 Fixed arm configuration — 173
    - 7.4.5.2 Adjusting the hand configuration — 174
- 7.5 Discussion — 175
  - 7.5.1 Evaluation of the proposed algorithm — 175
  - 7.5.2 Proposed improvements — 175
  - 7.5.3 Trajectory planning — 176
- 7.6 Summary — 177

8 Grasp planning — 178
- 8.1 Introduction — 178
- 8.2 Background — 179
  - 8.2.1 General model of grasping — 179
    - 8.2.1.1 Approaches to parallel jaw grasping — 179
  - 8.2.2 Choosing grasps and grasp parameters — 180
  - 8.2.3 Building integrated systems — 181
- 8.3 World spatial relationships in grasping — 181
- 8.4 Grasping concepts — 181

	8.4.1	Task requirement	182
		8.4.1.1 Basic task attributes	182
		8.4.1.2 Suitable task description	183
	8.4.2	Feature reasoning for grasping	184
	8.4.3	Geometric constraints in grasping	184
		8.4.3.1 Parallelity and exterior grasp condition	184
		8.4.3.2 Local accessibility	186
		8.4.3.3 Mutual visibility	188
		8.4.3.4 Finding the grasp point and approach direction	188
	8.4.4	Grasp force and grasp evaluation	190
		8.4.4.1 Analysis for a rigid gripper jaw surface	190
	8.4.5	Analysis for a soft contact	192
8.5	Design and implementation		195
	8.5.1	Task requirement	195
		8.5.1.1 Manipulability	196
		8.5.1.2 Torquability	198
		8.5.1.3 Rotatability	198
		8.5.1.4 Stability	199
		8.5.1.5 Format for task description	199
	8.5.2	Feature reasoning	200
	8.5.3	Geometrical constraints	201
	8.5.4	An example	205
8.6	Summary		205

9 Trajectory planning and control — 207
- 9.1 Introduction — 207
  - 9.1.1 Cartesian space control — 208
    - 9.1.1.1 Joint space control — 209
    - 9.1.1.2 Joint interpolated control — 211
- 9.2 Evaluation of trajectories — 214
- 9.3 Other trajectory evaluation approaches — 215
- 9.4 Background material — 216
- 9.5 Robots with more than 3 degrees of freedom — 223
- 9.6 Evaluation and analysis — 223
  - 9.6.1 Discussion — 226
- 9.7 Summary — 231

10 Considerations for generic kinematic structures — 233
- 10.1 Introduction — 233
- 10.2 Kinematic structures — 234
  - 10.2.1 Inverse kinematic solution — 234
  - 10.2.2 Jacobian — 235
  - 10.2.3 Degeneracy — 237
  - 10.2.4 Singularity — 237
  - 10.2.5 An example — 238
  - 10.2.6 Quartenion representation of rotations — 244

	10.3	Kinematic implementation	244
	10.4	Kinematic analysis	253
	10.5	Example	255
		10.5.1 Link parameters for cylindrical robots	255
		10.5.2 Inverse kinematics results	257
		10.5.3 Singular points for RPP (cylindrical) robots	258
	10.6	Pattern of kinematic behavior	263
	10.7	Summary	264
11	**Program synthesis and other planners**		**266**
	11.1	Introduction	266
	11.2	Spanning vector for assembly directions and other applications	266
		11.2.1 Determining control faces	269
		11.2.2 Finding mating faces	269
		11.2.3 Spanning vector	271
		11.2.3.1 Mathematical background	271
		11.2.3.2 Algorithm for finding spanning vector	273
		11.2.4 Representation of spanning vector	275
		11.2.5 Some examples of spanning vector	275
	11.3	Precedence generation	276
		11.3.1 Concept of assembly precedence	278
		11.3.2 Assumptions for precedence	279
		11.3.3 Spatial relationships and precedence	279
		11.3.4 Find all mating faces	283
		11.3.5 Determining the disassembly direction	284
		11.3.6 Precedence algorithm	285
		11.3.7 Interpretation of results	289
	11.4	Fine motion planning	289
	11.5	Program synthesis	290
References			**292**
Index			**300**

List of tables

1.1	Prominent robot languages	4
2.1	Notation of symbols	39
4.1	The structure of RALPH task-level commands	74
5.1	Table of varieties of noncontact sensors	105
5.2	Table of varieties of contact sensors	105
5.3	Table of ranked generic sensor combinations	111
5.4	Table of coordinate frames for generic sensor types	119
6.1	Typical values for the parameters used in the dynamic simulation of the two-block system	140
7.1	The joint categories for four robot body types	165
8.1	Task attributes for various tasks; on a scale of 0 to 2, where 0 means no requirement and 2 means strong requirement	182
8.2	Task attributes for mid level commands	194
8.3	Weightage of factors influencing manipulability	196
8.4	Weightage of factors influencing torquability	196
8.5	Weightage of factors influencing rotatability	198
8.6	Weightage of factors influencing stability	199
10.1	Link parameters for the GE-A4 robot	238

Acknowledgements

I wish to thank my students at the Automation and Robotics Laboratory both past and present who worked with me on a number of research projects which resulted in the development of various chapters of this book. In particular, I thank Dr Tzong-Shyan Kang and Dr Jang-Ping Chen for working with me on the *Feature Reasoning* project; Shuchieh Yeh and Mehran Kamran (*CAD Data and Symmetry* project); Hsu-Chang Liu (*Product Modeling* project); Ashok Vishnu (*Stability* project); Aditi Dubey (*Grasp Planner* project); Jagtap Prashant (*Precedence Generation* project); Ellen Lin (*Sensor Representation* project); Sven Haberer (*Gross motion planning* project); and Andy Rist (*Simulator* project). Many of our laboratory staff members helped to proof-read and comment on the book. In particular, I thank Mehran Kamran, Jyh-Haw Kang, Hsu-Chang Liu, and Shuchieh Yeh for providing this kind of needed support.

Finally, I wish to give special thanks to my family for their patience for all those hours I had to spend away from them developing the material for this manuscript. I appreciate their support.

This material is based partially upon research supported by the National Science Foundation, under grant number DMC-8610417 and the North Atlantic Treaty Organization. Any opinions, findings, and conclusions or recommendations expressed in this publication are those of the author and do not necessarily reflect the views of the National Science Foundation or the North Atlantic Treaty Organization.

B.O. Nnaji

Preface

Machines will gradually become programmed using computers which have the knowledge of how the objects in the world relate to one another. This book capitalizes on the fact that products which are manufactured can be designed on the computer and that information about the product such as its physical shape provide powerful information to reason about how to develop the process plan for their manufacture. This book explores the whole aspect of using the principles of how parts behave naturally to automatically generate programs that govern how to produce them.

The last decade saw tremendous work on how machines can be programmed to perform a variety of tasks automatically. Robotics has witnessed the most work on programming techniques. But it was not until the emergence of the advanced CAD system as a proper source of information representation about objects which are to be manipulated by the robot that it became viable for automated processors to generate robot programs without human interface. It became possible for objects to be described and for principles about how they interact in the world to be developed. The functions which the *features* designed into the objects serve for the objects can be adequately represented and used in reasoning about the manufacturing of the parts using the robot.

This book describes the necessary principles which must be developed for a robot to generate its own programs with the knowledge of the world in the CAD system. The reader will be taken through the basic theory of automatic robot programming; and representation of product information in a CAD system suitable for making inference about the product planning. Issues of how a task is represented to the machine are addressed in detail including spatial relationships and bill of materials. This task specification is based on a more contemporary view of the CAD system. In addition, the planning at all stages of robot motion is also presented.

This work represents over a decade of work in robotics and automated manufacturing. All the material presented has been tested and many of the principles are being used in many major industries in the world.

This text should be suitable for advanced undergraduate and graduate students in industrial, mechanical, manufacturing, and electrical engineering, as well as computer science. The book should also be appropriate for a course on robot programming, machine programming, and robotic assembly. It can also serve as a major part of a course in robotics, or any other automated manufacturing course.

Theory of Automatic Robot Assembly and Programming is also designed for practising engineers, computer scientists or managers of production function who desire to keep abreast of the technology or to use robots in their facility.

The principles which govern automatic robot programming are well explained within the application area of assembly throughout the book.

CHAPTER 1

Machine programming

1.1 INTRODUCTION

As we approach the end of the century, the traditional method of programming machines by manual method will be replaced for most machines by automatic methods. But this automatic method requires that the machines have information regarding the product design and representation as well as the process information which is necessary to generate the required process plan (or program). In advanced machines such as robots this problem is further complicated by the spatial dexterity of the machine.

The lack of widespread use of robots in industry today can be attributed in large measure to people's conception of what a robot should be to users. Many people still think that a robot should have all the attributes of a human being—dexterity, senses, vision, balance, response, computation, reasoning; except emotion. Emotion is the only allowance people give as to what should separate a robot from a human. It is expected that robots should have all the above qualities with greater performance in reasoning, computation and response. Science fiction movies have only served to exacerbate the situation with portrayals of man-made machines such as R2-D2, in the *Star Wars* movies, which is "capable" of most of the above qualities including emotion.

There is, therefore, great disappointment when people come face to face with the limitations of today's robots. These robots have no sense of vision, balance, dexterity, and worst of all no reasoning capability for interaction with the world. It is actually the robot's inability to deal with a generalized world that is the key problem. Certainly, this maps into the other problem areas. The robot's inability to interact with the world in a meaningful way is hinged particularly on understanding the world of geometry and spatial relationships; and how geometry relates to the function which objects and their attributes (features) serve for given situations (applications).

Interestingly, the problems which robots have in reasoning about the world extend to other machines. Certainly, a numerical control (NC) machine cannot be automatically programmed to drill two holes on a part unless the machine is made somehow to "comprehend" that the geometric form specified by the designer actually implies holes on the part and perhaps even that these holes imply certain manufacturing operations which the NC machine is capable of performing.

But what do we really mean when we say that a machine understands the world? Philosophically, can one really say that there is understanding of the world if the machine can read the information but cannot make use of such information? Would it not merely constitute a monitoring system—which takes information from the world, puts it (perhaps) in a form a human being can comprehend, and expects the actual decisions or procedures to be manually made?

Understanding means that inference can be made from information presented to the medium. A lack of inference capability implies that the medium cannot understand and further implies that the medium cannot reason. It is reasoning that enables a medium to make plans of actions in order to accomplish tasks.

We know that today's machine cannot make inferences which means that they cannot reason. In fact, what may appear to be planning capability on the part of some advanced machines is actually due to action protocols established by human programmers. There is no deductive reasoning vis-a-vis planning. Programming of machines entails the use of human understanding of the world and how the world relates to tasks which a machine performs within that world, to plan a sequence of related activities which enable the achievement of a task objective. When programming is left to a machine, as must be the case in automated manufacturing environments, the machine must have some of the elements of the human decision logic for inference.

This book is intended to present the various problems that have hindered the empowerment of inference ability on the programmable machines and the principles which will resolve these various problems.

1.2 PROBLEMS OF MACHINE REASONING

The good news is that while human beings often encounter unpredictable and highly complex situations in real life, machines, particularly those in manufacturing environments, will mostly encounter worlds which are ordered, planned and representable using a computer. In fact, the objects in a machine's world can be created using the CAD system – including the parts which the machines will perform operations on. But the bad news is that even within that limited world of CAD representable objects, clear behavior of objects and their features under certain machine operations have not been well understood and documented. Thus, little knowledge is available to an automatic machine planner regarding the methods of manipulating parts, what features on parts mean in terms of operation, the assemblability of parts, etc. There is therefore a host of related problems which we should examine to understand the nature of machine programming. The problems of robot programming are encapsulated in the wide range of problems pervading machine programming. In this book, we will discuss in some detail the elements of robot programming since it serves as an adequate denotation of the machine programming system.

1.3 ROBOT PROGRAMMING

Robot programming is the act of sequencing robot motions and activities in an order that will accomplish a desired set of tasks and teaching the robot those motions or activities. The activities can include interpretation of sensor data, sending and receiving of signals and data from other devices or actuation of the joints and end-effector. The teaching of the motion paths and activities can be accomplished in two major ways: (a) lead-through techniques (or *teach by showing*); and (b) language-based techniques. The lead-through methods are accomplished in powered or in non-powered modes. The

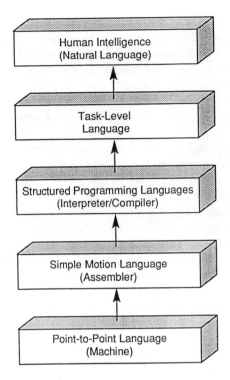

Fig. 1.1 Levels of programming languages.

powered lead-through method essentially allows for the robot to be guided to its destinations using a teach pendant to control the various joint actuators. Language-based programming techniques are quite similar to regular computer programming. Programs are normally specified in some type of language which can be low level or high level. There are two levels of language-based techniques: (a) *explicit robot programming languages*; and (b) *task-level programming*.

The explicit robot programming languages are *robot level* languages and require the robot programmer to be comfortable in computer programming and in the design of sensor-based motion strategies, which is a skill not usually possessed by workers on the manufacturing floor. Task-level programming, on the other hand, allows the user to specify the effects of the actions on objects rather than the sequence of manipulator motions needed to achieve the task as is the case in explicit robot programming languages. Since *task-level* programming is object-oriented, not robot-oriented, and deals with the effects on objects, not the motions of the robot, it is certainly an easier programming method. Figure 1.1 shows the various levels of programming languages available.

1.3.1 Explicit robot programming

Explicit robot programming is the act of specifying robot executable instructions in some robot-dependent textual language which can be translated by a system which is either an interpreter or a compiler. This programming scheme requires the instructions to be specified as a series of motion instructions which, when combined in some order, will accomplish the desired task. When the translation system is an interpreter, a program can be decoded instruction by instruction, and the robot control will also execute it in a stepwise manner. In cases when a compiler is the translation system, an intermediate set of codes is generated which is executed by an interpreter program.

Although the compiler method is more powerful, it often requires its own hardware and software. Therefore, the more prevalent explicit robot programming method is the interpreter method.

With the explicit programming method, it is possible to program robots both off-line and on-line. The off-line programming is accomplished by specification of trajectories and sensor calls which must be made in textual mode. Conditional statements are possible in this programming scheme.

There are two methods of specifying the trajectories as points of movement. In the first approach, the trajectory coordinates are defined explicitly in the program as numerical variables or constants. These values can be either measured from CAD drawing or read from an existing database into the program. Since this method can be very cumbersome, the second approach appears to be more promising. In the second method, symbols are used to specify the trajectory. For instance, the symbolic variables can be for the start, intermediate, and endpoints, in addition to orientation. The actual values can be obtained through teaching and inserted as text. A typical robot program can be supplemented by non-motion commands including, spatial relationships of objects in the world, representation language for sensors, sensor fusion, and other peripherals, real-time computational elements, etc.

As can be seen from Figure 1.1, the versatility of a robot language is dependent on the complexity of the programming scheme. The easiest approach is task-level but it is the most difficult to develop as will be seen later. The most verbose is machine language.

Table 1.1 Prominent robot languages

Developer	Language	Robot of implementation
Cincinnati Milacron	T3	T3
Unimation	RPL, VALII	Puma
Adept	VALII	Adept1
Sheinmann	AL, PAL	Stanford arm
IBM	AML, Funky, Emily, Maple Autopass	IBM arm
Bendix	RCL	PACS arm
General Electric	Help	Allegro
Anorad	Anorad	Anomatic
Olivetti	Sigla	Sigma
Stanford	WAVE	Stanford arm, Robovision
Automatix	RAIL	Autovision, Cybervision

There are currently over 100 explicit robot programming languages, almost as many as the varieties of robots that exist. Some of the languages include Algol, Pascal, Basic and VALII. The development of robot languages started with the extension of the standard computer languages to include some robot control commands. Table 10.1 contains some of the more prominent robot languages which have been implemented.

We will now illustrate the structure of explicit robot programming using VALII The VAL language was released in 1979 by Unimation, Inc. of Danbury, Connecticut, USA for programming of its PUMA series robots. It was developed for Unimation by Victor Sheinmann and Bruce Simons. VAL stands for Victor's Assembly Language. The VALII language released in 1984 is an enhancement of the VAL language. It is a computer-based control system and language that provides the capability to easily define a robot's task. It also allows the use of sensor information and working in unpredictable situations. The following include some of the important aspects of the language:

(a) Locations definition. Several methods can be used to define locations and for determining the current location of the robot. For instance

   HERE P1

given in mode defines the variable P1 to be the current location of the robot. Similarly, the command

   WHERE

requests the robot system to display the current location of the robot. The command TEACH is used to lead the robot through and record for the robot a series of locations using the teach pendant. A third method for defining locations for the robot is by using the POINT command. For instance,

   POINT PX = P1

sets the value of PX, which is a location variable, to be equal to the value of P1.

(b) Editing and control of programs. In order to initiate the creation or editing of a program, the following commands may be issued:

   EDIT PICKUP1
   ⋮
   E

The EDIT command opens a program named PICKUP1 and the E command, which stands for EXIT, enables for exiting from the programming mode to the monitor mode.

Before the execution of a program, the speed is normally specified, and this can be accomplished using the SPEED command. For instance,

   SPEED 70

specifies that the speed of the manipulator is 70 units on a scale which may range from 0.39 (very slow) to 12800 (very fast), with 100 as "normal" speed. With the value of 70, one would be setting the robot speed below normal speed. However, the actual speed that is used for programs is obtained from the product of the speed specified in the monitor mode (such as 70 above) and the speed specified within the text of the program. For instance, if 50 was set as the speed within the program, it means that the actual speed will be (0.7 × 50 = 35) 35 units for the "normal" speed. Execution of a specified program can be accomplished using the command:

EXECUTIVE PICKUP1

In cases where the program needs to be executed several iterations, the EXECUTE command may read as follows:

EXECUTE PICKUP1, n

where n is the number of iterations of execution.

There are many additional monitor commands which can be used in VALII, they include, STORE, COPY, LOAD, FLIST, RENAME, DELETE, etc.

(c) Motion commands. VALII possesses commands which are used for translational movements by the robot. The motion instruction,

MOVE PX1

enables the robot to move by a joint-interpolated motion to the point PX1. Another command related to this is

MOVES PX1

which causes the robot to move along a straight line from its current position to the point PX1.

Similarly, the commands APRO and DEPART are used to move the tool towards and away from some specified position and orientation, respectively. For instance,

APRO PX1 70
MOVE PX1
DEPART 70

means that the tool should move to the location (position and orientation) specified by PX1, but offset from the point along the tool z-axis by a distance of 70 mm (distances can be measured also in inches or other units). The DEPART command returns the tool to its original position by moving it away from the new point PX1 70 mm along the tool z-axis. Although both the APPRO and DEPART commands are carried out in joint-interpolated mode, it is possible to carry them out in straight line motions by using APPROS and DEPARTS. Other commands that apply to motion include, SPEED, DRIVE, and ALIGN. The DRIVE command can be used to change a single joint from its current location to another. For instance, if a joint is rotary, a command might issue as follows:

DRIVE 5, 55, 60

This implies that the angle of joint 5 should be changed by driving the joint 55° in the positive direction at a speed of 60 percent of the monitor speed.

(d) Control of the hand. The gripper of the hand can be controlled using some gripper control commands such as OPEN, CLOSE, GRASP. The commands can be specified as follows:

OPEN 60

⋮

CLOSE1 60

The open statement causes the robot gripper to open to 60 mm and the CLOSE1 command causes the robot to close immediately to 60 mm. GRASP is another command that causes the robot to close immediately. A GRASP command

GRASP 10.5, 100

causes the gripper to close immediately while checking to assure that the final opening is less than the 10.5 mm specified. After closing, the program should go to statement 100 in the program.

(e) Configuration control and interlock commands. There are configuration commands that assure that the manipulator is spatially configured to approach its task like the human arm of shoulder, elbow, wrist, etc. The command of RIGHTY or LEFTY will imply that the elbow is pointed to the right-hand side or left-hand side, respectively. Other such commands include, ABOVE, BELOW, etc.

Interlock commands are used to communicate about input or output signals. The RESET command turns off all external output signals and is typically used for initialization. The SIGNAL command is used to turn output signals on or off. A positive signal number is used to turn the corresponding signal on and a negative number turns it off. For instance,

SIGNAL 2, −5

turns output signal 2 on and output signal 5 off. There is also the REACT command which is used to interrupt regularly the program due to external sensor signals. For example,

REACT VAR2, SUBR5

Would cause the continuous monitoring of the external binary signal specified as variable VAR2. When the signal expected occurs depending on the signal sign, control is transferred to subroutine SUBR5. There are many other VALII commands that space will not allow to be adequately discussed in this book. There are INPUT/OUTPUT control commands such as PROMPT and TYPE; program control instruction sets such as IF...THEN...ELSE...END, WHILE...DO, DO...UNTIL, etc.

This section is intended to show the structure of commercially available explicit programming language. A complete set of VALII commands can be obtained from Unimation [Inc84].

1.4 MACHINE TASK-LEVEL PROGRAMMING

Developing a task-level language is difficult. Many researchers have worked on aspects of the language but few have been able to produce reasonable language programs. Even those that have been developed operate within very restricted domains. The more prominent ones include: AUTOPASS [LW77], RAPT [ACP83], [PAB78] and LAMA

[LPW77]. The elements of task-planning include, *world modeling, task specification, and manipulator program synthesis* [LP83a], [BHJ⁺83]. World modeling task must include the following information:

1. geometric description of all objects and machines in the task environment;
2. physical description of all objects, e.g. mass and inertia;
3. kinematic description of all linkages;
4. description of manipulator characteristics, e.g. joint limits, acceleration bounds, and sensor characteristics [BHJ⁺83].

In addition to the configurations of all objects in the machine world, the models should include uncertainty associated with each configuration. Geometric description of the objects in the machine world is the major element of a world model. The geometric data can be acquired from vision or from the CAD system. Since parts to be manufactured are modeled in the CAD system, it seems appropriate that the information should come from the CAD. In addition, the data so represented are more accurate and easier to reason with than the data which can be acquired in today's vision system. The impasse in complete development and implementation of a good automated process planner is due to the range of knowledge needed by the machine to interact with its world, the problems associated with acquiring the world knowledge, and the difficulty in developing a robust reasoning paradigm. Research in the 1980s [Nna88], [LP83a], [Pop87] has shown that it is best to represent world knowledge using the CAD system and to deal with the dynamic world of the robot using sensors such as the vision system. One method which has proven useful in world modeling is the Octree encoding scheme [IRF89]. The octree scheme is similar to both spatial enumeration and cell decomposition approaches. It is a hierarchical representation of finite 3-D space and is a natural extension of the quad tree concept which is used for 2-D shape representation.

Task specification deals with the specification of the sequence of configuration states from the initial state of manufacturing parts to the final state of the product. The specification can be accomplished in three ways [BHJ⁺83]:

1. using a CAD system to model and locate a sequence of objects configurations [NC90];
2. specifying the machine configurations and locating the features of the objects using the machine itself [GT78]; and
3. using symbolic spatial relationships among object features to constrain the configurations of objects [ACP83], [NC90], [PAB80a].

The first and second methods are prone to inaccuracy since the light pen or a mouse is often used in the first case and the machine (especially the robot) has limited accuracy. The third method holds out more promise. It involves the description of the machine configurations by a set of symbolic spatial relationships that are required to hold between objects in that configuration. Some advantages can be seen from this approach. For instance, the method does not depend on the accuracy of the light pen or the machine. Also, the families of configurations such as those on a surface or along an edge can be expressed. Although using symbolic spatial relationships is a more elegant way of specifying tasks, the fact that they do not specify tasks directly is a drawback. They require conversion into numbers and equations before use [LP83a].

In this book, we will exploit the powers of the CAD specification approach and the symbolic spatial relationship to provide the designer with a robust task specification system which can be used to capture the designer's intent and to propagate this intent to the process planner. In summary, task specification can be achieved by using a sequence of operations and/or symbolic spatial relationships to specify the sequence of states of the objects in the realization of the task. But rather than have a sequence of intermediate model states on a CAD system, an initial state on the CAD system coupled with the symbolic spatial relationships can completely specify the task. A product modeler is needed which can allow for the use of a mouse or light pen to point to features which mate and which can provide the designer with the syntax and protocol to specify orientation by using symbolic spatial relationship. It is also possible to derive operations by examining binary sets of parts relationships on the bill of material tree. The product modeler can build the bill of material automatically.

The typical components of manipulator program synthesis are grasp planning, motion planning and error detection. The output is normally a sequence of grasp commands, motion specifications and error tests. It may be instructive to note that the product modeler which will be described in this book produces the relationships including functions of features of parts which are essential in program synthesis. For instance, grasp planning is well constrained if spatial relationships have been specified which must exist between features of parts in an assembly. The search for grasp surfaces will therefore be constrained by such relationships.

This book is therefore concerned with machine task-level programming where the world of the machine such as a robot can be specified in a CAD environment. Since automatic machine programming can be for a range of applications, we restrict our illustrations in this book to the assembly domain, specifically where the robot is the executor of the assembly task.

1.5 ORGANIZATION OF THIS BOOK

An introduction to the concepts of machine programming has been made in this chapter. We will present the role of the CAD system and data in automated robot programming and assembly in the second chapter. In Chapter 3 we present the concept of spatial relationships which provides the mechanism to relate different bodies in assembly situations. In order to provide a good picture of the general problems needed in automatic robot programming, we present the structure of an automatic programming system in Chapter 4. Since dynamic world reasoning is largely due to sensor information, the representation of such sensor information to the manipulator is presented in Chapter 5. In Chapter 6, we present the concepts of *world modeling* and *task specification*. Starting from Chapter 7, we present the various issues associated with robot planning. Chapter 7 contains *gross motion planning*, Chapter 8 deals with *grasp planning* and *fine motion planning*, and *trajectory planning* is presented in Chapter 9. In order to ensure that a chosen robot will be able to execute a generated robot-level program, we present a method to transform the general robot-level commands into generic robot-level commands in Chapter 10. The very important issue of program synthesis including grasp planning, fine motion planning, and precedence generation in assembly are discussed in Chapter 11.

CHAPTER 2

CAD in automatic machine programming

2.1 INTRODUCTION

As discussed in Chapter 1, automatic robot programming requires that tasks which the robot must perform be specified to it in some task-level language. This method of specification implies that the robot must reason about its work environment in order to plan how to execute the task. One of the principal problems to be overcome in such planning is reasoning about the world of geometric entities whose shape define the configuration of the world. Although the world information can be captured using a variety of approaches including vision, the understanding of the world implies understanding geometric shapes. Since the CAD system normally processes knowledge of parts geometric composition, obtaining and representing the robot's world information in the CAD system is the most viable approach.

By means of acquiring the world knowledge from the CAD system and generating these commands, a robot is able to automatically perform the task. The major drawback in the system so far developed (as well as in RAPT, and AUTOPASS) is that those systems have not been capable of interpreting geometry to recognize geometric features and inferring function which the features are intended to serve. Even for the most simple geometries – the polyhedral objects, this has proven to be very cumbersome. The problem of inference of function from form (geometry) of form functional specifications is particularly difficult because it involves the regeneration of the designer's intent for the part or product. It invariably requires an experienced process planner to do this and amounts to a "reinvention of the wheel."

Representing information to the robot (task specification) using the CAD system is in itself fraught with problems. This is largely due to the limited capacity of today's CAD systems. This problem is not limited to assembly but rather spans all areas of applications of automatic process planning applications. Geometric data currently obtained from CAD systems are CAD system dependent. In addition, current CAD systems are incapable of carrying manufacturing process information other than geometry. What is needed is a new type of CAD system; a CAD system which is capable of capturing the designer's intent for the part or product and can transfer the product definition information to a process planner. This new CAD system should ideally be capable of translating the data so acquired at design stage into a neutral graphics medium in order to eliminate CAD system dependence of the geometric data. Function can be captured by the designer and transferred to the process planner, thus eliminating the huge stumbling block of today's automatic process planner – the inference of function from form. A system with these capabilities – ProMod – is described in Chapter 5. In the next section, we describe the type of information which a desired CAD system must provide for the robot task planning system.

2.2 DESIRED CAD DATA

Many of the concepts and algorithms used in this chapter come from the earlier works of the author and the PhD thesis of T.S. Kang which shed much light on geometric reasoning.

We have already mentioned that the future CAD system must produce both geometric and functional data which are relevant to the process planning function. In order to ensure that future planning systems deal with data which is easy to manipulate, a study of the type of data appropriate for reasoning about geometry has been conducted [NP90]. The following observations were made: since the essential geometric information in object reasoning is the boundary information, a good modeler must be able to present boundary information in the form of *vertices, edges, loops,* and *faces*, from which features may be extracted. Also, a good modeler must be capable of translating its 3-D geometric data into a neutral format such as IGES (Initial Graphics Exchange Specification). In addition, it seems a waste of very valuable time to design parts using the current approach of *free design* where models of parts are created and the process planner must "tease out" the feature embedded on the part using some type of feature recognition system. It seems inevitable that at times a designer will necessarily be reduced to "raw geometry," or that elaborations of features will take them outside classes whose manufacturing properties are explicitly present in a database [NP90]. The capability of recognizing instances of feature classes in an object description which does not have them explicitly present will be a continuing need. It is however obvious that a more reliable and efficient technique will be to employ the concepts of *design with features* [LD86, Pop87]. This means the existence of a particular form feature has been directly expressed by the user of the system, or inferred as being part of a module so expressed. In contrast, the concept of *free design* merely requires the designer to describe with explicit geometry the form of the features in a part without symbolic reference to the feature.

Although there are many available modeling schemes which can be used for geometric model representation on the CAD systems (e.g. sweep representation, primitive instances, spatial occupancy enumeration, cell decomposition) [Nna88], we are concerned only with the boundary representation method because they possess the best attributes of all the representation schemes for geometric reasoning. The boundary representation is important because it is close to computer graphics, unambiguous, and available to computing algorithms. Disadvantages include: verbosity, complicated validity checking and difficulty in creating objects. CSG is useful in the creation process because it has features such as simple validity checking, unambiguity, conciseness, ease of object creation and high descriptive power. Its disadvantages include: inefficiency in producing line drawings and inefficiency in graphic interaction operations. In addition to these benefits, at least 95 percent of manufacturing parts do not require multicurved surfaces found on ships' hulls, car bodies, and aircrafts and hence can be represented in CSG [VRH⁺78]. It is however possible to produce models using CSG and then transform the data to boundary data. These schemes essentially fit well in design with features, reasoning about geometry, transmutation of form into function, and the consequent product modeler; and result in a synergistic effect on the automatic process planning system.

It is well known that just as robot programming is currently manipulator dependent, product modeling is still CAD system dependent. In other words, there is no standard internal data representation across CAD system manufacturers. However, the advent of IGES, which is a neutral graphics data representation medium is the first step in this regard. For a CAD-based formalism to be truly beneficial to the manufacturing world, there must also be a standard CAD representation. IGES is one of the most popular neutral graphics system but is severely limited in the types of data which it can communicate. It can only manage with geometric information. Even this geometric information is quite restricted. New standards which are intended to carry functional information as well as geometric information are under development at the time of this writing. The most prominent among them is PDES/STEP [PDE88]. It is however possible for CAD data translated into a neutral form such as IGES to be restructured into a form understandable by a reasoning system [NC89]. This is useful in using such data for automatic process planning.

2.3 FEATURE, FEATURE CLASSIFICATION AND REPRESENTATION

Webster's dictionary defines a feature to be a prominent part or characteristic of an entity of interest. From an engineering point of view, a feature can be regarded as a portion of an object with a prominent characteristic. A universal definition of a feature has not yet been accepted. In this chapter, a generalized feature definition is presented. We will then present a procedure for classification and representation of features for geometric reasoning in automated machine programming.

2.3.1 Feature concept

A universally acceptable definition of a manufacturing feature is yet to be obtained. CAM-I described "form feature" as a specific geometric configuration formed on the surfaces, edges, or corners of a workpiece intended to modify or to aid in achieving a given function [PW85]. This definition of "form feature" has a very strong connection between "form feature" and the material removal processes. Examples such as holes, grooves, slots, bosses, walls, ribs, and pockets are the shapes or volumes associated with the surfaces of a part. However, considering the intrinsic nature of features in the production industry, the definition of features should not focus on a particular process. Dixon defined feature as "A feature is any geometric form or entity uniquely defined by its boundaries, or any uniquely defined geomeric attribute of a part, that is meaningful to any life cycle issue" [DLL⁺87]. A more recent definition of a feature is, "A feature is any named entity with attributes of both form and function" [Dix88]. Nnaji and Kang defined feature as, "A set of faces grouped together to perform a functional purpose"

[NK90]. In PDES, form feature is defined in the conceptual model of feature information for the Integrated Product Information Model (IPIM) as a stereotypical portion of a shape or a portion of a shape that fits a pattern or stereotype [PDE88]. The feature model is called "Form Features Information Model" [FFIM]. The FFIM is intended to be independent of product classes and user applications. However, the FFIM does not force user applications to implement specific form features nor application-independent nomenclature [PDE88]. A feature cannot just be standalone with specifications related only to "form." Other information such as shape-oriented (tolerances, surface finishes, dimensional items) and process (machining, assembly, sheet metal forming) information should also be included. There is a consensus among researchers in the field that features contain information with some bearing on both form and function. In this book therefore, feature is defined as "A set of geometric entities* (surfaces, edges, and vertices) together with specifications of the bounding relationships between them and which imply an engineering function on an object." The characteristics of a generalized feature processing scheme are described as follows:

- A feature can be a combination of a group of relative geometric entities (face, edge, loop, vertex) or just a geometric entity itself.
- Non-geometric and manufacturing information should be combined with the form of a feature itself.
- Representation scheme and data structure of features should be pertinent to particular fields of applications.
- A mapping function should be defined for an application-dependent system to map the form of features into a standardized format (PDES form feature representation,) and vice-versa (Figure 2.1). This will result in the possibility of exchange of product information from a CAD system to another or other advanced application systems.

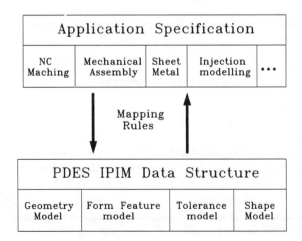

Fig. 2.1 Mapping an application-dependent feature onto a standardized format.

*Application-dependent.

In mechanically assembly or machining processes a feature is, usually, defined as a set of constituent faces. The geometric information related to a feature is obviously a subset of the object although some geometric entities (such as the edges located on the feature boundary) should be ignored in accomplishing the recognition process. In addition to the geometric information, some non-geometric information associated with a feature is also essential for process planning.

2.3.2 The Euler formula for features

Features of a 3-D object can be considered as an open object. Open object is defined as an object which contains single faces and single edges. From a topological point of view, the graph of any feature is 2-manifold and can be imbedded in a Möbius bend. The Möbius bend is a non-orientable surface and it is homomorphic to a disk. The Euler formula for features can be derived from the invariance of Euler characteristic of non-orientable surfaces. In this work, the graph used to represent features of a mechanical component are face-based. The Euler formula for the features with this representation is as follows:

$$v + f - e = s - h + r \tag{2.1}$$

where s, v, f, e are the number of shells, vertices, faces, and edges of the feature; r is the number of rings inside the feature faces; h is the genus of the graph.

Due to the different data representation of features, some modification of Euler formula is necessary. For example, in sheet metal working, if face-oriented representation is used to represent the geometric data of the part, the region inside a loop may be empty. In this situation, a face is not sufficient to represent a feature. Instead of using faces to build a graph, loops are used as the basis to construct the graph for features. The Euler formula is then modified as follows:

$$v + l_s - e = (s - h) + r \tag{2.2}$$

where e, s, h, r and v are the same as those in equation (2.1), l_s represents the loop with solid region.

2.3.3 Feature classification

Since manufacturing features are application dependent, the classification of features is also needed to be domain dependent. For example, a protrusion does not exist for a machining process when the raw material to produce the part is not in a pre-formed shape. However, protrusion is an important feature in an assembly process. It can be seen that for a variety application, features are interpreted in a different manner. Figure 2.2 shows an example of the different interpretation of given CAD data in two application domains. Features $\mathscr{AF}_1$ and $\mathscr{AF}_2$ are interpreted as a through hole and a blind hole respectively in mechanical assembly. While in machining the feature may be interpreted as a blind hole $\mathscr{NF}$ associated with $\mathscr{AF}_1$ and $\mathscr{AF}_2$.

In this research, the purpose of feature classification is to provide a search procedure for the feature recognition process. In order to classify features into several categories for a specified application, one must understand the geometric and topological variances

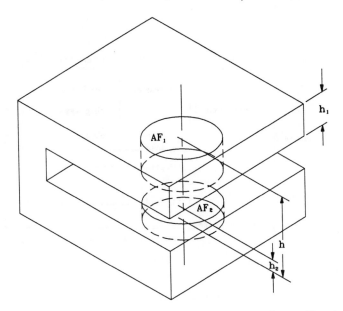

Fig. 2.2 A feature may be interpreted differently based on its Application Domain.

and invariances among the categories. The classification scheme should contain several layers in a hierarchical manner. For purposes of automation, the scheme to detect variances in a certain domain must be developed and can be computerized.

A generalized level for feature classification is shown in Figure 2.3. The classification scheme is based on the following criteria:

1. The Euler formula for features:
   From the value of genus (h) a feature is a passage that can or cannot be detected.
2. Feature boundary loop:
   Based on the type of the edges on a feature boundary, a feature is a depression, protrusion, or hybrid that can be detected.
3. Topological configuration of feature graph:
   Based on the configuration of the feature graph, the features whose graphs are homomorphic would be put into the same category.

Examples for describing the features belonging to single hole depression are shown in Figure 2.4. The further classification for a specified application can be followed to build the feature pattern database for the purpose of recognition.

To illustrate the contrast between assembly and other domains in feature classification, we present a quick summary of feature classification for sheet metal forming as shown in Figure 2.5. The criteria for the classification are as follows:

1. The Euler formula for features is used to detect if a feature is a passage or not.
2. Loop type: A loop can be one of the following type:
   (a) empty loop: the loop with no interior.
   (b) solid loop: a loop with interior.

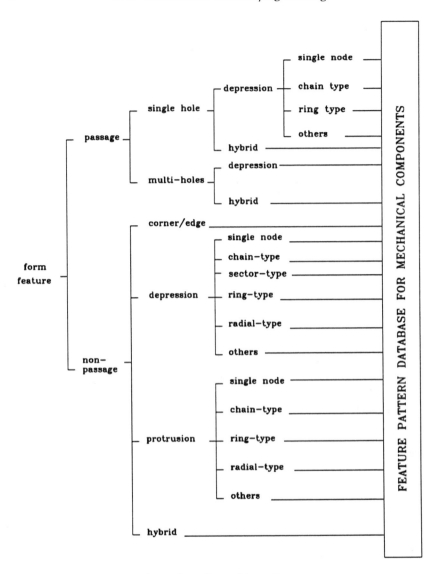

Fig. 2.3 Feature classification for mechanical assembly application.

The raw material of the sheet is considered to be a flat planar solid region. Therefore, the existence of any empty loop or any solid loop other than that of the planar face implies that a feature exists.
3. Cutting edge: A cutting edge is a convex edge that is connected to one or more thickness edges. Once cutting edges exist, they imply the existence of a hole or a slit.
4. Associated entity group: The existence of the associated entity group of a feature entity group establishes some characteristics of the feature. In addition, the distance

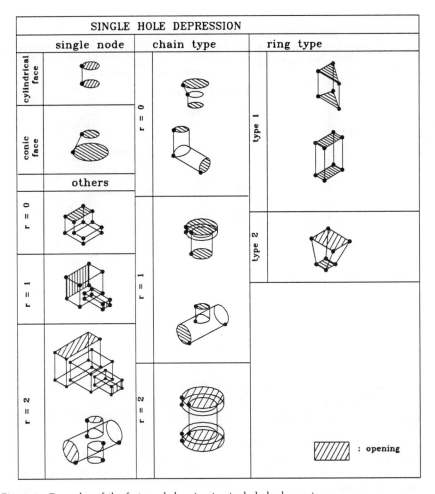

Fig. 2.4 Examples of the features belonging to single hole depression.

between a feature entity group and its associated entity group can be used to accomplish the task of classification.

A careful observation of properties of each application must be carried out for purposes of achieving unification and completeness (if possible) in a representation scheme.

2.3.4 Feature representation

The primary goal in developing a representation scheme for objects should be completeness in order to pave the way for further applications. The product model representation of a part should include both geometric data and non-geometric data. The geometric data can be directly derived from IGES through a geometric reasoning process. Some of the non-geometric information, such as rotational and reflective symmetries, of a part or a feature and pattern of features can also be obtained from topological and geometric

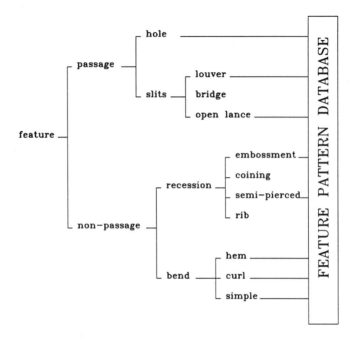

Fig. 2.5 Feature classification for sheet metal application.

reasoning based on graph and group theory. While the others (e.g. tolerances) must be acquired by the designer through a product model interface.

Two types of representations are generally used to represent form features:

1. *Explicit feature representation*: features are represented using a group of geometric entities and their relationship. This form of representation is shown in Figure 2.6.
2. *Implicit feature representation*: features are represented by a set of parameters as shown in Figure 2.7.

The representation scheme of an explicit feature representation is lower in level when compared to an implicit feature representation scheme. The latter can always be obtained by a deduction process from the former. Implicit feature representation can be regarded as a "macro view" of a feature and this representation may be useful in performing some operations in which the shape of a feature can be represented in a parametric way without ambiguity. This representation is used in a feature-based design systems for creating the part. However, the geometric information of features in this representation may not be complete if the boundary evaluation of the feature associated with the part is not followed. For example, a square hole in Figure 2.8 is difficult to represent by implicit approach. In addition, without explicitly presenting these features, the technological information, tolerances and surface finish, relative to faces cannot be specified. On the other hand, complex-shaped features (such as a thread on a cylindrical face) are hard to represent explicitly. In cases like these, the implicit feature representation is suitable to represent the feature. In this book, "explicit feature representation" accompanied by parameters associated with implicit feature representation is the scheme used to represent

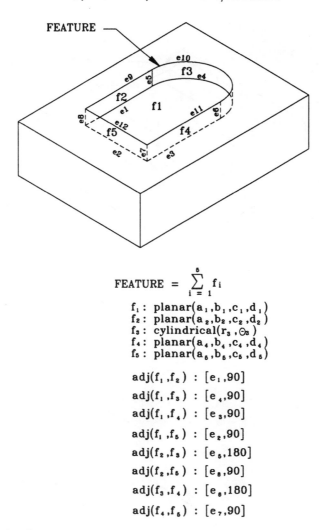

$$\text{FEATURE} = \sum_{i=1}^{5} f_i$$

f_1: planar(a_1,b_1,c_1,d_1)
f_2: planar(a_2,b_2,c_2,d_2)
f_3: cylindrical(r_3,Θ_3)
f_4: planar(a_4,b_4,c_4,d_4)
f_5: planar(a_5,b_5,c_5,d_5)

adj(f_1,f_2) : $[e_1,90]$
adj(f_1,f_3) : $[e_4,90]$
adj(f_1,f_4) : $[e_3,90]$
adj(f_1,f_5) : $[e_2,90]$
adj(f_2,f_3) : $[e_5,180]$
adj(f_2,f_5) : $[e_8,90]$
adj(f_3,f_4) : $[e_6,180]$
adj(f_4,f_5) : $[e_7,90]$

Fig. 2.6 Explicit feature representation.

features. This hybrid representation can reduce the complexity during the part creation process. For example, a threaded rod is hard and tedious both to create on the current CAD systems and to interpret in a feature extraction system. To represent a thread the following list of parameters should be provided: type, major diameter (d), mean diameter (d_m), minor diameter (d_r), pitch (p), thread angle (2α), chamfer (some parameters can be omitted if any standard, such as ANSI, DIM, and JIS, is applied). Complex features, such as thread, teeth, helical groove, instead of being generated in the modeling processes are assigned to a particular face of the part through a product modeler.

The data structure of each entity is designed to cover the geometric and non-geometric information in a hierarchical manner. For each feature graph the adjacent relationship among the nodes and the incident relationships between the nodes and linkages are

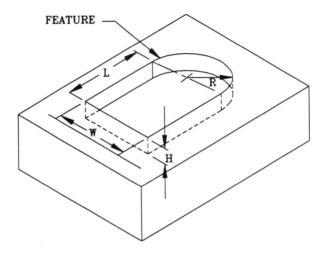

Fig. 2.7 Implicit feature representation.

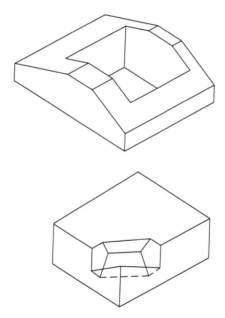

Fig. 2.8 A square hole which is difficult to represent using implicit feature representation.

explicitly represented. This information is helpful for identifying the features in a pattern process and finding the symmetries for features. For representing mechanical components, face is the basic geometric entity used in this representation, since the faces of a part can capture a great portion of the associated information of a part. The relational data structure, which captures both geometric and non-geometric information of the features, for representing a feature is shown in Figure 2.9.

Feature, feature classification and representation

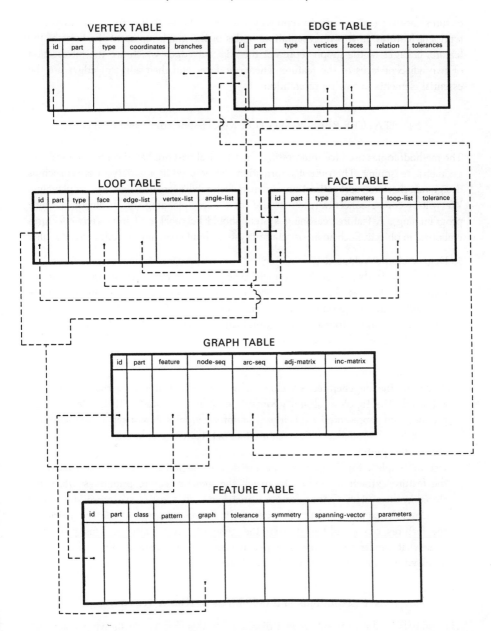

Fig. 2.9 Data structure for representing features.

2.3.5 Feature representation for mechanical components

In this book, relational boundary information is used to represent an object. Feature in this application can be regarded as a set of connected faces which perform an engineering function. Therefore, a data structure which contains both topology and geometry of features' face set is necessary to represent features. A face-based graph is developed to represent features. In this representation, nodes of the graph represent the faces of the feature, linkages of the graph represent the adjacent edges which are the intersection of two adjacent faces of the feature. Therefore, faces and their adjacent edges are the essential elements in this representation.

2.4 FEATURE REASONING FOR MECHANICAL COMPONENTS

The methodologies used for interpreting a mechanical part are based on topological and geometric reasoning. The output information can be used in a variety of areas such as design rule checking, cost estimation, and manufacturability analysis and process planning. We will of course concentrate on its use for process planning (or automatic machine programming). A feature reasoning system should be capable of interpreting features embodied in objects. Such an interpretation system should contain the following modules:

1. Degenerated envelope determination.
   Find the basic shapes (such as parallel-piped, right pyramid, right prism, cylinder, etc.) that match the shape of the object. Each basic shape has a coordinate frame attached to it. The geometric information of the object is then transformed based on the object frame to establish a coordinate-independent information module from the IGES input. This is essential for the subsequent steps involved in process planning.
2. Feature extraction.
   Based on the degenerated envelope of a part, the features of the part can be extracted. The topological and geometric entities associated with the features are grouped and represented explicitly. This information is comprised of external and internal feature bounded loops, feature face set, feature graph, etc.
3. Feature recognition.
   Use the explicit topological and geometric information of features extracted from the feature extraction module to match the generic feature primitives which are stored in the system. Once a feature is matched with a feature pattern in the feature database, the parameters and other geometric information such as the symmetries of the features can be obtained from the generic feature patterns. This process is somewhat similar to the feature-based design system, however; the tedious steps involved in human interface would be eliminated.

2.4.1 Notations and definitions in feature reasoning

The symbols used for the subsequent discussion in this chapter are shown in Table 2.1. The definition of geometric entities used for reasoning about features of mechanical components follow.

Definitions

Envelope
The smallest convex hull that can enclose an object.

Degenerate envelope
The *basic shape* of an object obtained by neglecting the small protrusions on the object.

For purposes of interpretation, features are classified into the two main categories of external and internal features:

1. *external feature:*
   At least one of the feature's faces intersects with the object's envelope.
2. *internal feature:*
   The intersections of the entire feature's faces is completely located inside the object's envelope. There is no intersection between any feature's face and the envelope. In this case, an inner loop, formed by the feature's faces, is used to represent the border of the feature. This type of feature is always a sub-feature of another feature.
   $F_{i,j}$ of object i possessing a total of N faces, then

$$F_{i,j} = \{f_{i,j,1}, f_{i,j,2}, \ldots, f_{i,j,N}\} \tag{2.3}$$

3. $F_{i,j}$ is an external feature, if

$$F_{i,j} \cap Ev_{i,j} = FBL_{i,j} \tag{2.4}$$

or it is an internal feature, if

$$F_{i,j} \cap Ev_{i,j} = \phi \tag{2.5}$$

An internal feature cannot exist alone; it must be a sub-feature of an external feature.

Edge
An edge is the intersection of *two* adjacent faces of a solid, which is one of the characteristics of adjacent topology. The type of an edge can be a line segment, arc, or circle. The attribute of an edge can be convex, concave, or neutral. An edge is said to be convex if the angle between its intersection surfaces is less than $180°$; it is said to be concave if the angle between the intersection surfaces is greater than $180°$; and it is said to be neutral if the angle between the intersection surfaces is equal to $180°$.

For any $e_{i,l} \in \varepsilon_i$ let $e_{i,l}$ be the intersection of faces $f_{i,l,m}$ and $f_{i,l,n}$ then

$$\text{attribute } (e_{i,l}) = \begin{cases} \text{CONVEX} & \text{if } \angle(f_{i,l,m} \cap f_{i,l,n}) < 180° \\ \text{CONCAVE} & \text{if } \angle(f_{i,l,m} \cap f_{i,l,n}) > 180° \\ \text{NEUTRAL} & \text{if } \angle(f_{i,l,m} \cap f_{i,l,n}) = 180° \end{cases}$$

A neutral edge is not the physical edge of a solid, but is used to separate a composite face into two types of faces. For boundary information based representation, the existence of neutral edges is necessary. Neutral edges are used to specify the boundary of faces with different types (for example, a planar and cylindrical faces) which are tangent to each other. Since every face of the object is bounded by a set of edges, neutral edges should be included in the boundary information of solids. A neutral edge can, moreover, be classified into a neutral-convex or neutral-concave edge depending upon the attributes of its adjacent faces. An example of an edge's attributes is shown in Figure 2.10.

Feature faces set (FSS)
A group or set of connected faces which is used to represent a feature.

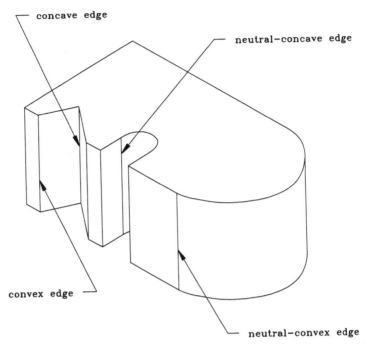

Fig. 2.10 Edge attributes.

External feature border edge (EFBE)
An edge is called *EFBE* if one, and only one, of its adjacent faces is an envelope face.

Feature bounding loop (FBL)
A *FBL* is a closed loop(s) used to specify the boundary of a feature and it can be one of the following types:

1. external feature bounding loop (EFBL): an *EFBL* is a closed loop which is composed of consecutive *EFBEs*. For each external feature, there is exactly one *EFBL* except where the external feature is a through hole or a *neck*.
2. internal feature bounding loop (IFBL): an *IFBL* is a closed loop formed by the edges of feature faces.

The attributes of edges in *FBL* is a very important characteristic in the process of interpretation of a feature. For example, if a *FBL* contains nothing but concave edges, this feature must belong to the protrusion category; while if a *FBL* is composed of concave edges this feature would belong to the depression category.

Figure 2.11 shows several types of external and internal features and presents their *FBLs*.

Face bounded loop (SBL)
A closed loop which specifies the boundary of a face. This loop may be composed of both convex and concave edges. The combination of different types of edges will be an important criteria in decomposing complex features.

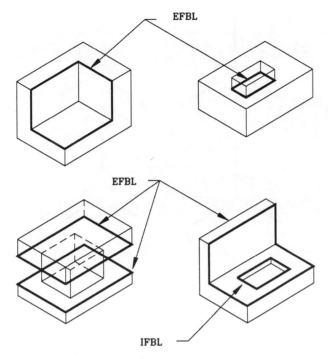

Fig. 2.11 External/internal feature bounding loop.

Co-face loop level (CFLL)
CFLL is the property of *face bounded loop* (SBL). The CFLL of the outermost SBL of co-surface bounded faces is said to be zero, while the CFLL of inner loops (or rings) is said to be one. A face is bounded by exactly one loop with its $CFLL = 0$ and none or several loops with their $CFLL = 1$.

Feature neighboring faces set (FNF)
A face set that contains all the faces adjacent to any face in a *feature face set* (FFS). The intersection between a FFS and its FNF is an empty set.

2.4.2 Envelope determination

Basically, an envelope is the smallest convex hull that can enclose an object. However, convex hulls may have a complex or irregular shape. We use primitive types of convex hulls, such as ppd, right cylinder, cone, or prism, which are of interest. An object's envelope is essential because it is the basis upon which the object coordinate frame is determined. The coordinate frames for the typical envelopes are shown in Figure 2.12.

For a geometric reasoning based system, it is helpful in finding the degenerate envelope [BPM82] instead of the envelope for an object. A degenerate envelope of an object is, geometrically, the basic shape of the object and can be obtained by neglecting the protrusions on the object. Using the degenerate envelope of the parts is useful for a

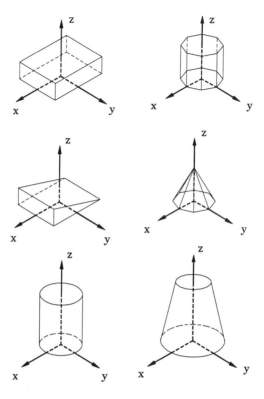

Fig. 2.12 Object coordinate frames.

feature reasoning process associated with tooling procedures specially in assembly processes. For an assembly operation, the concept of a degenerate envelope is essential, while for a machining operation the material envelope is useful. However, the envelope of an object can always be derived from the degenerated envelope of the object. Using the geometric information from IGES, the envelope can be found using the concepts of half-space boundary to find the envelope for an object. Since the normal of each face of an object has already been found, it would not be difficult to determine whether a certain face is part of an envelope surface or not. For example, a planar face's surface equation is

$$\lambda x + \mu y + \nu z - \rho = 0 \qquad (2.6)$$

where $\lambda^2 + \mu^2 + \nu^2 = 1$.

Let V be the vertex set of a convex object, $v_i(x_i, y_i, z_i) \in V$. v_i is located outside the object, if

$$\lambda x_i + \mu y_i + \nu z_i - \rho_i \geqslant 0. \qquad (2.7)$$

On the other hand, the plane surface which satisfies equation (2.8) is called a *supporting plane*.

$$\lambda x_i + \mu y_i + \nu z_i \leqslant 0 \qquad (2.8)$$

Smart [Sma88] defined a supporting plane π of a three-dimensional set as S if and only if π contains at least one boundary point of S and S lies entirely in one side of the closed half-spaces determined by π. In this book, instead of using the supporting plane, the support surface is used, since this procedure can be applied to determine the envelope surface for the cylindrical and conic surfaces as well. An object contains faces which are the partition of cylindrical or conic faces, the auxiliary planes would be generated according to the following procedures:

Let f_i be a surface revolutionary face of an object $V_{gs} = \{v_{gs1}, v_{gs2}, \ldots, v_{gsn}\}$ and $V_{ge} = \{v_{ge1}, v_{ge2}, \ldots, v_{gem}\}$ be the ordered sets of vertices located on the generatrix of f_i at start position and end position, $V_{gs} \neq \phi$ and $V_{ge} \neq \phi$.

$$e_{gs} = \begin{cases} (v_{gs1}, v_{gsn}) & \text{if } n > 1 \\ \phi & \text{otherwise} \end{cases}$$

$$e_{ge} = \begin{cases} (v_{ge1}, v_{gem}) & \text{if } m > 1 \\ \phi & \text{otherwise} \end{cases}$$

An auxiliary plane P_{aux} is needed to replace f_i in determining the envelope of the object. P_{aux} can be established using the following procedure:

1. If attribute of f_i is concave then P_{aux} can be determined by two parallel lines which are located on the start and end positions of the generatrix.
2. If attribute of f_i is convex and either attribute of e_{gs} or e_{ge} is not concave then P_{aux} can be determined by a point v_m located on the generatrix at the half of the entire revolution angle. In this case P_{aux} is tangent to f_i at v_m.
3. If attribute of f_i is concave and the attributes of e_{gs} and e_{ge} are concave then P_{aux} is determined using the same procedure as that described in (1).

The methodology for obtaining the envelope surfaces is described as follows:

1. Extract all the faces bounded by one or more level 0 loops from the translated neutral geometric data file of an object, and check them individually:

    If the ratio, γ, of the occupied area of the inner loop over the total area of the outer loop for a face of an object is less than a certain value which is assigned when using a particular application, where

$$\gamma = \frac{\text{Area}_{innerloop}}{\text{Area}_{outerloop}} \quad (2.9)$$

then find all the vertices located on the positive side, the normal direction of this face. Put these vertices in a set called *exclude-vertex-set* (EVS) and their associated faces in an *exclude-face-set* (EFS). Faces in EFS will not be considered in finding an envelope. They show the existence of a protruded feature.

2. A face, $f \in FNF$, if

$$\forall v_i \in V - EVS \ni v_i \in \mathcal{H}^- \quad (2.10)$$

where V is the set of vertices for the object, and $\mathcal{H}^-$ is the negative partition of the 3-D space. A plane, $P_j: \lambda x + \mu y + vz - \rho = 0$, for example, $P_j \in FNF$ if and only if

$$\lambda x_i + \mu y_i + v z_i \leq \rho \, \forall v_i(x_i, y_i, z_i) \in V - EVS, i = 1, 2, 3, \ldots \quad (2.11)$$

An example of this is shown in Figure 2.13.

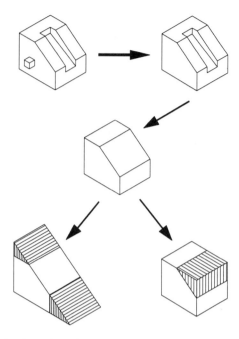

Fig. 2.13 An example of envelope determination.

3. Use the outer face set to match the *envelope rule sets* and decide what the envelope should be for a particular object. For example, if all the "envelope surfaces" are planar, and six of them are either parallel or perpendicular to one another then the envelope of this object would be a *ppd*. The dimension of the envelope can also be calculated from *EFS*. At the same time, a homogeneous transformation must be applied to all the vertices in order for them to coincide with the CAD coordinate frame used with the pre-defined object coordinate frame. This is essential in building a coordinate-independent system.

2.4.3 Coordinate frame mapping

Different modeling systems create the coordinate frame for each envelope type differently. For example, in I-DEAS GEOMOD [SDR86] the origin of a ppd(block) is located on the center of its volume, while in PADL-2 [PAD85] the origin is located on the bottom left corner vertex. In addition, if a homogeneous transformation is applied to the object on a CAD system, the location of the object may be changed and could be difficult to trace. Here, a strategy based on mapping a CAD-based object onto our pre-set local coordinate frame is developed. A parallel-piped envelope is used as an example to illustrate the procedure of coordinate frame mapping.

The local object coordinate frame can be assigned, according to the criteria of:

$$L \geqslant W \geqslant H$$

where L, W, H are the dimensions of a *ppd* along x-axis, y-axis, and z-axis of its object coordinate frame, respectively. An object coordinate frame generated in this fashion would satisfy the natural orientation of the object. It would be convenient for a vision system to determine where the object is located in the world. Since for neutral data such as IGES database, geometrical information is referred to with respect to the CAD coordinate frame where the model was created, it is essential to map the CAD coordinate frame onto our pre-defined local coordinate frames. This relation can be described by a three by three rotational matrix, for the orientation relationship between the CAD frame and our object frame, followed by a one by three translational matrix. This transformation can be done using the following processes

1. Find an "orientation face set" (OFS), $OFS \subseteq FNF$, such that

$$\forall f_i, f_j \in OFS, \quad \mathbf{n}_{f_i} \cdot \mathbf{n}_{f_j} = 0 \tag{2.12}$$

where $\mathbf{n}_{f_i}$ is the normal vector of $f_i \in OFS$. The number of elements in OFS is three. The normal vectors of the faces in OFS determine the orientation of the object.

2. $\forall f_k \in OFS$

$$f_k : \lambda_k x_k + \mu_k y_k + \nu_k z_k - \rho_k = 0 \tag{2.13}$$

find the length along each normal direction

$$L_k = \max\{|\rho_i - \rho_k|\} \forall v_i(x_i, y_i, z_i) \in V - EVS \tag{2.14}$$

where $\rho_i = \lambda_k x_i + \mu_k y_i + \nu_k z_i$

3. Sort $\{OFS\}$ by L_k, and put the normals of the three faces into "orientation direction set" (ODS). Let

$$ODS = \{\mathbf{n}_1, \mathbf{n}_2, \mathbf{n}_3\} \tag{2.15}$$

To satisfy the right-hand-screw rule

$$\mathbf{n}_1 \times \mathbf{n}_2 = \mathbf{n}_3 \tag{2.16}$$

4. Find the orientation of the object frame with respect to the CAD frame.
   Let $ODS = \{\mathbf{A}_1, \mathbf{A}_2, \mathbf{A}_3\}$, where

$$\begin{aligned} \mathbf{A}_1 &= [a_{11}, a_{21}, a_{31}] \\ \mathbf{A}_2 &= [a_{12}, a_{22}, a_{32}] \\ \mathbf{A}_3 &= [a_{13}, a_{23}, a_{33}] \end{aligned} \tag{2.17}$$

then the orientation relationship between the object coordinate frame and CAD coordinate frame can be described by matrix $\mathbf{A}$, where

$$\mathbf{A} = \begin{bmatrix} a_{11} & a_{12} & a_{13} \\ a_{21} & a_{22} & a_{23} \\ a_{31} & a_{32} & a_{33} \end{bmatrix}$$

for $v_i(x_i, y_i, z_i) \in V$ it is clear that

$$\mathbf{A}^{-1} \begin{bmatrix} x_i \\ y_i \\ z_i \end{bmatrix} = \begin{bmatrix} \hat{x}_i \\ \hat{y}_i \\ \hat{z}_i \end{bmatrix}$$

where $(\hat{x}_i \hat{y}_i \hat{z}_i)^T$ are the coordinates of v_i after an orientation modification.
It is also obvious that $\mathbf{A}^{-1} = \mathbf{A}^T$.

5. Find the origin of the object frame with respect to the CAD frame.
   The dimension, length(L), width(W), and height(H), can be obtained by:

$$L = \max(\text{coord}_x) - \min(\text{coord}_x)$$
$$W = \max(\text{coord}_y) - \min(\text{coord}_y)$$
$$H = \max(\text{coord}_z) - \min(\text{coord}_z)$$

and the origin of the coordinate frame in the IGES file can also be found as follows:

$$O_x = \frac{\max(\text{coord}_x) + \min(\text{coord}_x)}{2}$$

$$O_y = \frac{\max(\text{coord}_y) + \min(\text{coord}_y)}{2}$$

$$O_z = \min(\text{coord}_z)$$

The origin of the CAD frame then needs to be translated to (O_x, O_y, O_z), with the coordinates of each vertex changed according to the following equation:

$$\begin{bmatrix} \bar{x}_i \\ \bar{y}_i \\ \bar{z}_i \end{bmatrix} = \begin{bmatrix} \hat{x}_i - O_x \\ \hat{y}_i - O_y \\ \hat{z}_i - O_z \end{bmatrix}$$

where $(\bar{x}_i \bar{y}_i \bar{z}_i)^T$ are the coordinates for v_i in our object coordinate frame.

2.4.4 Extraction of form features

The form feature extraction process is based solely on geometric reasoning. The relationships between the features are represented in a hierarchical manner. A root feature is a feature that contains all the edges of an external feature bounding loop(s) (EFBL). A child feature of a feature is the feature which may contain none or some but not all of the edges located on EFBL. Obviously, if a child feature of a feature does not contain any edge in EFBL, then the implication is that an internal feature bounding loop would exist. The existence of an IFBL implies the existence of a sub-feature, while the reverse is not necessarily true. The scheme for obtaining the hierarchical relationships between the features of an object will be discussed in "Feature Decomposition". In this section, features are extracted based on the EFBLs and connectivity of faces. The feature extraction process groups sets of geometric entities based on the connectivity of the faces of an object. The purpose of the feature extractor is to provide sufficient geometric information for the subsequent feature reasoning process. The prioritized sequence of extraction process is as follows:

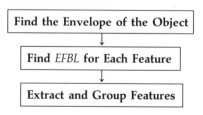

The algorithm to find the feature face set is described as follows:

1. Find an external feature entity including:
   - external feature bounding edges.
   - external feature bounding loops.
   - set of vertices which are located on both feature bounding loops and edges of the envelope.
2. Find all the connected faces using back tracking method.

2.4.5 Feature recognition

The most important principle to abide by in recognizing a feature is avoiding ambiguity. The criteria used for a feature recognition process is based on the feature classification described earlier in this chapter. Then generic feature primitives or feature families which have different geometric characteristics under a specified category in feature classification must be developed. A feature may be decomposed during the recognition process. In general, the recognition process starts with topological checking. If necessary, this is followed by geometric checking. The schema for a feature recognition process is shown in Figure 2.14. The pattern matching process begins to match the extracted form feature with the generic feature primitives stored in the feature pattern database. If no feature primitives can match the extracted feature, the process would continue to decompose the feature into several partitions according to the rules set in the decomposition process and return to the pattern matching process to match these decomposed features individually. The pattern matching and decomposition processes would be executed repeatedly, until either all the decomposed features are recognized or any of the decomposed features cannot be decomposed any further. If a feature is not recognized, it would be put on a blackboard. The features on the blackboard will be eventually inserted into the generic feature primitive database manually [Kan91].

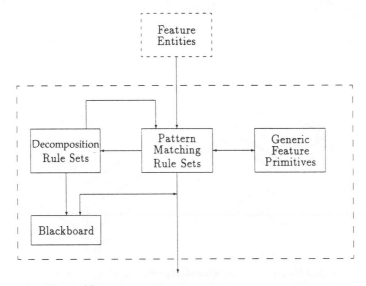

Fig. 2.14 The schema of feature recognition process.

2.4.6 Criteria used for feature recognition

The feature recognition problem involves determining if there exists a one-to-one correspondence between the nodes of a face-based feature graph and a generic feature primitive stored in feature database. Based on the feature classification scheme discussed earlier, a feature can be classified into a category, according to the geometric and topological characteristics of the feature. The identical checking between the pattern and feature is then followed. A feature carries the information of both form and function, the form of a feature can be affected by the boundary condition while the function of the feature may not (Figure 2.15(a, b)). Therefore, some information about feature entities should be ignored during the matching process. The feature boundary loop (*FBL*) of a feature is used to represent the relationship between the feature and the other part of an object. Using *FBL* as a criteria to do the feature recognition would be considered weak from a feature recognition standpoint. The length of a curve of a feature should also necessarily be ignored, since the size of the feature should be considered when the patterns are represented by primitives. The information of the feature entities employed in the recognition process would be addressed at the nodes and the entities that specified the relations between the nodes of the feature graph. The face-face adjacency and the face-edge incidence matrices of a feature capture sufficient information about the configuration of the feature for the recognition process. The adjacency matrix of a face-based feature graph, G, with n nodes and no parallel edges is an n by n symmetric binary matrix $\mathbf{A}(G) = [x_{ij}]$ such that [Kan91]

$$x_{ij} = x_{ji} = \begin{cases} 1 & \text{there is an edge between } i\text{th and } j\text{th nodes,} \\ 0 & \text{otherwise} \end{cases}$$

The incidence matrix of a face-based feature graph, G, with n nodes, m edges, and no self-loop is a m by n binary matrix, $\mathbf{I}(G) = [y_{ij}]$, such that

$$y_{ij} = \begin{cases} 1 & \text{edge } i \text{ is incident on node } j \\ 0 & \text{otherwise} \end{cases}$$

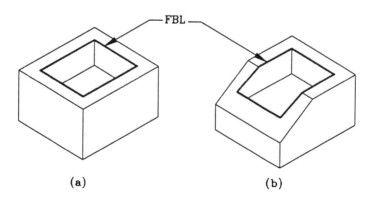

(a) (b)

Fig. 2.15 *FBL* of a feature should be ignored in pattern matching process.

It must be noted that the elements in the adjacency and incidence graphs are not just carrying topological information. Geometric information such as the angle between two adjacent faces and the types of nodes and linkages is also included. The reason to use both adjacency and incidence matrices for topological checking is because:

- Regardless of the existence of parallel edges and self-loop, both adjacency and incidence matrices of a feature can be used to represent the feature without any ambiguity.
- Parallel edges can be detected from the incidence matrix of a feature.
- Self-loops of a feature can be detected from the adjacency matrix of the feature.
- Permutation of any rows or columns in an incidence matrix will result in a graph isomorphic to the original one. Two graphs with no self-loops are isomorphic if

$$I(G_1) = M \cdot I(G_2)$$

where $I(G_1)$ and $I(G_2)$ are the incidence matrices of G_1 and G_2. M is a row or column interchanging matrix.
- Permutation of rows and its corresponding columns of an adjacency matrix imply reordering of the nodes. Two graphs G_1 and G_2 with no parallel edges are isomorphic if

$$A(G_1) = P^{-1} \cdot A(G_2) \cdot P$$

where $A(G_1)$ and $A(G_2)$ are the adjacency matrices of G_1 and G_2. P is a permutation matrix.

The information structure of adjacency and incidence matrices of the feature is as follows:

- feature face adjacency matrix
    1. adjacency matrix mask
    2. node sequence
    3. number of nodes
    4. nodes(faces)
        - type of face
        - set of linkage incident on the node
    5. adjacent information(edges)
        - types of adjacent edges
        - angular dimension of two adjacent faces
- feature edge-face incidence matrix
    1. incidence matrix mask
    2. node sequence
    3. number of nodes
    4. linkage sequence
    5. number of linkages
    6. nodes(faces)
        - type of face
        - set of linkage incident on the node

2.4.7 Generic feature primitives

Feature pattern database is used for the pattern matching process. In this database, feature external loop (*FBL*) of a feature is ignored. The characteristic of a feature is not affected by the boundary of the feature. The criteria to build a generic feature primitive is as follows:

1. Faces and their adjacent edges must be labeled.
2. The sequence of faces in the adjacency matrix is determined using the starting face and edge and the loop direction of faces.

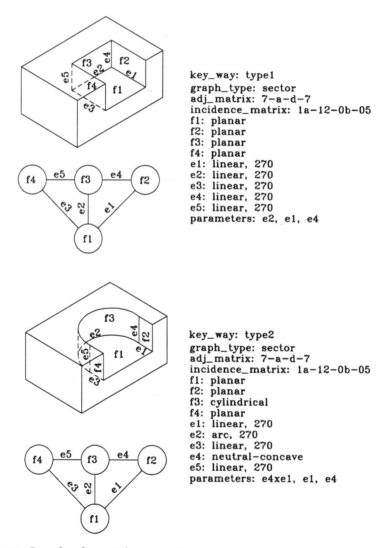

```
key_way: type1
graph_type: sector
adj_matrix: 7-a-d-7
incidence_matrix: 1a-12-0b-05
f1: planar
f2: planar
f3: planar
f4: planar
e1: linear, 270
e2: linear, 270
e3: linear, 270
e4: linear, 270
e5: linear, 270
parameters: e2, e1, e4
```

```
key_way: type2
graph_type: sector
adj_matrix: 7-a-d-7
incidence_matrix: 1a-12-0b-05
f1: planar
f2: planar
f3: cylindrical
f4: planar
e1: linear, 270
e2: arc, 270
e3: linear, 270
e4: neutral-concave
e5: linear, 270
parameters: e4xe1, e1, e4
```

Fig. 2.16 Examples of generic feature primitives.

3. The edge sequence in the incidence matrix is determined by the face sequence in the adjacency matrix and the loop direction of the faces.

The methodology to determine the face and edge sequence for the adjacency matrix and incidence matrix of a generic feature primitive is:

1. Choose a face, f_i, which has the highest degree number in a face-based feature graph as the starting face.
2. Choose an adjacency edge, e_j, whose type has the least appearance number in the loop of f_i as the starting edge.
3. Start by traversing the faces in the feature graph using the backtracking search (depth first) according to the loop directions of the faces such that a face is visited only once.
4. Traverse the adjacent edges of the feature graph using the breadth-first search according to the face sequence determined in (3) and the direction of the faces such that every edge is visited only once.

The adjacency and incidence matrices of a feature graph are simply used for checking if a testing feature and a generic feature primitive are topologically isomorphic. The result will be used to determine whether further geometric checking is necessary. The mapping relationships can be captured from the topological checking. The geometric information needed to be carried by a generic feature primitive in order to perform the recognition process is as follows:

1. type of faces in the feature graph.
2. type of adjacent edges in the feature graph.

For example, the instances of a keyway and a groove as shown in Figure 2.16 show the information structure of two types of generic keyways. The items shown in the adjacency and incidence matrices indicate the hexadecimal values of each row of the matrices [Kan91].

face	1	2	3	4	value
1	0	1	1	1	7
2	1	0	1	0	a
3	1	1	0	1	d
4	1	0	1	0	a

face/edge	1	2	3	4	5	value
1	1	1	1	0	0	1a
2	1	0	0	1	0	12
3	0	1	0	1	1	0b
4	0	0	1	0	1	05

It must be pointed out that this feature pattern is fully application dependent. The domain discussed here is mechanical assembly.

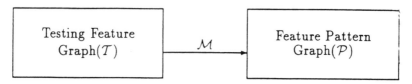

Fig. 2.17 Mapping a testing feature graph onto a feature pattern graph.

2.4.8 Pattern matching

The recognition strategy used is "learning before recognition." The decision rules used to classify features into several categories are implemented to find the location of appropriate patterns quickly. The matching rules can be applied to locate the right primitive in feature pattern database for a specified testing form of a feature. Once a primitive is matched with the testing feature, the associated parameters and attributes of the primitive will be attached to this testing feature.

The matching rules are built according to the criteria described in section 4.6. In order to detect that two face-based graphs are matched to each other, isomorphic checking must be performed. The isomorphism $\mathcal{M}$ maps a testing feature graph, $\mathcal{T}$, onto a pre-set feature pattern graph, $\mathcal{P}$, as shown in Figure 2.17.

Obviously, $\mathcal{M}$ is a bijective mapping and it should be coordinates independent. Not only must the adjacent topology of the feature face graphs (FFGs) of the testing feature and generic primitive be the same, but the information carried on each linkage of them should be identical. The entities of interest in $\mathcal{M}$ are the faces(nodes) and the adjacent edges(linkages). Since the information carried in faces are different, the nodes are not indistinguishable. Faces may be distinguished by their types. The angular dimension between any two faces should also be checked.

The rules used in form feature classification are used as the decision rules to locate the feature primitive for a particular test feature. The procedure for the pattern matching process is described as follows:

1. Find the face sequence and edge sequence according to the procedure described in section 4.5.
2. Check topological equivalence by comparing the corresponding values in the adjacency and incidence matrices of test feature and the generic feature primitive.
3. If the test feature is isomorphic to the generic feature primitive, check the geometric contents of each node and linkage of the test feature graph to its corresponding nodes and linkage in the generic feature graph; otherwise change the starting condition and go to (1).
4. If the geometric checking is passed then the feature is recognized; otherwise, change the starting conditions and go to (1).
5. If there is no match between the test feature and the current feature primitive, pick the following feature primitive in the feature pattern database, repeat the whole process.

The geometric information is used to generate the face and edge sequences of the adjacency and incidence matrices for the topological checking. For an n nodes, m linkages feature graph, the complexity to check topologically identity is reduced from $n!$ to $2m$.

2.4.9 Feature decomposition

After a feature, represented by a feature face set, is extracted using the procedure described above, it may be necessary to decompose this feature into two or more features if the feature is not recognized. Then all of the decomposed features should be tested individually.

The decomposition process will result in several independent co-level sub-features. The strategy used to decompose a feature is:

1. The type of *FFG* can be classified into chain-type graph and non-chain-type graph. Chain-type graph is one in which each linkage has one endpoint in common. A chain-type graph is decomposable if two or more faces in the graph are co-surface.

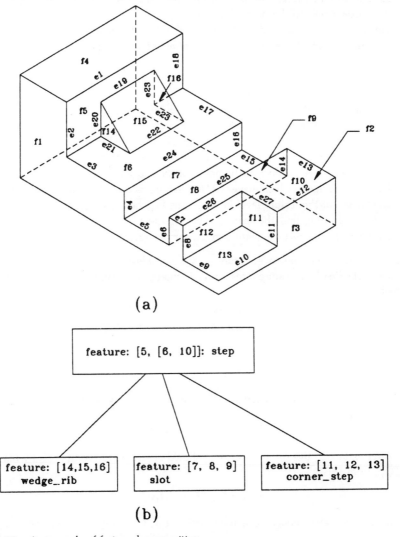

Fig. 2.18 An example of feature decomposition.

The sub-graphs can be found by extracting all the faces between every two co-surfaces. For any non-chain-type FFG, if it is decomposable, there must exist one or more articulation nodes(faces). A node is said to be an articulation node, if the degree of this node is greater than one and the removal of a node from a connected graph will result in the generation of two or more disjoint sub-graphs.

2. Find the first level FFG by extracting the faces from the envelope. Using the extracted FFG as FNF to find all the internal feature bounding loops (IFBLs), if any FBL exist then the sub-feature of the current level FFG can be extracted by traversing the current FBL.
3. Find the co-surface faces in FFG. If they exist, extract all the faces in between as its sub-feature.
4. Find the articulation nodes of the FFG. The FFG, hence, can be separated into several disjoint sub-graphs by splitting the articulation nodes. These sub-graphs are 1-isomorphic to the original graph.
5. For each articulation node, $\mathscr{A}$, update the FNF by removing each face $F_i \in FNF$ such that $\mathscr{A}$ and F_i are disjoint. Then adding $\mathscr{A}$ to the updated FNF.

$$FNF' = FNF \cup \{\mathscr{A}\} - \{\mathscr{B}\}$$

where $\mathscr{B} \in FNF$, $\mathscr{A}$ and $\mathscr{B}$ are disjoint.

6. Consider every disjoint sub-graph as an individual graph.
   (a) If any IFBL is formed, then a sub-feature is generated. The face set of the sub-feature is a face set of this sub-graph subtracted from the articulation node.
   (b) If no sub-FBLs exist, then the feature is a major feature in this level. The feature face set is the product of the union of the sub-graph and the articulation face.
7. Find all the faces which have CFLLs and extract all the connected faces as a sub-feature of the feature level in which the face belongs.
8. For each feature, find the feature type for the feature itself and its sub-features. Meanwhile, construct the feature graph for this feature.

An example of this shown in Figure 2.18(a) and illustrates how this strategy can be applied. The first level feature faces can be found by traversing the first level FBL which yields

$$FFG_1 = \{f_5, f_6, f_7, f_8, f_9, f_{10}, f_{11}, f_{12}, f_{13}\}$$
$$\text{with } FNF_1 = \{f_1, f_2, f_3, f_4\}$$

Let the node FFG_1 be the FNF; a sub-FBL can then be found

$$FFG_{1,1} = \{f_{14}, f_{15}, f_{16}\}$$

From FFG_1 f_6 and f_{10} are co-surfaces, then the faces in between should be extracted as a sub-feature.

$$FFG_{1,2} = \{f_7, f_8, f_9\}$$

After extracting this sub-feature, FFG_1 should be updated to form FFG_2 by removing $FFG_{1,2}$ from FFG_1:

$$FFG_2 = \{f_5, \{f_6, f_{10}\}, f_{11}, f_{12}, f_{13}\}$$

Now, an articulation node, f_{10}, is found from FFG_1 by splitting f_{10} from FFG_1 to yield two disjoint sub-graphs. These are defined as FFG_3 and FFG_4, respectively

$$FFG_3 = \{f_5, \{f_6, f_{10}\}\}, \quad FFG_4 = \{f_{10}, f_{11}, f_{12}, f_{13}\}$$

Based on FNF the feature boundary loop (FBL) of FFG_3 can be found:

$$FBL_{FFG_3} = \{e_1, e_2, \{e_3, e_7\}, e_{26}, e_{27}, e_{12}, \{e_{13}, e_{17}\}, e_{18}\}$$

The feature boundary edges for FFG_4 cannot be found based on the current FNF. Therefore, the articulation node, f_{10}, needs to be removed from FFG_4 and acts as a feature neighboring face

$$FNF' = FNF \cup \{f_{10}\} = \{f_1, f_2, f_3, f_{10}, f_{10}\}$$

Based on FNF', for $FFG_4 - \{f_{10}\}$ there is a sub-FBL formed

$$FBL_{FFG_4} = \{e_8, e_9, e_11, e_27, e_26\}$$

Therefore, a sub-feature $FFG_{1,3} = \{f_{11}, f_{12}, f_{13}\}$ is extracted from FFG_4. For FFG_3; there is no sub-FBL formed under FNF', so the main feature can be found as:

$$FFG_{1,0} = \{f_5, f_6, f_{10}\}$$

The next step is to find the major features as well as the sub-features and build a hierarchical feature graph for this combination feature. By matching the generic feature in the database, the results are shown as follows:

$FFG_{1,0}$: step,

$FFG_{1,1}$: wedge-rib,

$FFG_{1,2}$: groove,

$FFG_{1,3}$: corner-step.

The feature graph of this feature can be built as shown in Figure 2.18(b).

In this chapter, we have shown the role of the computer-aided design system in providing necessary information for an automatic machine programming system. It is

Table 2.1 Notation of symbols

OBJ_i	an object
v_i	the vertex set of Obj_i
$v_{i,j}$	the vertex j in v_i
$\mathcal{E}_i$	the edge set of Obj_i
$e_{i,j}$	the edge j in ε_i
$\mathcal{F}_i$	the face set of Obj_i
$f_{i,j}$	the face j in $\mathcal{F}_i$
Ev_i	the set of envelope's faces of Obj_i
$FE_{i,j}$	the set of faces constructing feature j
$f_{i,j,k}$	the kth face in $F_{i,j}$
$FBL_{i,j}$	the feature bounding loop of feature j
$FNF_{i,j}$	the neighboring face set of feature j

obvious that for manufacturing operations such as assembly that the robotic system must possess the product data, and that this data should necessarily provide feature information. The reason for concentrating on features should be obvious. Assembly of components implies mating of features thus reducing the assembly reasoning on the level of mating to feature reasoning. In addition, understanding the configuration of an object permits the possibility of reasoning about other aspects of the geometry, for instance grasping certain faces at points that will provide for stable grasping and manipulation.

In the next chapter, we will explore how to relate mating features and interacting bodies using spatial relationship concepts.

CHAPTER 3

Spatial relationships

3.1 INTRODUCTION

In order to link design and manufacturing processes automatically, a product's specifications must be carried from the beginning of design to the final stage of inspection. For example, Figure 3.1(a) shows a product with two parts, one is a plate P, with two holes, and the other one is a bent shaft, S. Figure 3.1(b) shows the features of the two parts in Figure 3.1(a). The features of part P are extracted and classified as a *Cvex_block_base*, *C_hole*$_1$, *C_hole*$_2$, *Fillet*$_1$, and *Fillet*$_2$. We have *C_shaft*$_1$, *C_shaft*$_2$, and *Elbow* features for part S. In Figure 3.1(c) and (d), the product is made by snapping *C_shaft*$_1$ of S into the *C_hole*$_1$ of P. Traditionally, these mating criteria are annotated in engineering drawings but this requires trained humans to create and to interpret the notes. In order to represent these criteria and other aspects of a designer's intention in a computer, two approaches have been proposed. Features may be extracted from the geometric information associated with parts modeled on a CAD system [NL90a, NK90]. One then specifies the design specifications (tolerances, surface finish, etc.) for features. The other alternative is to use a CAD system that is *feature-based* [Dix88]. In that case, the system would have a feature database. Designers choose the appropriate features and decide upon feature attributes based on design specifications. One then models the parts by combining these features together. These kinds of schemes have been explained in Chapter 2.

Spatial relationships were proposed by Popplestone *et al.* [AP75, PAB80b, PAB80a, Pop87] in 1975 to express the relative positions of parts in a product's final state by specifying spatial relationships between features. The types of spatial relationships described there include against, coplanar, fits, parax, lin, rot, and fix. A mathematical model is applied to transform a system of spatial relationships into equations which are solved to determine the location of components in a product. In our development here, we follow this idea with certain modifications and expansions. We define *design with spatial relationships* not only for inferring the assembly position of parts but also as a model to capture designers' intent. In our application, the design procedure is as follows [NL90b, LN91, BNRew]:

1. Develop product specifications.
2. Design the nominal form of components.
3. Assign product specifications and functions between assembly features of parts.
4. Distribute the product specifications among the parts automatically.
5. Perform mechanical analysis of each component.

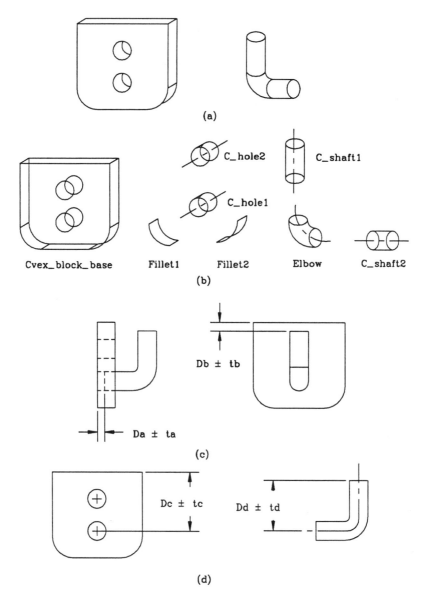

Fig. 3.1 The relationships between features and specifications.

Note that all aspects of this procedure are subject to iteration. In this product design process, part representation in the form of *boundary representations* is required. However, the models can be created with some other modeling techniques (such as CSG) provided that a conversion to boundary representation is possible. The parts are created and represented in a nominal form first, then the generic data is transferred to a neutral graphics format. Using an interactive process, designers assign product specifications and functions to features of parts.

3.2 BACKGROUND

Much current work in assembly planning focuses on the representation of all the possible sequences of an assembly. De Fazio and Whitney [FD87] generate all the possible assembly sequences from a liaison diagram through a question–answer process which establishes precedence constraints on the liaisons; this work is an improved version of Bourtjault [Bou84b] in terms of the number of questions being asked. The algorithm described in [FD87] asks $2l$ questions where l is the number of liaisons as opposed to the exponential procedure used in [Bou84b]. Homem de Mello and Sanderson [dM89] present a representation for assembly plans using AND/OR graphs derived by finding recursively the feasible subsets of an assembly (subassembly). Wolter [Wol88] proposed an implicit representation *constraint graph* from which one can derive the feasible assembly plans when necessary.

One must realize that the ordering of assembly operations depends strongly on the geometry of assembly components and the kinematic constraints imposed among these components. Fundamentally, the *spatial legality constraint* – no two bodies occupy the same volume of Euclidean space at the same time – has to be obeyed. In order to check legality one has to be able to compute the relative positions of assembly bodies, not only when the bodies are in their final configuration but also while they are being moved into those configurations.

3.2.1 Importance of a solid modeler in spatial reasoning

The geometric data of a part output from a solid modeler is non-variable. This body is considered as a rigid body with a fixed shape and size unless further attributes that deal with flexibility are attached. In the real world, both rigid and non-rigid parts play necessary roles in making a product. A close look at the non-rigid parts will reveal that for them to perform functions, boundary conditions at their attached ends have to be initialized, e.g. a spring with two ends hooked on fixed pins, a wire with its two ends plugged into sockets, a round belt which forms two or more arcs at its ends and which are mounted on pulleys, etc. Therefore, in our application, the non-rigid parts are modeled as a rigid body with its assembly end(s) fixed in the shapes as shown in an assembly, and the rest of them are flagged with certain properties of flexibility.

When modeling a part, some common features such as helical features which are widely used can be parametrically abstracted. However, these features are computationally expensive to represent. In this case, they are reduced into a simple form and the parametric properties are employed when the mating spatial relationship is assigned. The advantage of assigning detailed information in the assembly mode instead of the part modeling mode is that most of the mating features are formed in the same manner. For example, in the assembly of gears, screws and screw holes, all the features that mate have compatible forms. With this approach, redundant parameter assignment and human errors are reduced. Some additional examples of simplification of features are: (1) a spring might have its round wire represented as square and the winding part is unified to be a cylindrical shell, (2) a gear becoming a disk, (3) a rack becoming a polyhedral shaft, (4) a spring washer looks like a regular washer.

3.2.2 Spatial relationships developments

We will describe a fully product assembly modeling system (PAM) which embodies the attributes enunciated later in Chapter 6. The importance of PAM is that a designer can now directly specify intentions or functions for a product without writing notes and several process planning information which the planner would normally expend large amounts of time deriving. They can be captured automatically and propagated using PAM.

The RAPT system developed by Popplestone, Ambler and Bellos [AP75, ACP83] infers the positions of bodies from specified symbolic spatial relationships between features of bodies. The original implementation of RAPT mapped these relationships into a set of algebraic equations via constraint propagation. Thus, the final task became that of simplifying and solving symbolically a set of algebraic equations – a time-consuming process. The slowness of the first system led to the development of a second one which dealt with constraints on the degrees of freedom of the relative motions using methodologies for chaining and merging constraints, involving a system of rewrite rules and a table of look-up procedure. The main shortcoming of the second system was the non-closedness of the set of constraints on the degrees of freedom considered under the composition operation whereby not all combinations of constraints can be tabulated.

Hervé [Her78] showed how to apply the theory of continuous groups to the kinematic analysis of mechanisms. Popplestone [Pop84] showed how group theory could be used in the treatment of the systems of spatial relations occurring in RAPT. Thomas and Torras [TT88a] implemented in software the use of Hervé's group-theoretic approach to treat such systems of spatial relationships, using symbolic methods.

In more recent research, a geometrical representation of symmetry groups was developed that allows us to compute relative positions of bodies in assemblies without requiring symbolic computation [Liu90]. They classified spatial relationships into TR subgroups of the Euclidean group in terms of a coset, with their intersections resulting in either a coset or null of a symmetry group. Thus the conceptual elegance of group theory has been given a computationally tractable realization.

However, in their work, the spatial relationships adopted are lower pair couplings, i.e. two bodies are in contact along a surface (plane or curved). Thus the mating features have a common symmetry group. Lower pair couplings are crucial for the analysis of kinematic motion of a mechanism, but, for product modeling or assembly, the types of spatial relationships required are beyond this scope. For instance, a cylinder which is placed on top of a planar surface. The possible motions which allow the two bodies to remain in the same spatial relationship, the type of planar-cylindrical surfaces contact in this case, include moving along the planar surface and rotating about the centerline of the cylinder, simultaneously sometimes. There will be no common symmetry group when the mating body pair does not have a surface contact.

In the next two sections, a brief overview of spatial relationships in [AP75, PAB80b, PAB80a] is given.

3.2.2.1 Mathematical representation

The post-multiplied transformation matrices used as tools to describe the spatial relationships are shown as follows:

- Translating a distance of (a, b, c):

$$trans(a, b, c) = \begin{pmatrix} 1 & 0 & 0 & 0 \\ 0 & 1 & 0 & 0 \\ 0 & 0 & 1 & 0 \\ a & b & c & 1 \end{pmatrix}$$

- Rotating about the X-axis by θ:

$$twix(\theta) = \begin{pmatrix} 1 & 0 & 0 & 0 \\ 0 & \cos(\theta) & \sin(\theta) & 0 \\ 0 & -\sin(\theta) & \cos(\theta) & 0 \\ 0 & 0 & 0 & 1 \end{pmatrix}$$

- Rotating about the Y-axis which brings the positive X-axis to the negative X-axis:

$$M = \begin{pmatrix} -1 & 0 & 0 & 0 \\ 0 & 1 & 0 & 0 \\ 0 & 0 & -1 & 0 \\ 0 & 0 & 0 & 1 \end{pmatrix}$$

- The standard or world axes:

$$W = \begin{pmatrix} X \\ Y \\ Z \\ O \end{pmatrix} = \begin{pmatrix} 1 & 0 & 0 & 1 \\ 0 & 1 & 0 & 1 \\ 0 & 0 & 1 & 1 \\ 0 & 0 & 0 & 1 \end{pmatrix}$$

If W_1 is a set of axes and p is a position, then W_{1p} is the set of axes W_1 transformed by p.

In their early work, feature is defined as either a plane face or a cylindrical shaft or a hole. Each body B has a set of features, *featsof*(B), associated with it. Each feature F belongs to only one body, *bodyof*(F), and has a set of axes embedded in it. The position of a feature of a body is defined to be that position which will transform the body axes into the feature axes. There is a function *posfeat* from features to positions. If F is a face, then the axes represented by Wf have their origin lying in F and the X-axis of Wf is normal to F, and is pointing out from it. If F is a shaft, then Wf has its origin at the

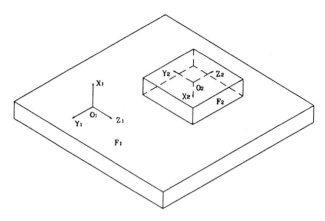

Fig. 3.2 Face F_1 against face F_2.

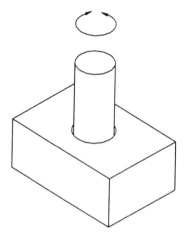

Fig. 3.3 Shaft in hole, with one degree of freedom.

tip of the shaft and the X-axis of Wf points outwards along the axis of the shaft. If F is a hole, then Wf points outwards along the axis of the hole. The features of a body will be changed when the body is changed in location. Let S be some world state, and let $p = posbod(B, S)$. Then the axes of the feature F in the state S are represented by Wfp.

3.2.2.2 Expressions for positions of bodies in terms of relations between features

If B_1 and B_2 are two bodies with faces F_1 and F_2. Let $posfeat\,(F_1) = f_1$ and $posfeat\,(F_2) = f_2$. Suppose S is some world state in which F_1 is against F_2. Let the bodies have positions p_1 and p_2 in S. The face axes O_1X_1, O_1Y_1, O_1Z_1 and O_2X_2, O_2Y_2, O_2Z_2 are represented by Wf_1p_1 and Wf_2p_2 (see Figure 3.2). There is a relationship between these sets of axes, namely that the X-axes of each set are in opposite directions and the origin of each lies in the YZ plane of the other. Let O_1 be a new origin of coordinates with the new axes which are O_1X_1, O_1Y_1, and O_1Z_1. In order to do this we have to post-multiply all the vectors and axis-matrices by $p_1^{-1}f_1^{-1}$, so that while W will be the new axis-matrix for F_1, the new axis-matrix for F_2 is $Wf_2p_2p_1^{-1}f_1^{-1}$. The relation between the axes of F_1 and F_2 can be expressed as

$$Wf_2p_2p_1^{-1}f_1^{-1} = W\,Mtwix(\theta)trans(0, y, z) \qquad (3.1)$$

where y, z and θ express the displacement and orientation of the F_2 axes with respect to the F_1 axes, and M turns the X-axis through π radians. Since all the matrices are non-singular, equation (3.1) can be rewritten as an equation for p_2 in the form

$$p_2 = f_2^{-1}Mtwix(\theta)trans(0, y, z)f_1p_1 \qquad (3.2)$$

An example is taken from [AP75] which illustrates the formulations and solutions in their work.

Example:

Given a fixed block of height 20, with *posbod I* and with a hole of depth 8 drilled into its top surface at $(50, 50, 20)$, i.e.

$$posfeat(\text{hole feature}) = XTOZtrans(50, 50, 20),$$
$$posfeat(\text{face at bottom of hole}) = XTOZtrans(0, 0, 12)$$

put a shaft into the hole so that the shaft ($posfeat(\text{shaft}) = M$) *fits* the hole, and the end face of the post is *against* the bottom of the hole (see Figure 3.3). Equating the position of the shaft derived through the *fits* relation to the fixed world

$$(M^{-1}Mtwix(\theta_1)trans(x_1, 0, 0)XTOZtrans(50, 50, 20)I)$$

with the position derived through the *against* relation

$$(M^{-1}Mtwix(\theta_2)trans(0, y_1, z_1)XTOZtrans(0, 0, 12)I)$$

produces the equation

$$twix(\theta_1)trans(x_1, 0, 0)XTOZtrans(50, 50, 20)$$
$$= twix(\theta_2)trans(0, y_1, z_1)XTOZtrans(0, 0, 12)$$

with

$$G(\text{shaft}) = twix(\theta_2)trans(0, y_1, z_1)XTOZtrans(0, 0, 12)$$

Solving the rotation equations produces the real equation

$$\theta_2 - \theta_1 = 0$$

Now substituting θ_2 for θ_1 in the equation, and solving the translation equation gives

$$x_1 = -8, y_1 = 50, z_1 = 50$$

and $G(\text{shaft})$ becomes

$$twix(\theta_2)trans(0, 50, 50)XTOZtrans(0, 0, 12)$$

i.e. the shaft fitted into the hole has only one degree of freedom-rotation about its own axis.

3.2.3 Features and spatial relationships

An *assembly* is a collection of bodies which bear some spatial relationship to each other. Two related bodies in an assembly do not make contact over their whole surface; instead *features* of each body are in contact. We may consider a feature as being some part of the surface of a body. Thus for example, a *journal* may be a feature of a shaft, and an *interior surface* may be a feature of a bearing, and we may say that the journal *fits* the interior surface. Similarly, we may say that the teeth of gears *mesh* or that the threaded portion of the shank of a bolt *fits* a hole tapped in some body.

We have defined in Chapter 2 that a feature is: *a set of surfaces together with specifications of the bounding relationships between them and which imply an engineering function of an object* [Nna88]. We note that there are other definitions for features including [Dix88, Sha88, CIas], etc. But this definition embodies the various definitions and at the same time is computationally tractable. Features are recognized as being application dependent, such as for machining, assembly, grasping in robotic assembly, etc. From this definition of a feature, it can be seen that surfaces are the basic primitives from which a feature is composed. But engineering specifications, such as tolerance and surface finish, are referenced to a single planar or cylindrical surface. So, in assigning spatial relationships, we consider features including planar faces and any features with principal axes, e.g. cylinder and sphere (either patches or curvilinear surfaces in *boundary representation* are applicable). This simplifies the feature extraction problem since these features are embedded implicitly in a part's geometric representation.

One can determine a feature's functionality from its shape or vice-versa [NL90a]. Strategies developed from these principles, in design as well as in assembly planning, have great power for reasoning about an assembly configuration from local information. For example, one typical rule of thumb is to mate a concave feature to a convex one: a rod is inserted into a hole or a block is put on the step of the 'L' part, for instance. *Feasible approach directions* are derived based on such concave mating features. Mating features can be derived from contacting surfaces of assembled components.

3.3 SPATIAL RELATIONSHIPS

In Popplestone *et al.*'s works, spatial relationships are focused on the configuration of a product. From his definition of spatial relationships such as, against, coplanar, fits, parax, lin, rot, and fix, comes the assumption that all the parts are located by assigning features that relate to some basis (a surface or a line); for instance, (1) against is two faces physically touching one another, (2) coplanar and parax deals with the faces not having physically contacted but lying on the same plane and (3) fit is two features contacting in a certain portion of faces. Lin, rot, and fix are derived from the combination of against, parallel, parax, and fit. There is no relationship that can handle the assembly specifications such as that shown in Figure 3.1(c), to insert a bent shaft to a hole with two planar faces away in the distance of D_a. In this section, we introduced a revised form of spatial relationships which can be generally applied to assembly and capable of accepting the design specifications which can be used by a robot in automatically planning and assembly.

3.3.1 Assembly locations

The representation of an assembly in some CAD systems uses the transformation matrix relating the parts coordinate frame to an assembly position. This requires that the designers know the detailed dimensions associated with each part in order to perform a precise transformation. The efficiency of defining an assembly depends mainly on the

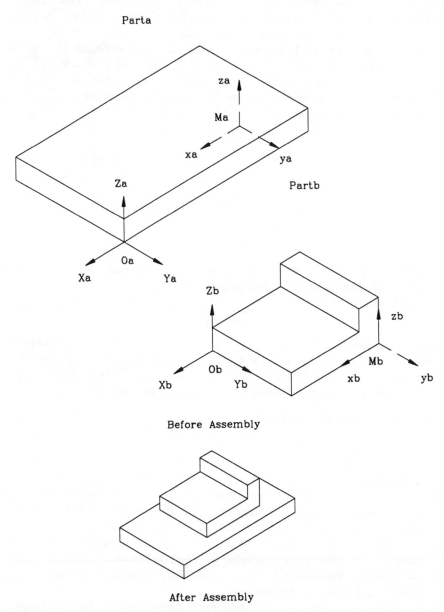

Fig. 3.4 Assembly with mating frames.

designer's knowledge about the parts' geometrical dimensions. This requires a great deal of memorization and calculation, especially in 3D modeling where the dimensions are difficult to label for visual comprehension.

In this chapter, we express an assembly in terms of mating frames. In each part or semi-assembly, a mating frame is attached on the body derived from the spatial relationships assigned by the designer. The mating frame is in the form of a homogenous transformation matrix with respect to the 3D model origin of parts or semi-assembly. In Figure 3.4, which shows the assembly relations between $PART_a$ and $PART_b$, the origin coordinate frame of $PART_a$ is O_a and the mating frame is M_a, also O_b and M_b are the origin and mating coordinate frames for $PART_b$. The original coordinate frame is automatically created by the CAD modeler which may lie inside or outside the body, and all the vertices are located relative to the origin. In order to assemble $PART_a$ with $PART_b$ one has to match the mating frame M_a exactly with M_b. If the transformation matrix for $PART_b$ with respect to $PART_a$ is $^{O_a}T_{O_b}$, them

$$^{O_a}T_{O_b} = {}^{O_a}T_{M_a} \cdot {}^{O_b}T_{M_b}^{-1}$$

In the following sections, we describe the possibility of relative motion using degree of freedom representations defined as a geometric formalism, and the type of spatial relationships. The resultant degree of freedom thus found through intersecting the assigned spatial relationships and mating frames are decided upon subsequently.

3.3.2 Types of spatial relationship

Six types of spatial relationships for assembly are shown below. Each of them is explained in terms of physical configuration and applications.

Against

Against relationship means that the faces touch at some point. Against relationship is the most basic spatial relationship and applies to any parts assembly. Any combintion of two features can possess this property. We provide examples of nine different types of against (Figure 3.5), which are widely used inside for assembly as follows:

1. Two planar faces are against one another only if the direction of the normals are in opposite directions and they touch over an area. This relationship holds true in numerous occasions where parts have planar faces.
2. A cylindrical feature touches a planar face along a line. Application examples: gear meshes with a rack, a roller contacts with a plane, a shaft fixed at a plane etc.
3. A spherical feature touches a planar feature at a point. Application examples: one end of a fixture holding a part, a ball fitting into the planar site of a shell, locating the probe of a coordinate inspection machine over a plane, etc.
4. Two cylindrical features touch over a point. Application examples: locating the shafts before welding or the positioning rollers inside a bearing.
5. Two spherical features touch over a point. Application examples: ball bearing assembly and coordinate measurement.
6. A spherical feature touches a cylindrical feature over a point. Application examples: coordinate measurement, bearings.

Spatial relationships

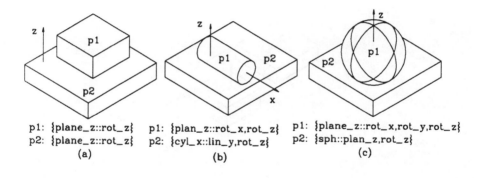

p1: {plane_z::rot_z}   p1: {plan_z::rot_x,rot_z}   p1: {plane_z::rot_x,rot_y,rot_z}
p2: {plane_z::rot_z}   p2: {cyl_x::lin_y,rot_z}   p2: {sph::plan_z,rot_z}
(a)   (b)   (c)

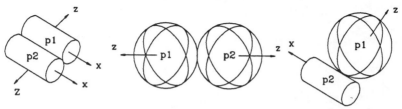

p1: {cyl_x::rot_x,rot_z}   p1: {sph::rot_x,rot_y,rot_z}   p1: {cyl_x::rot_x,rot_y,rot_z}
p2: {cyl_x::rot_x,rot_z}   p2: {sph::rot_x,rot_y,rot_z}   p2: {sph::lin_x,rot_x,rot_z}
(d)   (e)   (f)

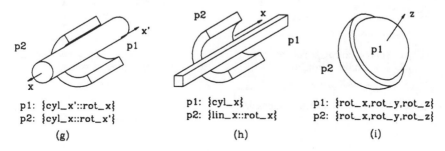

p1: {cyl_x'::rot_x}   p1: {cyl_x}   p1: {rot_x,rot_y,rot_z}
p2: {cyl_x::rot_x'}   p2: {lin_x::rot_x}   p2: {rot_x,rot_y,rot_z}
(g)   (h)   (i)

Fig. 3.5 Against spatial relationships.

7. A cylindrical feature touches a concave cylindrical feature over a line. Application examples: off-center rotation assembly, bearing, and hooking a spring end on a shaft.
8. A planar face touches a concave cylindrical feature over two lines. Application examples: hooking a spring end over a square shaft.
9. A spherical feature touches a concave spherical feature over a sphere contact area when these two spheres are of the same size. Application example: ball joint, bearing assembly, and coordinate measurement.

52   *Spatial relationships*

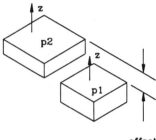

p1: {plane_z::rot_z}
p2: {plane_z::rot_z}
(a)

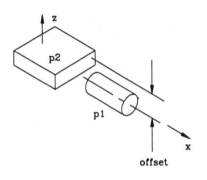 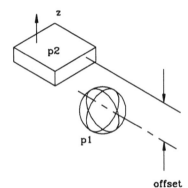

p1: {plane_z::rot_x,rot_z}   p1: {plane_z::rot_x,rot_y,rot_z}
p2: {cyl_x::lin_y,rot_z}   p2: {sph::plan_z,rot_z}
(b)   (c)

Fig. 3.6   Parallel-offset spatial relationships.

Parallel-offset

In Figure 3.6, the parallel relation holds between planar faces, cylindrical and spherical features. In two parallel planar faces, the outward normals are pointing in the same direction. This relationship exists without the physical contact of two features with an offset distance. Application examples: locating a welding part and indicating the depth of insertion operation.

Parax-offset

In Figure 3.7(a), this relationship is similar to *parallel*, but the outward normals of the parallel planar faces are in opposite directions.

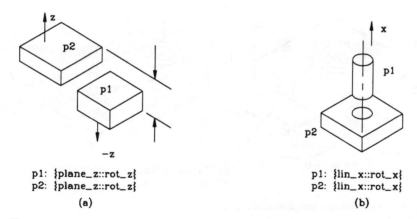

Fig. 3.7 Spatial relationships: (a) parax-offset, and (b) aligned.

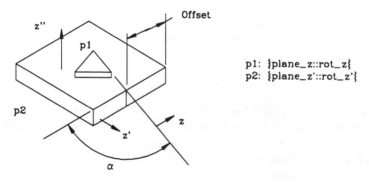

Fig. 3.8 Incline-offset spatial relationships.

Aligned

Two features are *aligned* if their centerlines are collinear. In Figure 3.7(b), the centerlines of cylindrical or spherical features are collinear. Application examples: insertion and any assembly requiring an alignment with cylindrical shafts or holes.

Incline-offset

The *inclination* relation holds for an angle between two planar faces. The *offset* describes the distance from a planar face of a part to the intersection line of two faces which makes the included angle (Figure 3.8). The intersection line is always parallel to the offset face. The include angle is rotated along the intersection line and the direction of the rotation is clockwise based on the direction of this line which is chosen from a face normal by picking a face. Application examples: welding, jigs and fixtures.

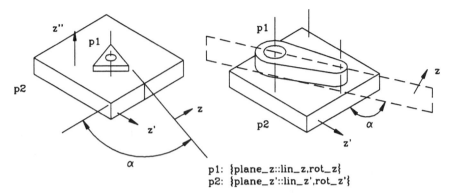

Fig. 3.9 Include-angle spatial relationships.

Include-angle

Include-angle, Figure 3.9, is similar to *incline-offset*. To have an include angle between two planar faces in their positive normal direction. The rotation is clockwise with respect to a normal of a picking face. The rotational axis has to be parallel to the normals of above two planar faces. This relationship is mainly designed for the assemblies having rotational degrees of freedom. Therefore, designers can assign a temporary or a fixed orientation to them when the assembly is taking place.

3.3.3 Degrees of freedom

In kinematics, a free rigid body has six degrees of freedom [Hun78]. To establish these six degrees of freedoms, if a point O can be marked in a rigid body, then a directed line Ox is pointed out of the body. An xy-plane can then be chosen through O and containing Ox. These three geometrical elements, the point O, the line Ox, and the xy-plane, uniquely determine a right-hand xyz-system of orthogonal axes in the body. Now, take another system, XYZ, as a fixed reference; in it, the point O is located at (X_O, Y_O, Z_O). The line Ox cuts the XY-plane in a point, say (X_l, Y_l). The normal to the xy-plane makes an angle v with the Z-axis. If the body is placed in an arbitrary position relative to the reference system, the five distances X_O, Y_O, Z_O, X_l, Y_l, and the angle v can then all be measured. In general the body cannot move in any manner without altering at least one of these six quantities. Therefore, we can consider these six elements as coordinates which locate the body. In the above example, the six degrees of freedom are five translations and one rotation. Alternatively, one could have chosen two direction-angles for the line O_x; or, in some other system of coordinates for the point O, other three lengths X_O, Y_O, Z_O could be replaced by two lengths and one angle, or by one length and two angles. For whatever system is used, no fewer than six coordinates are required to locate the body; if there are more than six, say $6 + m$, then it is always true that there are m identities which inter-relate the six coordinates leaving only six independent coordinates. Therefore, the six degrees of freedom are not necessarily 3 translations and 3 rotations; one may have a wide choice to define them.

A spatial relationship can be interpreted as a constraint imposed on the degrees of freedom between mating or interacting features. In every mating feature, a feature-coordinate frame is attached (see Figure 3.5(a), (b) and (c)). A virtual-coordinate frame is derived to represent the coordinate frame needed for expressing degrees of freedom. The axes of the virtual-coordinate frame are a selection of the axes of a feature-coordinate frame of the mating features. The origin of a virtual-coordinate frame cannot be determined and is not necessary at this point. This is due to the fact that the fixed position of the mating parts is not determined. In mating of two planar features, the virtual-coordinate frame can be either one of the feature-coordinate frames (Figure 3.5(a)). In mating with a cylindrical feature, the X-axis of the virtual-coordinate frame is the same as the Z-axis of the feature-coordinate frame, the Z-axis of the virtual-coordinate frame has its direction defined as the vector starting from a point in the contact line and pointing towards its features center or center line (Figure 3.5(b)), and the X-axis of the virtual-coordinate frame can be either the X- or Y-axis of the feature-coordinate frame. In mating with a spherical feature, the Z-axis of the virtual-coordinate frame has its direction defined as a vector starting from a contact point and pointing towards its feature center. The X- and Y-axes are selected from any two axes of the feature-coordinate frame (Figure 3.5(c)). We classify the types of degrees of freedom based on virtual-coordinate frames as follows:

1. *lin_n*: linear translation along *n*-axis, where *n* contains a fixed point and a vector;
2. *rot_n*: rotation about *n*-axis, where *n* contains a fixed point and a vector;
3. *cir_n*: translating along a circle with *n*-axis, where the fixed point of *n* is the center of the circle and the vector of *n* is perpendicular to the circle;
4. *plane_n, cyl_n, sph*: translating along a planar, cylindrical, spherical surface etc. For a planar surface, a fixed point of *n* can be any point and the vector of *n* is the normal vector of the face; for a cylindrical surface, a fixed point of *n* can be any point on the centerline of the cylindrical feature and the vector of *n* is in either direction of the centerline; and for a spherical surface, a fixed point of *n* is the center of the sphere and the vector of *n* can be in any direction. If $body_1$ is translating along a surface of $body_2$, the contact area for $body_1$ is constant and the orientation of these two bodies relative to the contact area remains the same. This implies that $body_1$ keeps its orientation when moving along $body_2$ and $body_2$ is fixed. Conversely, if $body_1$ is fixed, $body_2$ can translate and rotate itself to have every portion of that surface in contact with $body_1$; for example, a planar surface of $body_1$ against a cylindrical surface of $body_2$. In the case when $body_2$ is fixed, the translation path of $body_1$ is *cyl_n*; when $body_1$ is fixed $body_2$ may *rot_x* and *trans_x* to achieve the same result;
5. *roll-lin_n*: rolling along a corner. As shown in Figure 3.10, a line connecting two round features of a body may roll around the corner and maintain the two round features in contact with two faces. It is difficult to derive this type of motion with respect to any of these two parts, so, a virtual line is adopted. The line function describes one of the degrees of freedom.

The degrees of freedom of a part can now be expressed as {degrees of freedom moving along the feature of the relative mating part :: degrees of freedom moving itself}, where the relative mating part is fixed. For example, in Figure 3.5(b), part *p1* is against with part *p2* over a line. The degree of freedom for *p1* with respect to *p2* describes the relative motion which maintains a line contact. *p1* may move along the mating planar face of *p2* in the path of keeping the contact line in parallel. This can be

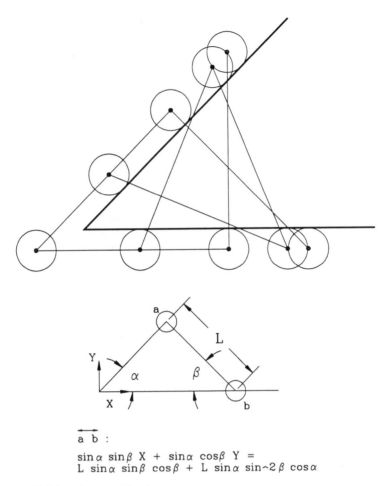

$\overrightarrow{a\ b}$:

$\sin\alpha \sin\beta\ X + \sin\alpha \cos\beta\ Y =$
$L \sin\alpha \sin\beta \cos\beta + L \sin\alpha \sin^2\beta \cos\alpha$

Fig. 3.10 Roll-line degrees of freedom.

described as the equation for an infinite plane, $Ax + By + Cz = D$. For $p1$ itself, $p1$ can rotate about the z-axes which gives a new orientation about the contact line, and $p1$ can rotate about x-axes to generate a new contact line with respect to itself. So, the degree of freedom for $p1$ with respect to $p2$ is $\{plane\_z :: rot\_x, rot\_z\}$. For $p2$ with respect to $p1$, $p2$ can translate along the cylindrical surface of $p1$, and $p2$ can move itself along the planar face and rotate about z-axis. The degree of freedom for $p2$ is $\{cyl\_x :: lin\_y, rot\_z\}$.

3.3.4 Intersection of degrees of freedom

A rule-based system can be developed for assisting the selection of an appropriate set of spatial relationships and for inferring the final state of degrees of freedom. From the previous section, surface is a type of degree of freedom. For a $plane\_z_a$ degree of freedom, a body may move on a planar surface along two lin. When a new $plane\_z_b$ is introduced, the remaining degree of freedom is derived by intersecting these two $planes$. For example,

Spatial relationships

if $plane\_z_a$ is parallel to $plane\_z_b$, then the degree of freedom may be either $plane\_z_a$ or $plane\_z_b$; if they are not parallel, then lin_n is the degree of freedom obtained from the intersection of $plane\_z_a$ and $plane\_z_b$. The intersection of surface degrees of freedom are available only when a *plane* is involved, such as, $circle\_n$ is the result of intersecting $plane\_z_a$ with cyl_n (or sph_n) together. In the intersection of two rotational degrees of freedom, say rot_z_a and rot_z_b, if they share the same rotational axis then rot_z_a or rot_z_b remains, if not, they cancel each other. In the above examples, we demonstrate some simple rules for the reduction of degrees of freedom, but in some cases, such as combining spatial relationships with the numbers of rotational degrees of freedom, the solution becomes complicated. Fortunately, the number of spatial relationships is fixed, so the number of rules is finite. We show some general reduction rules as follows:

1. $\{\{plane\_z_a :: rot\_z_a\} \cap \{plane\_z_b :: rot\_z_b\} \parallel vec\_of(z_a) \cdot vec\_of(z_b)| = 1\}$
   $= \{plane\_z_a :: rot\_z_a\}$ or $\{plane\_z_b :: rot\_z_b\}$
2. $\{\{plane\_z_a :: rot\_z_a\} \cap \{plane\_z_b :: rot\_z_b\} \parallel vec\_of(z_a) \cdot vec\_of(z_b)| \neq 1\}$
   $= \{lin\_n_c | lin\_n_c = intersect(plane\_z_a, plane\_z_b)\}$
3. $\{\{plane\_z_a :: rot\_z_a\} \cap \{lin\_n_b\} \parallel vec\_of(z_a) \cdot vec\_of(n_b)| = 1\}$
   $= \{lin\_n_b\}$
4. $\{\{plane\_z_a :: rot\_z_a\} \cap \{lin\_v_b\} \parallel vec\_of(z_a) \cdot vec\_of(z_a)| \neq 1\}$
   $= \{fix_c\}$
5. $\{\{plane\_z_a :: rot\_x_a, rot\_z_a\} \cap \{plane\_z_b :: rot\_x_b, rot\_z_b\} \mid$
   $x_a = x_b, |vec\_of(z_a) \cdot vec\_of(z_b)| = 1\}$
   $= \{plane\_z_a :: rot\_x_a, rot\_z_a\}$ or $\{plane\_z_b :: rot\_x_b, rot\_z_b\}$
6. $\{\{plane\_z_a :: rot\_x_a, rot\_z_a\} \cap \{plane\_z_b :: rot\_x_b, rot\_z_b\} \mid$
   $x_a = x_b, |vec\_of(z_a) \cdot vec\_of(z_b)| \neq 1\}$
   $= \{lin\_x_c :: rot\_x_c | lin\_x_c = cent\_of(x_a) = cent\_of(x_b),$
   $rot\_x_c = rot\_x_a = rot\_x_b\}$
7. $\{\{plane\_z_a :: rot\_z_a\} \cap \{lin\_x_b :: rot\_x_b\} \parallel vec\_of(z_a) \cdot vec\_of(z_b)| < 1\}$
   $= \parallel \{fix_c :: rot\_x_c | rot\_x_c = rot\_x_b\}$
8. $\{\{plane\_z_a :: lin\_z_a, rot\_z_a\} \cap \{plane\_z_b :: lin\_z_b, rot\_z_b\} \mid$
   $|vec\_of(z_a) \cdot vec\_of(z_b)| = 1\}$
   $= \{plane\_z_a :: lin\_z_a, rot\_z_a\}$ or $\{plane\_z_b :: lin\_z_b, rot\_z_b\}$
9. $\{\{plane\_z_a :: lin\_z_a, rot\_z_a\} \cap \{plane\_z_b :: lin\_z_b, rot\_z_b\} \mid$
   $|vec\_of(z_a) \cdot vec\_of(z_b)| < 1\}$
   $= \{lin\_n_c, plane\_z_c | lin\_n_c = intersect(plane\_z_a, plane\_z_b)$
   $plane\_z_c = plane\_of(lin\_z_a, lin\_z_b)\},$
10. $\{\{plane\_z_a :: rot\_x_a, rot\_y_a, rot\_z_a\} \cap \{plane\_z_b :: rot\_x_b, rot\_y_b, rot\_z_b\} \mid$
    $cent\_of(rot\_x_a, rot\_y_b, rot\_z_c) = cent\_of(rot\_x_b, rot\_y_b, rot\_z_b),$
    $|vec\_of(z_a) \cdot vec\_of(z_b)| < 1\}$
    $= \{lin\_v_c :: rot\_x_c, rot\_y_c, rot\_z_c |$
    $lin\_v_c = intersect(plane\_z_a, plane\_z_b), rot\_x_c = rot\_x_a | \text{ or } |rot\_x_b,$
    $rot\_y_c = rot\_y_a | \text{ or } |rot\_y_b, rot\_z_c = rot\_z_a | \text{ or } |rot\_z_b\},$
11. $\{\{plane\_z_a :: rot\_x_a, rot\_y_a, rot\_z_a\} \cap \{plane\_z_b :: rot\_x_b, rot\_y_b, rot\_z_b\} \mid$
    $cent\_of(rot\_x_a, rot\_y_b, rot\_z_c) = cent\_of(rot\_x_b, rot\_y_b, rot\_z_b),$
    $|vec\_of(z_a) \cdot vec\_of(z_b)| = 1\}$
    $= \{plane\_z_a :: rot\_x_a, rot\_y_a, rot\_z_a\} | \text{or} | \{plane\_z_b :: rot\_x_b, rot\_y_b, rot\_z_b\}$
12. $\{\{plane\_z_a :: rot\_x_a, rot\_y_a, rot\_z_a\} \cap \{plane\_z_b :: rot\_x_b, rot\_y_b, rot\_z_b\} \mid$

$cent\_of(x_a, y_z, z_a) \neq cent\_of(x_b, y_b, z_b), vec\_of(z_a) \cdot vec\_of(z_b) = -1\}$
$= \{plane\_z_c :: rot\_x_c | plane\_z_c = plane\_z_a$ or
$plane\_z_b, x_c = axis(cent\_of(x_a, y_a, z_a)cent\_of(x_b, y_b, z_b))\}$

13. $\{\{plane\_z_a :: rot\_x_a, rot\_y_a, rot\_z_a\} \cap \{plane\_z_b :: rot\_x_b, rot\_y_b, rot\_z_b\} |$
$cent\_of(x_a, y_a, z_a) \neq cent\_of(x_b, y_b, z_b),$
$vec\_of(z_a) = vec\_of(z_b), \| vec\_of(rot\_z_a) \times vec\_of(rot\_z_b) \| = 0\}$
$= \{plane\_z_c :: rot\_z_c, rot\_x_c | plane\_z_c = plane\_z_a$ or $plane\_z_b,$
$rot\_z_c = rot\_z_a$ or $rot\_z_b, x_c = axis(cent\_of(rot\_x_a)cent\_of(rot\_x_b))\}$

14. $\{\{plane\_z_a :: rot\_x_a, rot\_y_a, rot\_z_a\} \cap \{plane\_z_b :: rot\_x_b, rot\_y_b, rot\_z_b\} |$
$0 < |vec\_of(z_a) \cdot vec\_of(z_b)| < 1\}$
$= \{lin\_n1_c :: rot\_n2_c, rot\_n3_c, roll\_line_c |$
$point\_of(n2_c) = point\_of(x_a, y_a, z_a), vec\_of(n2_c) = vec\_of(z_b),$
$point\_of(n3_c) = point\_of(x_b, y_b, z_b), vec\_of(n3_c) = vec\_of(z_a),$
$roll\_line_c = roll\ cent\_of(x_a, y_a, z_a), cent\_of(x_b, y_b, z_b)\}$

15. $\{\{plane\_z_a :: rot\_x_a, rot\_z_a\} \cap \{plane\_z_b :: rot\_x_b, rot\_y_b, rot\_z_b\} |$
$vec\_of(z_a) = vec\_of(z_b), vec\_of(rot\_z_a) \times vec\_of(rot\_z_b) = 0\}$
$= \{plane\_z_c :: rot\_z_c, | plane\_z_c = plane\_z_a$ or $plane\_z_b, rot\_z_c = rot\_z_a$ or $rot\_z_b\}$

16. $\{\{plane\_z_a :: rot\_z_a\} \cap \{plane\_z_b :: rot\_x_b, rot\_y_b, rot\_z_b\} |$
$0 < |vec\_of(z_a) \cdot vec\_of(z_b)| < |1\}$
$= \{lin\_n1_c :: rot\_n2_c | lin\_n1_c = intersect(plane\_z_a, plane\_z_b),$
$point\_of(n2_c) = cent\_of(x_b, y_b, z_b), vec\_of(n2_c) = vec\_of(z_a)\}$

17. $\{\{lin\_n1_a :: rot\_n2_a\} \cap \{plane\_z_b :: rot\_x_b, rot\_y_b, rot\_z_b\} |$
$vec\_of(n2_a) \cdot vec\_of(z_b) = 0\}$
$= \{lin\_n1_c | lin\_n1_c = lin\_n1_a\}$

18. $\{\{lin\_n1_a\} \cap \{plane\_z_b :: rot\_x_b, rot\_y_b, rot\_z_b\} |$
$vec\_of(n1_a) \cdot vec\_of(z_b) = 0\}$
$= \{fix_c\}$

Notation:

- *Axis*: *Axis* is a translation or rotation axis which contains a start point and a vector.
- $point\_of(Axis) \Rightarrow$ Return start point of *Axis*.
- $vec\_of(Axis) \Rightarrow$ Return vector of *Axis*.
- $cent\_of(Axis\_x, Axis\text{-}y, Axis\text{-}z) \Rightarrow$ Return origin of the *Axis-x, Axis-y, Axis-z* coordinate frame. Also, this origin is a center point of a sphere.
- $axis(Point_a, Point_b) \Rightarrow$ Return an *Axis* with $Point_a$ as start point and $Point_b - Point_a$ as vector.
- $intersect(Dof_a, Dof_b) \Rightarrow$ Return intersection of two degrees of freedom.
- $roll(Point_a, Point_b) \Rightarrow$ Return a *roll_line* degree of freedom as described in (section 4.3).

Figure 3.11 and Figure 3.12 show two examples. In the first example, a product has four components, say $part_a, part_b, part_c$, and $part_d$. The spatial relationships between $part_a$ and $part_b$ are: (1) planar face f_1 *against* cylindrical face $c1$, (2) planar face f_2 *against* cylindrical face c_1, and (3) planar face f_4 *parallel* to planar face f_1 with *offset* 0. The degrees of freedom for them are: (1) $\{plane\_z_a :: rot\_x_a, rot\_z_a\}$, (2) $\{plane\_z_b :: rot\_x_b, rot\_z_b\}$, and (3) $\{plane\_z_c :: rot\_z_c\}$. In the first two sets of degrees of freedom, both x_a and x_b represent the centerline of the cylindrical face, and z_a is not parallel to z_b, so the intersection is

Spatial relationships

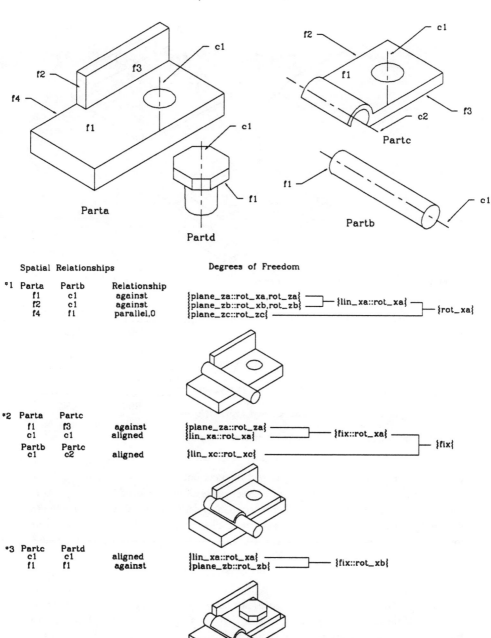

Fig. 3.11 Intersection of degrees of freedom, first example.

60   *Spatial relationships*

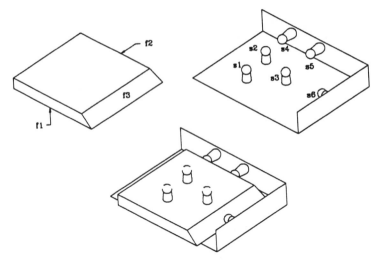

Spatial Relationships			Degrees of freedom		
Parta	Partb				
f1	s1	against	{plane_za::rot_xa,rot_ya,rot_za}	{plane_za::rot_xg,rot_za}	
f1	s2	against	{plane_zb::rot_xb,rot_yb,rot_zb}		
f1	s3	against	{plane_zc::rot_xc,rot_yc,rot_zc}	{plane_za::rot_za}	
f2	s4	against	{plane_zd::rot_xd,rot_yd,rot_zd}	{lin_n1::rot_n2}	
f2	s5	against	{plane_za::rot_xe,rot_ye,rot_ze}	{lin_n1}	
f3	s6	against	{plane_zf::rot_xf,roty_f,rot_zf}	{fix}	

Fig. 3.12 Intersection of degrees of freedom, second example.

$\{lin\_x_a :: rot\_x_a\}$ (rule 6). Next, we intersect the new one with $\{plane\_z_c :: rot\_z_c\}$. Since x_a is not parallel to z_c, the final state of degrees of freedom is reduced to $\{rot\_x_a\}$. After $part_a$ and $part_b$ are put together, another set of spatial relationships may be issued between $part_c$ and $\{part_a, part_b\}$. In the second example, a block is planned to be put on a fixture for machining or inspection operation. The spatial relationships are planar face f_1 *against* spherical faces s_1, s_2, s_3, s_4, s_5, and s_6. The set of degrees of freedom for each relationship is $\{plane_z :: rot_x, rot_y, rot_z\}$. From rule 13, the intersection of the first two degrees of freedom is $\{plane\_z_a :: rot\_x_g, rot\_z_a\}$ which is equal to a cylindrical face, by connecting two spheres (s_1 and s_2) together, *against* a planar face. For the second intersection, rule 15 is applied, we have $\{plane\_z_a :: rot\_z_a\}$ which is equal to *against* with two planar faces. After the last iteration, a $\{fix\}$ is reached which implies none of the degrees of freedom exist.

3.3.5 Inferring mating frames

A mating frame is a coordinate frame of a part relative to the origin coordinate frame of that part. Each component has its own mating frame and also for each subassembly. As described in section 4.1, to assemble two components is to match their mating frames

Spatial relationships

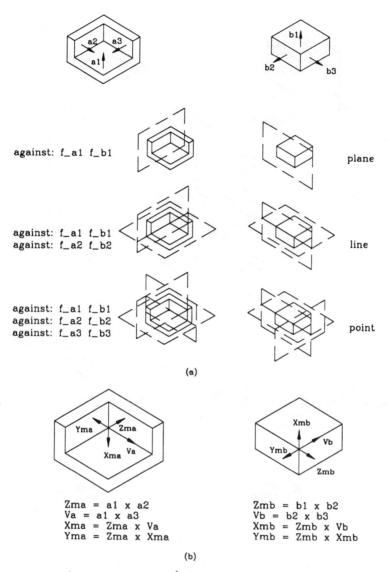

Fig. 3.13 Mating frame extraction, example 1.

together. The way to find mating frames is subdivided in two procedures where one is to locate the position of a mating frame and the other one is to find the orientation.

Positioning is derived from intersecting the features of spatial relationships, which is similar to the intersection of degrees of freedom. In Figure 3.13(a), the assigned mating features of $part_a$ are face $a_1, a_2,$ and a_3. In the first *against*, the position of the mating frame is located on that against face. When two *againsts* hold, the mating frame is positioned on a line which is the intersection of those two *againsts* faces. If a third *against* is assigned, the mating frame is fixed in the intersection of three *against* faces. This

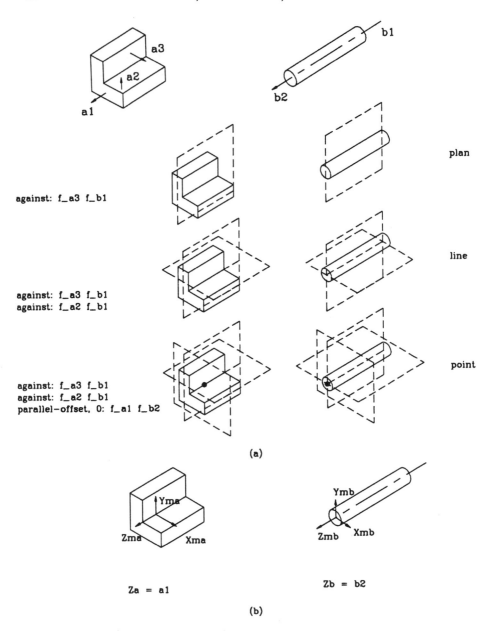

Fig. 3.14 Mating frame extraction, example 2.

method is also applied to $part_b$. Figure 3.14(a) shows a cylinder being assembled with an "L" shaped part. For the spatial relationships which contain cylindrical (sphere) features, the centerline (center) of one of the cylindrical (sphere) features takes the place of the line of the faces intersection, and the relative features are raised along the Z-axis of the feature-coordinate with distance r of the cylinder (sphere). In Figure 3.14(a), the faces of a_2 and a_3 are raised r distance, so the intersection point of the faces $a_1, a_2,$ and a_3 is allowable for placing a cylinder with the position of the mating frame in the centerline.

The orientation of a mating frame is derived from features with normal vectors. For three planar face against relationships, if the mating faces normal are $a_1, a_2,$ and a_3, and none of them are parallel, then the Xm_a, Ym_a, and Zm_a axes of a mating frame can be decided upon as follows:

$$Zm_a = a_1 \times a_2$$
$$V_a = a_1 \times a_3$$
$$Xm_a = Zm_a \times a_1$$
$$Ym_a = Zm_a \times Xm_a$$

With the same sequence of calculations, one can find the direction of Xm_b, Ym_b, and Zm_b mating axes of the mating part. This example is shown in Figure 3.13(b). The reason for both parts having a consistent orientation has to do with the basic property involved in crossing vectors, which is:

$$a \times b = -a \times -b$$

Relationships such as *parallel-offset, incline-offset,* and *include-angle* may fit into this method with reversing one of the parts face normal direction.

For a cylindrical (spherical) feature, its face cannot provide enough information for finding orientation, because the feature itself has rotational degree(s) of freedom about the centerline (center). In this case, the orientation can be decided if (1) combining with spatial relationships which has planar face with planar face relationships, or (2) spatial relationships with multiple cylindrical (spherical) features, e.g. insert a part with two cylindrical features into another with two hole features, and cylindrical or hole features are not aligned. In Figure 3.14(b), the mating frame with Z-axis aligns with a_1 and the X- and Y-axes are randomly selected which are perpendicular to each other.

3.4 PRODUCT SPECIFICATION ATTRIBUTES

Product specifications describe the constraints of form and function of a product. Tolerances are form specifications which specify the available range for position, orientation, and form of a product or components. Function specifications describe the functional performance of a product or its individual components. Form specifications influence the functional performance. Function is the product of interaction of forms. For example, in order to fasten two blocks with a screw, the bolt of the screw must have the same type of thread as the threaded hole of the block below. Based on design principles, the hole for the block above has to be a smooth cylindrical surface. The outside diameters should be the same for the threaded holes and the depth of thread of the bolt should move far enough to contact with the threaded hole of the bottom block. It can be seen that the form specifications needed are: thread type, thread length and the diameter tolerance. In assembling the forms, the *fasten* function can thus be achieved. In this example, if designers assigned the screw-fasten function between the threaded hole and the bolt with a specified load then they provide enough information for searching the form specifications for both the bolt and the threaded hole. The approach used in PAM is that the most appropriate time to assign the specifications to a product is in the assembly mode which saves human effort and clarifies the intention of the designer.

Product specification attributes such as functional descriptions and positional specifications can be specified by the designers, others can be automatically (or interactively, when the default information is insufficient) generated through the specifications assigned. These specification attributes contain the information about the methods of assembly (such as glue, weld, snap fit, etc.) and the tolerances associated with positioning the parts. Through assignment of the spatial relationships, these attributes are inferred. For example, in Figure 3.1, the positional tolerances $Da \pm ta$ and $Db \pm tb$ are assigned with association of *parallel-offset*, and then the tolerance $Dc \pm tc$ and $Dd \pm td$ of components are decided upon. The product specification attributes within spatial relationships are described as follows:

Against

1. Snap: shows the faces contact which will cause deformation of the components.
2. Gears mesh: indicates that the *against* features are gears with a certain type of gear parameters.
3. Welding: indicates the types of welding and joint with two faces *against* to each other.
4. Glue: indicates the type of glue processes.
5. Contact force: indicates the force required for making an assembly. The force reading may feed back to stop the approach motion of the manipulator.
6. Motion: indicates the relative motion parameters between two against faces.
7. Lubrication: indicates the type of lubricant required to put on before assembly.

Parallel-offset

1. Offset: gives the distance between two features.
2. Tolerance: indicates the tolerance associated with the dimensional offset.
3. Welding: indicates the types of welding and joint with the gap as the offset between two planar faces.

Parax-offset

1. Offset: gives the distance between two features.
2. Tolerance: indicates the tolerance of offset.

Incline-offset

1. Angle: indicates an include angle between two features.
2. Tolerance_angle: indicates the tolerance of the value of angle.
3. Offset: gives a distance between two features.
4. Tolerance_offset: indicates the tolerance of offset.

Include-angle

1. Angle: indicates an include angle between two features.
2. Tolerance-angle: indicates the tolerance of the value of angle.

Aligned

1. Snap: indicates the force required for snapping two features together.
2. Welding: indicates the types of welding and joint with two faces *against* to each other.
3. Screw: indicates the type of thread parameter.
4. Torque: indicates the torque required to screw two features together.
5. Glue: indicates the type of glues process used.
6. Motion: indicates the relative motion parameters between two against faces.
7. Lubrication: indicates the type of lubricant required to put on before assembly.

With complete specifications assigned, this information can then provide for post-processor such as tolerance propagation and process planning. The applications are described in the next section.

3.5 APPLICATIONS

In a system such as PAM the spatial relationships engine is a kernel which accepts neutral geometric information of parts as input and the output will be models that are useful for assembly process planning. Some of the applications such as bill of materials and precedence constraints, feasible approach directions, and world modeling are described briefly below.

3.5.1 Bill of materials and precedence constraints

A bill of materials is represented in a binary tree with parts or subassemblies as leaves which are sequentially generated when the spatial relationships of parts are assigned. The assembly precedence constraints can also be acquired from the bill of materials and the spatial relationships between parts. A part cannot be assembled if there is an object crossing the assembly directions. *Feasible approach directions* can be used to find the possibility of assembly as well as disassembly. In disassembling objects, once an object covers another object, the assembly direction of the object below will be covered until the object above is removed. In the bill of materials tree, if the parts have spatial relationships with the same ancestor(s) and do not block the assembly of their offspring(s), then these parts have the same precedence. For example, in surface mounting, chips are mounted on a PCB, the chips will always have the spatial relationships with PCB only. Consequently, the assembly precedence is the same for these chips. After this, the precedence of assembly operations is built for determining the operations sequence.

3.5.2 Feasible approach directions

Some researchers [Woo87] have developed a procedure to generate a sequence of motions for removing parts from a final assembly. For example, Woo uses the notion of *monotonicity* for generating *disassembly direction*. In general, disassembly directions are the same as assembly directions since the problem of finding how to assemble a set of given parts can be converted into an equivalent problem of finding how the same parts can be disassembled.

The determination of the feasible approach directions in a polyhedral representation is done as follows [BNYew, JN91]: wherever there is a planar contact between two parts, the parts can be approached in any direction in the half space created by the mating face. Thus if $F = f_1, f_2, \ldots, f_j$ represents a face set of all planar mating faces for a part then, the set of approach directions due to the ith mating face is given by:

$$R_i = \{r | r \cdot n_i \geq 0\}$$

Here n_i is the unit normal vector to the ith mating face. The set of resultant approach directions due to all the j mating faces is:

$$R = \cap_{i=1}^{j} R_i$$

3.5.3 World modeling

Assembly stations can be modeled with robot, conveyor, vision system, jigs and fixtures, etc. in static or dynamic form in terms of the spatial relationships characteristic attributes assigned. Lozano-Perez [LP83c] has pointed out that the attributes needed to model the world should include the following:

1. physical description of all objects, e.g. mass and inertia;
2. kinematic description of all linkages:
3. description of manipulator characteristics, e.g. joint limits, acceleration bounds, and sensor characteristics;
4. task specifications of all objects.

The first attribute, the physical description of a single object can be evaluated by a modeler (such as PAM) itself. When it is given, the physical properties of an assembly may then be calculated through combining the properties of individual ones. With regard to the second and third attributes, the kinematic description of linkages and their characteristics can be assigned and evaluated through modeling the manipulator with spatial relationships. Spatial relationships implicitly represent the operational requirements to achieve assembly which provide the information to satisfy task specifications.

CHAPTER 4

Structure of an automatic robot programmer

4.1 INTRODUCTION

Manipulator task planning is a difficult and broad topic. It contains many unsolved problems that are still in the research stage, such as obstacle avoidance, robot sensing and motion, redundant robots, etc.

A number of model-based manipulator systems have been proposed, but none so far has industrial application to date although some have been partially implemented [LP83a]. However, several of them do provide useful features that we can benefit from.

LAMA, designed and partially implemented at MIT [LPW77], is a system that takes obstacle avoidance, grasping, and error prediction into consideration. AU-TOPASS, developed and partially implemented at IBM [LW77], is a system with a certain degree of similarity to LAMA. It considers grasp points, trajectory, collision detection, and provides a set of general operations, in which the most general subset, PLACE statement was implemented. The emphasis in RAPT, implemented at University of Edinburgh, Edinburgh, Scotland [AP75] [ACP83] [PAB78] [PAB80b] [PAB80a] [Pop87], is on specifying tasks geometrically. Though it has no features of obstacle avoidance, automatic grasping, or sensory operations, the embedded spatial relationships concept provides us with a very useful tool in task planning. It provides the ability of considering an assembly task as fulfilling a spatial relationship which can be expressed and solved by mathematical equations.

So far, no system can deal with all assembly operations or adequately deal with all assembly situations. Most of the existing systems expand a task-level description into an executable program for a specific manipulator.

Existing research on task planning mainly concentrates on grasp planning and path planning. Generally speaking, geometric reasoning plays a very important role for task planning, especially for obstacle avoidance problems and deciding grasp points and approach locations.

Finding a stable grasp and a safe approach are heavily dependent on geometrical calculation, and are not simple tasks. For grasp planning, most methods focus on finding safe grasp configurations for initially grasping the object, and can only deal with limited object models. Also, a simple gripper, usually parallel-jaw gripper, is assumed, and very little attention is paid to stability and uncertainty. Often, these methods are not general enough for industrial application [LP83c].

It is possible to model the working environment and find a collision-free path under certain restrictions and assumptions. Yet the computational load it takes to find these paths is enormous and hard to apply in real industrial problems. For example, the *configuration space* (to be discussed in detail in Chapter 7) approach is employed by Lozano-Perez [LP80] for an obstacle avoidance problem, which can be applied in 2D but would be a very difficult task to extend to 3D. While Brooks [Bro83a] [Bro83b] uses *generalized cones* to represent the free space, and an algorithm has been developed to find the collision-free path, again, these approaches are computationally intensive.

Some groups intended to integrate both pure AI's strategic planning and robot task-level planning. In the work of Liu and Arbib [LA86], the property of an object is represented in a hierarchical structure. The planner generates a sequence of feature-level assembly commands in rearranged correct order if the given task-level command order is not suitable. *Planning with constraints* method is employed for goals' propagation and rearrangement. This approach emphasizes semantic level strategic control of task planning instead of dealing with the physical and geometric characteristics of the task. A similar approach in strategic control was utilized in [VA86], where a planning system with several components in charge of strategic planning, tactical planning, path planning, sensory feedback, and lower level control for the robot was outlined. The first phase refinement of goals up to abstract feature level, and part of the geometric reasoning component have been implemented. Basically, these approaches cover only limited object types and assembly operators.

This chapter describes a manipulator assembly task planner, RALPH (Robot Assembly Language Planner in Hierarchy), that processes the knowledge of the working environment and generates a sequence of general, manipulator independent commands. The planner takes a very high level command, such as "insert PEG into HOLE" without further specifications, reasons about the involved object features using the information from the CAD system, and generates a process plan for the manipulator to automatically perform the task. RALPH will be used to illustrate the general concept of task-level automatic programming since it processes most of the attributes needed in such a planner. RALPH, as a task planner that can accept commands which are as human-oriented as possible, such as *insert peg1 into hole1*, and generate the corresponding general manipulator-level commands automatically to drive the manipulator performing the specified task.

In order to see how convenient a task-level language could be, let us take a simple pick-and-place operation as an example. Imagine that there are two objects on a working table. Point_B and Point_C are the points right above the two objects, from where the gripper can start approaching the targets respectively. Point_A is the location where the robot hand rested originally. Following is the step-by-step command sequence.

Robot-level commands:
move from Point_A to Point_B
approach D
grasp
depart D
move from Point_B to Point_C
approach D
ungrasp
move from Point_C to Point_A

where Point_A, Point_B, Point_C, and distance D must be specified previously. The user might find this tedious and it requires language programming skill to program in this level. It would be much easier if the user could specify the same operation simply by:

place Object 1 Face 6 against Object 2 Face 1

However, this expression requires knowledge of the objects to be able to reason about the appropriate surfaces for which mating can take place. We desire to leave such reasoning to the robot planner and go to a level where we can express our task simply. Here we present a robot assembly task planner which could perform the operation by issuing a command similar to its human-oriented formalism equivalent:

Task-level command:
place Object 1 onto Object 2

Without specifying which faces of the objects will be mating. By means of analyzing the form of the objects and inferring their functions [NA88], sets of feasible grasping and mating faces can be generated. The *task-level* planner then searches for the optimal solution by finding the sets of grasping and mating faces that require the least number of rotations and shortest translations. This can be done by the algebra engine [Nna88] [Pop87], which can simplify the algebraic expressions and solve the equations. After the optimal solution has been selected, the task planner passes the newly generated mid-level command to the *mid-level* planner. Following is the mid-level command.

Mid-level command:
place Object 1 Face 6 (orientation specification) against Object 2 Face 1

The mid-level planner then decomposes the mid-level command into the *general robot-level* commands, which are robot independent and continue to be similar to the group of *robot-level commands* which we showed at the very beginning of this section.

After the general robot-level commands have been generated, the next step will be translating the robot independent commands into robot dependent commands to actually perform the task. Since there are several types of robots (cartesian, scara, anthropomorphic, etc.) that might be used in assembly work, direct decomposition of the general robot-level commands into the *specific robot-level* commands may be too large a task for the planner to deal with. There is a need for a general robot-level planner to translate the general robot-level commands into *generic robot-level* commands, which are robot type specific. These generic robot-level commands could be translated directly into robot-specific robot-level commands.

The overall command levels will be:

Tasklevel
→ *MidLevel*
　→ *GeneralRobotLevel*
　　→ *GenericRobotLevel*
　　　→ *SpecificRobotLevel*

The hierarchy of the planning is shown in Figure 4.1.

The naming of the planners follows the convention that the named type of command goes into that type of planner. For example, general robot-level commands, which are

70   *Structure of an automatic robot programmer*

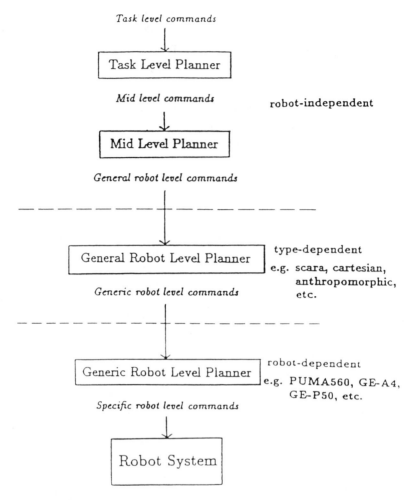

Fig. 4.1 Hierarchy of task planning.

generated by a mid-level planner, will become input to the general robot planner. The first two planners are robot independent, and thus we have not yet included the model of the robot being used for manipulation. The last and most important step is to check whether the generated path for the object being manipulated will constrain it to collide with any object. If so, the collision-free path generator will be invoked to plan for a new path to perform the task. At the beginning stage of planning the task, we assume that the path being generated is collision free. When it comes to the general and generic specific planning, the planner will check the configurations of the robot to ensure that the generated path will not be impossible for the robot to realize.

By using the robot specific command file, which is generated by the generic robot-level planner, the on-line robot motion planner can be activated to drive the robot performing the task. It is important to mention that the planning issues addressed by RALPH are for the assembly domain only.

4.2 AN OVERVIEW OF RALPH

As discussed in Chapter 1, there are two essential features that a task-level language should possess. The first is the geometric 3D modeling of the workspace. The second is the ability to process the world knowledge and generate the step-by-step robot-level commands to perform the specified task [Nna88]. For the geometric 3D modeling, some research groups have developed several useful task-level languages [LP83a] [AP75] [ACP83] [LPW77] [PAB78] that model the world at the same time that users are specifying the assembly tasks. Actually, at the design phase of an assembly, a designer has everything modeled on the CAD system. It would be both convenient and economical to extract

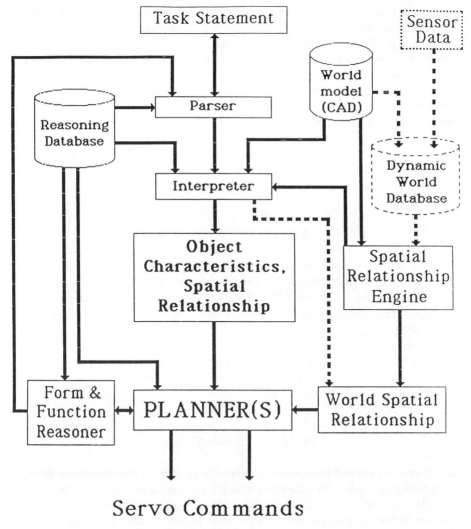

Fig. 4.2 The structure of RALPH.

the necessary data directly from the CAD system to build the world knowledge. This is the way RALPH [Robot Assembly Language Planner in Hierarchy) represents the knowledge of the 3D world.

As shown in Figure 4.2, the task statement first goes into the parser for grammatical checking. Then the object interpreter interprets the objects' and features' information in the current command, and passes the interpreted information to the planners. The hierarchy of planning has been shown in Figure 4.1, which is from the *task-level planner*, to the *mid-level planner*, the *general robot-level planner*, and then the *generic robot-level planner*. The task-level planner invokes the geometric and function reasoning engine to reason about mating and grasping information. In the current version of RALPH a task specification system to be discussed in Chapter 6 limits the mating and grasping problem space by providing the spatially related surfaces. Meanwhile, if an illogical command such as *insert hole into box* has been issued, an error message will be returned to the user. Following will be the choice of the optimal feasible grasping sets and mating surfaces and deciding the mating features orientation by the task-level planner. The mid-level planner then decomposes the mid-level internal commands into the general robot-level commands.

The rest of the work is translating the robot independent commands into type/robot dependent commands. This will be done by the subsequent work when getting into specific robot application. As long as different types of general robot-level planners check the path to ensure the generated sequences will not be in conflict with the robot configuration, the commands can be translated further into a special robot understandable commands. As a portability concern, Figure 4.3 shows this feature of RALPH. Thus, RALPH can be used by any manipulator which has a RALPH driver.

4.3 WORLD KNOWLEDGE DATABASE

One of RALPH's important features is directly extracting the information from the CAD system. Using CAD-based designs this way is very convenient. It makes the best use of the CAD information and saves RALPH users from having to model the assembly parts again in the reasoning world, since these models have already been built by the product designer.

When the planner is processing the commands, a working memory will be used to keep updating the current status of the environment, e.g. the current locations of objects, current relationships among objects, etc. Thus the planners can simulate the situations when the assembly is really happening.

4.4 RALPH COMMANDS

We will consider the first three levels of commands for RALPH. The first is task level, which is easy for humans to understand. This level is written by the user to input to the RALPH system. The second one consists of mid-level internal commands. This is generated by the task-level planner and is processed by the mid-level planner to expand into the next level's commands. The last one is the generated output, robot independent,

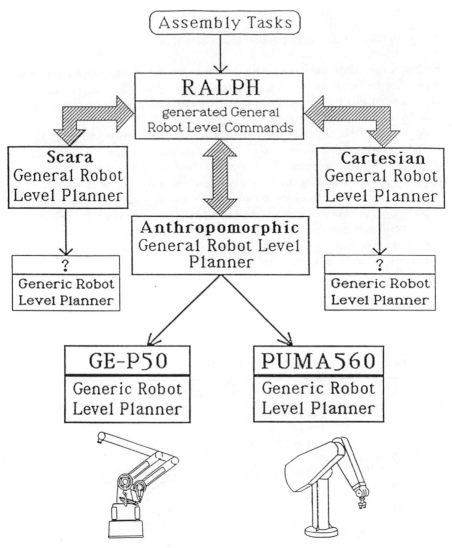

Fig. 4.3 The portability of RALPH.

general robot-level commands. This can be accepted by different types of lower level planners to further translate into generic robot-level, then specific robot-level commands.

4.4.1 Task-level commands

In examining those most frequently used commands given to human workers to perform assembly tasks, we come up with several sets of commands. These commands have a structure very close to their natural language equivalents. This is one of the main advantages that RALPH would provide.

Basically, RALPH task-level commands are of the format:

[Operator] [Loose object] [Preposition] [Fixed object]

For binary operations, the first object located between the operator and the preposition is movable, i.e. this is the object being operated upon. We call it the *loose* object. The object located after the preposition is considered immovable during this command, and is called the *fixed* object.

As seen in Table 4.1, there are four sets of commands, *ASSEMBLE, REORIENT, HOLD, RELEASE*, which have totally different functions. The command tree structure is from the left to the right. The higher the level number, the lower branch it is decomposed into. Following is a detailed explanation of each set of commands. Those commands with and without parentheses after them are binary and unary operators respectively. For binary operations, the first object located between the operator and the preposition is movable, i.e. this is the object being operated upon. We call it the *loose* object. The object located after the preposition is considered immovable during this command, and is called the *fixed* object.

Table 4.1 The structure of RALPH task-level commands

Level 1	Level 2	Level 3
ASSEMBLE (WITH)	PLACE(ONTO) PLACE(OVER) PLACE(BESIDES)	
	INSERT(INTO)	DROP(INTO) DROP_INSERT(INTO) PUSH_INSERT(INTO)
	SCREW(INTO)	SCREWR(INTO) SCREWL(INTO)
REORIENT	FLIP	FLIP x, $\chi \in N$ $\chi = 1 \rightarrow$ the easiest way to flip $\chi = 2 \rightarrow$ harder than $\chi = 1$ to flip $\chi = 3 \rightarrow$ harder than $\chi = 2$ to flip and so on...
	TURN	TURN x, χ in number of turns, e.g. 0.25
		TURN x, χ in number of degrees, e.g. 275
HOLD	HOLD HOLD(AGAINST)	
RELEASE		

(1) ASSEMBLE (WITH)

ASSEMBLE has five descendants: PLACE (ONTO), PLACE (OVER), PLACE (BESIDES), INSERT (INTO), and SCREW (INTO). Their functions are very trivial, except for the PLACE (OVER) command which may be applied in the case like *"PLACE box_with_hole OVER rod"*. The INSERT set can be further decomposed into DROP (INTO), DROP_INSERT (INTO), PUSH_INSERT (INTO), whereas, the sons of SCREW are SCREWR (INTO), which stands for right screw, and SCREWL (INTO) for left screw.

(2) REORIENT

IF REORIENT is employed alone, there is a need for further inference. For those objects that should not be flipped (e.g. there are some *loose* parts on the top of it), it will be decomposed into a TURN command, which is a rotation about the Z-axis of the world frame. Otherwise, it will be treated as FLIP. These commands can be specified in a more detailed way with a number after them. For example, *flip 1* means to flip the object by the easiest way according to the current location of the gripper, while *flip 2* will choose the next easiest way to flip the object, etc. The degree of "ease of flipping" is determined by the amount of rotation and translation it takes to achieve the goal. If no number is specified when using TURN, the system will infer the amount of rotation. This rotation will be calculated according to a convention that a human would consider "obvious". For example, a hexagonal nut would be turned 60 degrees, and a cube would be turned 90 degrees, etc. Otherwise it will be rotated by the specified amount.

(3) HOLD

If HOLD is issued without AGAINST, it is just a simple unary operation: move to the object, then hold it without changing the location of the object. There is only a grasping set, and no mating sets that will be generated by the geometric and function reasoning engine. The grasping set should contain those ways that expose the features. Thus other machines or robots can perform additional operations on the held object. When AGAINST is used along with HOLD, it means to pick up the movable object and hold it against the specified immovable object. Note that HOLD can also be used in multiple robots assembly situations.

(4) RELEASE

RELEASE is the simplest unary operation. If the held object is stable when opening the gripper, then it should perform: ungrasp the object, then depart from the object. In the case that the object is being held against a wall, it should move the object to a stable location first, then ungrasp. HOLD and RELEASE should be used as a pair.

Operators can employ these commands at each level, but the issued command will have to be inferred by the geometric and function reasoning engine to its executable level if it has many possibilities of inference when specified. One task may be specified in more than one way. For example, to insert A into B could be accomplished by issuing *DROP_INSERT A INTO B* (convention) or *INSERT A INTO B*, or even *ASSEMBLE A WITH B*. The difference is the inference load on the geometric and function reasoning engine. In the last case, it takes the geometric and function reasoning engine the most effort to infer the mating and grasping sets by analyzing the features of part A and B. Different part features will result in different internal mid-level commands even when they use the same command, ASSEMBLE A WITH B.

Notice the "power/generality trade off" in this command structure tree. The higher the level, the more general and less powerful the command could be, and vice versa. In efficiency aspect, the higher the level, the longer the time it takes to infer the mating and grasping sets. Nevertheless, task-level planning and decomposition into general robot-level commands is off-line processing. Running time is not crucial in this aspect. We believe that the command tree feature could provide the flexibility for both awkward and dexterous RALPH (or similar systems) users.

4.4.2 Mid-level internal commands

The format of the mid-level internal commands is very similar to the task-level commands. The only difference is the added specification in the commands. If a less inferred and non-executable command is issued, it has to be broken down into the more explicit, executable command of that group. For example, if the command ASSEMBLE (WITH) is issued, the right command to use when it comes to mid-level might be PLACE (ONTO), or PLACE (OVER), or PLACE (BESIDES), or DROP (INTO), ... or SCREWL (ONTO). In order to show the exact format of these commands, we use A and B to indicate mating objects.

PLACE A (feature, orientation) ONTO B (feature)
PLACE A (feature, orientation) OVER B (feature)
PLACE A (feature, orientation) BESIDES B (feature)
INSERT A (feature, orientation) INTO B (feature)
DROP A (feature, orientation) INTO B (feature)
DROP_INSERT A (feature, orientation) INTO B (feature)
PUSH_INSERT A (feature, orientation) INTO B (feature)
SCREWR A (feature) INTO B (feature)
SCREWL A (feature) INTO B (feature)
FLIP A (feature, direction)
TURN A (feature, direction)
HOLD A (feature, orientation)
HOLD A (feature, orientation) AGAINST B (feature)
RELEASE A (motion_flag, feature, orientation)

4.4.3 General robot-level commands

The generation of this level of commands is the expansion of an internal mid-level command into a sequence of robot motions, which are general, robot independent, and are not executable by a robot yet. They are standard output for the higher level planner and need further translation into type/robot specific commands if they are used to command the robot performing the task. Following are the basic components.

A RALPH task-level command is decomposed into a sequence of the following basic components.

RALPH commands

TRANS: translate in x, y, z, directions. → *trans*([P_x, P_y, P_z]).
where P_x, P_y, and P_z are real numbers.

ALIGN: change orientation (Euler angles or roll, pitch, yaw).
→ *align*(**X**_*vector*, **Y**_*vector*, **Z**_*vector*). where **X**_*vector*, **Y**_*vector*, **Z**_*vector* are three-tuple lists corresponding to X-, Y-, and Z-axes.

MOVE: change location (orientation and position).
→ *move*([**X**_*vector*, **Y**_*vector*, **Z**_*vector*, **P**_*vector*]).
where **X**_vector, **Y**_vector, **Z**_vector denote the orientation and **P**_vector is the position. They are all three-tuple lists.

MOVEC: take a coarse motion path with the specified points.
(If the coarse motion is not available for the robot being used to perform the task, MOVEC will be treated as MOVE.)
→ *movec*([P_1, P_2, ... P_n]). where P_1, P_2, ... P_n are the coarse motion points.

GRASP: close gripper so that the distance between the jaws is the specified amount.
→ *grasp*(AMOUNT). where amount is a real number in the system defined unit length.

UNGRASP: open gripper so that the distance between the jaws is the specified amount.
→ *ungrasp*(AMOUNT). where Amount is a real number in the system defined unit length.

APPROACH: move gripper along one positive axis of gripper's frame.
→ *approach*(AXIS, AMOUNT).
where AXIS is a three-tuple list, and Amount is a real number in the system defined length. (AXIS is *plus_x*, *minus_x*, or *z*.)

FAPPROACH: move gripper along one positive axis of gripper's frame with the specified maximum force (This is a force approach, which is used in PUSH_INSERT command.)
→ *fapproach* (AXIS, AMOUNT, FORCE).
where AXIS is a three-tuple list, AMOUNT is a real number in the system defined length, and FORCE is a real number in the system defined unit force. (AXIS is *plus_x*, *minus−x*, or *z*.)

DEPART: *move gripper along one negative axis of gripper's frame.*
→ *depart* (AXIS, AMOUNT). where AXIS is a three-tuple list, and amount is a real number in the system defined length (AXIS is *plus−x*, *minus_x*, or *z*.)

LINEAR: move the gripper to the location along a linear path.
→ *linear* ([**X**_*vector*, **Y**_*vector*, **Z**_*vector*, **P**_*vector*]).
where **X**_vector, **Y**_vector, **Z**_vector denote the orientation and **P**_vector is the position. They are all three-tuple lists.

FLINEAR: move the gripper to the location along a linear path with the specified maximum force. (This is a linear motion with force.)
→ *flinear*([**X**_*vector*, **Y**_*vector*, **Z**_*vector*, **P**_*vector*], FORCE).
where **X**_vector, **Y**_vector, **Z**_vector denote the orientation and **P**_vector is the position. They are all three-tuple lists.
FORCE is in the system defined unit force.

ROTR: rotate about positive Roll direction of current gripper's frame.
→ rotr(THETA).
where theta is in the unit of degree.
ROTL: rotate about negative roll direction of current gripper's frame.
→ rotl(THETA).
where theta is in the unit of degree.
WAIT_FOR: remain motionless for a specified amount of time.
→ wait_for(N). where N is a positive real number.
WAIT_UNTIL: remain motionless for a specific I/O signal.
→ wait_until(IO_SIGNAL).
SIGNAL: send a type of signal to a particular device.
→ signal(TYPE(DEVICE_NAME)).

The task-level commands are decomposed into sequences of combinations of basic components. For convenience, we introduce two notational short cut commands, PICK and PUT, in the following text since many task-level commands contain PICK and PUT sequences in their decomposed forms:

PICK: TRANS, ALIGN, APPROACH, GRASP PUT: TRANS, ALIGN, APPROACH, UNGRASP, DEPART
PLACE (ONTO): PICK, MOVEC, PUT
PLACE (OVER): PICK, MOVEC, PUT
PLACE (BESIDES): PICK, MOVEC, PUT
INSERT (INTO): PICK, MOVEC, PUT
DROP (INTO): PICK, MOVEC, TRANS, ALIGN, UNGRASP
DROP_INSERT (INTO): PICK, MOVEC, PUT
PUSH_INSERT (INTO): PICK, MOVEC, TRANS, ALIGN, FAPPROACH, UNGRASP, DEPART
SCREWR (INTO): PICK, MOVEC, TRANS, ALIGN, APPROACH, ROTR, UNGRASP, DEPART
SCREWL (INTO): PICK, MOVEC, TRANS, ALIGN, APPROACH, ROTL, UNGRASP, DEPART
SCREWR (OVER): PICK, MOVEC, TRANS, ALIGN, APPROACH, ROTR, UNGRASP, DEPART
SCREWL (OVER): PICK, MOVEC, TRANS, ALIGN, APPROACH, ROTL, UNGRASP, DEPART
ROLL: PICK, TRANS, ROTR, TRANS, UNGRASP, DEPART
YAW: PICK, TRANS, ROTR, TRANS, UNGRASP, DEPART
REORIENT: PICK, TRANS, ALIGN, TRANS, UNGRASP, DEPART
HOLD: PICK, WAIT, UNGRASP, DEPART
HOLD (AGAINST): PICK, MOVEC, TRANS, ALIGN, APPROACH, WAIT, DEPART, PUT
UNSCREWR (ONTO): PICK, ROTR, DEPART, PUT
UNSCREWL (ONTO): PICK, ROTL, DEPART, PUT
GETTOOL: PICK, DEPART
RETURNTOOL: PUT

If we take a good look at these decomposed forms, it is easy to see an interesting phenomenon: the sequences are very similar. Actually, the so called "task level" is for the human being's perception of the task. To a robot, it doesn't make any difference whether the command is PLACE (INTO) or INSERT (INTO). The only difference is the specified amount and the direction of the motions. Avoidance of the tedious motion specification (either by teaching or numerical data) is thus the purpose of task-level programming.

4.5 MATHEMATICAL CONSIDERATION

Since no motion of the robot is specified to a task planner (such as RALPH) at task level, the goal position and orientation of the robot will have to be inferred by the planner itself. It is not a trivial task obtaining the related information. There is data extraction, symbolic equation solving, and configuration checking included in this process. We have presented in Chapter 3 the concept of symbolic spatial relationships which drive the relationships in the robot's world. The main idea is: each object in the working environment has a *location* (or a set of *axes*) embedded in it, which is defined relative to the *world frame*. The *world frame* is a *coordinate frame* that is used for reference. An object is formed as a set of faces configured into some topology with geometric attributes, and is described by *features* for assembly purposes. In Chapter 2, we defined a *feature* as a set of surfaces together with a specification bounding relation between them. For example, a *shaft* or a *hole* is a *feature*. The shaft and the hole are both composed of two plane surfaces, and a convex and a concave cylindrical face, respectively. Each feature also has its own *location* (or a set of *axes*) defined with respect to the object it belongs to. That is why the *location* of an object can be transformed to anywhere in the world, but the *locations* of the features remain the same within the object. Figures 4.4 and 4.5 show some embedded axes of features and the objects they belong to.

When a command applies to objects, there will be one or more spatial relationships constraining the objects' features to achieve the goal. Each relationship can be described by an equation with transformation variables in it, which links the two objects together under the stated relation. By solving for the variables in the equations, the new location of the manipulated object with respect to the other object can be obtained.

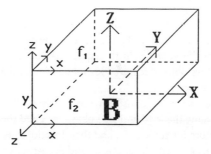

Fig. 4.4 The axes of plane faces (f_1, f_2) and the primitive object (B_2).

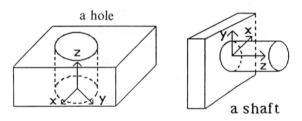

Fig. 4.5 The axes of a hole and a shaft.

By multiplying rotations and/or translations to one location, we can transform the old location to a new location. If we *postmultiply* the transformation matrix to a location, then the transformation matrix is made with respect to the frame of the position, not the reference frame (world frame) [Pau81]. Since a coordinate frame is actually a transformation matrix, the same technique can be used to obtain a new expression of a frame relative to other objects. As shown in Figure 4.4, assume the axis of the block is B which is defined relative to the *world frame*, and the axis (or location) of one of the two faces is f_1, which is defined with respect to the object frame B. Then, the location of the face relative to the *world frame* is represented by $B*f_1$, where "$*$" denotes the matrix multiplication.

Note that in our system the axes of primitives are as defined in Chapter 3. Generally speaking, the object frame is always chosen at the center of one of the largest faces, and has the Z-axis point through the object, and the Y-axis aligned with the longest edge of that face. When dealing with faces, we always choose the Z-axis as normal, and the south-west corner as the origin of a planar face frame. For example, the normal of a face is defined as the Z-axis, and the Y-axis is chosen in the direction that aligns with the face edge, where the X-axis is equal to the cross product of Y and Z.

The spatial relations we employ are defined in Chapter 3. *Against* and *coplanar* hold between two planar faces, while *fits* holds between a pair of convex and concave features. A convex feature is a projection with a convex contour, such as a peg, and a concave feature is a hollow portion with a concave contour, such as a hole. *Aligned* also holds between cylindrical features, but unlike *fits*, it specifies the Z-axes of the two features pointing to the same direction. In the case that a through hole is aligned with another hole, the directions of the Z-axes could be the same or opposite, which is dependent on the way of mating and grasping.

The main advantage of spatial relationships is to make abstract ideas concrete. Using mathematical expressions to describe the relations is the way to quantify the abstract concepts. By solving for the variables in the equations, the new object location can be expressed in terms of other objects' feature locations and thus chains the related features together for planning's concern.

We define **M** as the matrix to transform the Z-axis to its opposite direction. As shown in Figure 4.6, assume the left-hand side object is A and the right-hand side object is B. Object A has a feature (or surface) F_1, and their locations are p_1 and f_1, while object B has a feature (or surface) F_2 with locations p_2 and f_2 respectively. Following is a brief summary of the frequently used mathematical equations that represent the spatial relationships. See reference [PAB80b] for detailed discussion.

Mathematical consideration

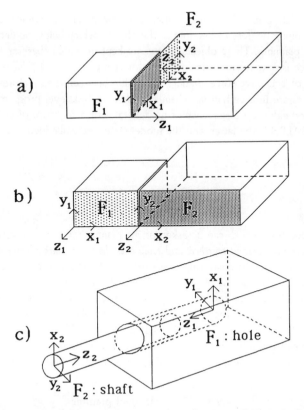

Fig. 4.6 (a) Face F_2 *against* face F_1 (b) Face F_2 *coplanar* face F_1 (c) Shaft F_2 *fits* hole F_1.

1. If F_1 is a plane face and F_2 is also a plane face, and F_1 and F_2 are *against* each other (see Figure 4.6a), then there are three remaining degrees of freedom (θ, y, and z) for the objects. The equation is

$$p_2 = p_1 * f_1 * trans(x, y, 0) * rot(z, \theta) * \mathbf{M} * f_2^{-1}$$

2. If F_1 is a plane face and F_2 is a plane face (see Figure 4.6b), and F_1 is *coplanar* with F_2, then there are three remaining degrees of freedom (θ, x, and y) for the objects. The equation is

$$p_2 = p_1 * f_1 * trans(x, y, 0) * rot(z, \theta) * f_2^{-1}$$

3. If F_1 is a shaft and F_2 is a hole, and F_1 *fits* F_2 or conversely (see Figure 4.6c), then there are two remaining degrees of freedom (θ, and z) for the objects. The equation is

$$p_2 = p_1 * f_1 * trans(0, 0, z) * rot(z, \theta) * \mathbf{M} * f_2^{-1}$$

4. If F_1 is a through hole and F_2 is a hole, and F_1 *aligned* F_2 with their Z-axes pointing to the same direction, then there are two remaining degrees of freedom (θ, and z) for the objects. The equation is

$$p_2 = p_1 * f_1 * trans(0, 0, z) * rot(z, \theta) * f_2^{-1}$$

Besides calculating new locations, spatial relationships also provide the information about the remaining degrees of freedom of the objects, which helps the decision making of robot path planning. These objects' status are kept in the database and are updated whenever there is new relationship formation or change.

When two objects have some spatial relationship between their features, we can say that the objects are in a certain *model state* or *state*. In task-level programming, a task can be specified either by a sequence of model states or a sequence of operations. The approach of RALPH is the latter, and the model states are embedded in each command implicitly.

4.6 TASK PLANNER

The type of questions which can be asked of the planner for assembling two parts will be: *How do these two parts fit together?* and *How does the manipulatior perform the assembly operation?* The approach to answering these two questions are discussed below.

We will examine two levels of planners. The first one is the *task-level planner*, which decides the target locations for the manipulator to reach, and find the path points for the manipulator to execute the command. The second planner is the *mid-level planner*, which accepts the output of the task-level planner (i.e. mid-level internal commands) and decomposes them into the general manipulator-level commands.

We assume the target manipulator has 6 degrees of freedom with a jaw-type gripper.

4.6.1 The task-level planner

The process from receiving the task-level commands to generating the mid-level internal commands, involves tremendous reasoning and computational information. In order for two parts to assemble together, the system starts with reasoning about the mating surfaces by the geometric and function reasoning engine. For grasping the *loose* object, the geometric and function reasoning engine also finds a set of possible grasps. Then, the spatial relationship engine will be invoked to calculate the final location of the *loose* object. From there, the task-level planner begins to calculate the target locations for the manipulator to reach and plans a path for it to follow.

1. Algorithm

An algorithm for the task-level planner is described as follows.

1. Read in one task command, prepare the needed information for the geometric and function reasoning engine.
2. Invoke the geometric and function reasoning engine to reason about mating surfaces and grasping set using the interpreted objects information.
3. Pass the set of mating surfaces to the spatial relationship engine to solve for the final locations of the *loose* object.
4. Search for the optimal grasp that allows for the most efficient path with the fewest translations and rotations.

5. Calculate the *grasp, approach_grasp, ungrasp,* and *approach_ungrasp* locations for the manipulator to reach.
6. Calculate the intermediate locations to avoid intrinsic collision.
7. Generate the corresponding mid-level internal command.
8. Pass the generated mid-level internal command to the mid-level planner for decomposition.
9. Wait for the message from the general manipulator-level planner. If the generated path is in conflict with certain type(s) of manipulator, then repeat step 1 through 8 until no conflict message has been returned from the general manipulator-level planner, which means that a feasible path has been generated.
10. Update necessary information to keep track of current status of the *world*.
11. Repeat 1 through 10 until the whole set of task commands have been processed.

The generation of mating faces and grasping sets is inferred by the geometric and function reasoning engine [NA88]. As shown in Table 4.1, each task-level command will be inferred to its executable command if it is issued by a non-executable command. When the task-level planner is invoked, not only is the command in the most inferred, executable form (this may not be the same one as the user issued), but also the sets {*mating_surfaces*}, {*ordered_feasible*}, will be part of the command. The set {*ordered_feasible*} is the prioritized list of the set {*feasible_grasps*}, which contains possible grasping vectors that point in the directions of grasping corresponding to the decided mating faces. The priority ranking at this phase of analysis is based on a single criteria: the amount of contact surface area available for a gripper.

In the case of mating symmetric features, there might be several ways to accomplish the goal. Based on the criterion of choosing the most convenient way of mating, the geometric and function reasoning engine is able to choose the right set of mating faces.

The first element in {*ordered_feasible*} may not necessarily be the right one to choose. This is because the "best grasp" may not be the most convenient one for the manipulator to grasp according to current orientation and position of the gripper. The task-level planner first seeks the optimal grasp that allows for the most efficient path with the fewest translations and rotations. Then the planner starts to calculate the exact locations that the manipulator should go to, which are defined relative to the reference *world frame*.

2. Goals

There are two goals that the task-level planner must achieve:

i. Calculates the target locations, which include the *grasp, approach_grasp, ungrasp,* and *approach_ungrasp* locations, that the manipulator (gripper) should reach.
ii. Plans an "intrinsic" collision-free path, that is, to find the intermediate locations for the manipulator to follow, which will help the manipulator in avoiding the collision between the mating objects when assembly is happening.

i. Calculate locations

Generally speaking, there are four locations which should be calculated:

GRASP

APPROACH_GRASP
UNGRASP
APPROACH_UNGRASP

These locations are represented by 4 × 4 homogeneous transformation matrices, which contain a 3 × 3 orientation matrix and a position vector.

- *Grasp location*

The information contained in the grasping set is actually a vector that could be used as the gripper Y direction, where the gripper frame is defined in Figure 4.8. Since the locations generated by the task planner are defined relative to the world coordinate system, we refer to the gripper frame as X, Y, Z, and P vectors, which correspond to the n, o, a, and p vectors in Paul's kinematic notations [Pau81].

From the interpreted objects information, we decide an approaching surface, $S_{approach}$, on the *loose* object, which will allow for vertical gripper approaches to the object during grasping. The gripper Z direction can be chosen as the opposite of the normal of $S_{approach}$, then the gripper X direction is equal to $Y \times Z$. An algorithm is needed to find the center of $S_{approach}$ so that the P vector of the *grasp* location can be easily calculated.

- *Approach_grasp location*

This location is calculated from the *grasp* location. Assume the given approach distance is D, and the position vector of the *grasp* location is P_1. The orientation part of the matrix, which contains the three X, Y, and Z vectors, is the same as that of the *grasp* location. The position vector P can be obtained by:

$$P = P_1 - D * Z$$

- *Ungrasp location*

Assume the *grasp* location is $T^{loose\_object}_{grasp}$, which means the *grasp* location is defined relative to the *loose* object, and the final location of the *loose* object is T^{world}_{final}, which is defined relative to the *world frame*. Then the *ungrasp* location, UNGRASP, is calculated by:

$$UNGRASP = T^{world}_{final} * T^{loose\_object}_{grasp}$$

- *Approach_ungrasp location*

This location is also calculated from the *ungrasp* location. Assume the given approach distance is D, and the position vector of the *ungrasp* location is P_2. The orientation part of the matrix is the same as that of the *ungrasp* location. From the information of the *fixed* object, we can reason about the *orthogonal assembly direction* to be the vector OAD. Then the position vector P can be obtained by:

$$P = P_2 - D * OAD$$

Note that a simple constraint is made in deciding the OAD vector: there are only two directions for the *loose* object to approach the *fixed* object when there are multiple ways of approaching. They are: "approaching from the top" and "approaching from the side." Only vertical and horizontal approaches are considered at this stage of the research. In grasp planning, we employ a common choice: place the gripper in contact with a pair of parallel surfaces of the object, or on the diameter of a cylindrical surface if that's the only choice. We also constrain the directions of grasping to "grasp from the top" and "grasp from the side". Figure 4.8 shows the grasp directions for a block.

ii. Intrinsic collision-free path

This goal can be achieved by choosing the path as a sequence of vertical and horizontal movements.

Besides the two approaching locations, *approach_grasp*, and *approach_ungrasp*, from where the gripper starts to linearly approach the targets, there are at least two intermediate locations in-between. The first location is found by lifting the grasping location up to a suitable height, which must be a certain amount higher than that of the ungrasp location. The second one is found by horizontally moving the first location to the location above the ungrasp location.

For a typical *place_onto* operation as shown in Figure 4.9, where $\mathbf{Q}_1$, $\mathbf{Q}_2$ are target locations, A_1, A_2 are approaching locations, and M_1, M_2 are intermediate path points, the paths will be.

- *Grasp path*

Move from current location to the *approach_grasp* location A_1, then approach to the *grasp location* Q_1 by linear motion, then grasp the *loose* object. We assume that this is a crowded working environment, so there is a need to move to A_1 again by linear motion to avoid collisions with other objects.

- *Intermediate path*

Move to M_1 then M_2 by *coarse motion*, as there is no need to stop at these intermediate points. If the target manipulator doesn't have this ability, then this path could be treated as ordinary gross motion.

- *Ungrasp path*

Move to the *approach_ungrasp* location A_2, then approach to the *ungrasp* location Q_2 by linear motion, then ungrasp the *loose* object. After linearly moving to A_2 again, the robot returns to the robot rest location.

The path that we have planned here is surely not the most efficient one, because we haven't incorporated any sophisticated obstacle avoidance algorithm. Yet it is reasonable and safe enough for most of the assembly situations.

4.6.2 The mid-level planner

So far, the task-level planner has paved the road to the last step, which is to generate the output commands. The main function of the mid-level planner is to decompose the commands into the general manipulator-level commands. According to the decision made by the task-level planner, the information that comes with each command to this stage should be enough for the mid-level planner to expand each command.

The information which comes with each mid-level internal command is the first, intermediate, and the last locations the manipulator should go to, which are grasp, intermediate, and ungrasp locations in general. The compositions are highly dependent on different operators we use, especially for the ungrasp path. These locations are represented by 4×4 homogeneous transformation matrices defined relative to the *world frame*. When it comes to the lower level planning which follows the mid-level planner, there is no "world frame" any more, and only the manipulator coordinate system is recognizable. The specific manipulator interface will operator on the 4×4 homogeneous transformation matrices and drive the manipulator.

1. Algorithm

Based on the decomposition we described in section 4.4.3, the mid-level planner has to calculate or extract the necessary numerical data that should be specified in each basic component. The general idea is:

1. Initiate necessary options specified for this particular task.
2. Decompose the grasp and the intermediate path in general.
3. Decompose the ungrasp path according to different operators.

Note that the sequence proposed above is the usual case for binary operation commands. It might be implemented a little bit differently for the unary operations, especially for the intermediate path and the ungrasp part, but the main idea is the same.

2. Example

For the typical *place_onto* operation as shown in Figure 4.9, the paths will be decomposed into the following basic components. Assume that the robot is at *Robot_rest_location* before executing the task, and W_1, W_2 stands for gripper width which opens and closes the gripper, respectively.

- *Grasp path*

MOVE (A_1).   (*approach_grasp* location)
APPROACH (Q_1).   (*grasp* location)
GRASP (W_1).
DEPART (A_1).   (*approach_grasp* location)

- *Intermediate path*

MOVEC ([M_1, M_2]). (intermediate path points)

Note that the MOVEC command is to take M_1 to M_2 as coarse motion points so that the robot will travel through or bypass these points smoothly. If the robot doesn't have the capability of coarse motion, this command will be treated as a series of MOVE commands:

MOVE (M_1).
MOVE (M_2).

- *Ungrasp path*

MOVE (A_2). (*approach_ungrasp* location)
APPROACH (Q_2). (*ungrasp* location)
UNGRASP (W_2).
DEPART (A_2). (*approach_ungrasp* location)
MOVE (*Robot_rest_location*).

4.7 AN EXAMPLE OF ASSEMBLY TASK

Here we present an example to briefly show how the processes of RALPH can collaborate together to accomplish the task.

As shown in Figure 4.7, the RALPH task-level command is in the compound format, *assemble A with (B with C)*, which is to put object B on the step of object C with the holes

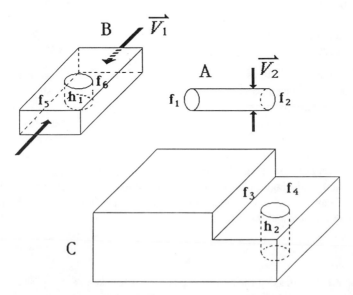

Fig. 4.7 An example of assembly task.

aligned, then insert A into the holes on B and C. We list the duty of each process sequentially for the purpose of this demonstration.

1. Parser

Checks the syntax of the command and breaks the compound command into single commands:

1. assemble B with C
2. assemble A with B_C

and then translates the commands into the internal forms:

1. assemble_with (B, C)
2. assemble_with (A, B_C)

2. Object-interpreter

Interprets objects A, B, C, and passes their information to the task-level planner which invokes the geometric and function reasoning engine for mating features information.

3. Geometric and function reasoning engine

After analyzing the form of objects and inferring their functions, the geometric and function reasoning engine would be able to generate the necessary information for the planners.

1. place_onto (B, C)
   mating features: $\{f_5, f_3\}, \{f_6, f_4\}, (h_1, h_2)$
   grasping set: $\{V_1\}$
2. insert_into (A, B)
   mating features: $\{f_1, [h_1, h_2]\}$
   grasping set: $\{V_1\}$

Note that the mating features for command (2) are $\{f_1, [h_1, h_2]\}$ instead of $\{f_1, h_1\}$. This means that f_1 should be inserted into both h_1 and h_2 in one assembly step.

For some operations like *place (onto)*, *place (over)*, and *place (besides)*, when specified in the less-inferred form, it is not necessary for the geometric and function reasoning engine to figure out what the actual command should be. This is because these types of operations only differ in target locations. But for others such as *insert (into)*, *drop_insert (into)* and *push_insert (into)* that have big differences at the terminal actions (e.g. *drop* versus *push*), it is important to distinguish between *insert*, *drop_insert*, and *push_insert*. Which one to choose depends on the assembly situations, for example, the depth of the hole, the diameters of the hole and the part being inserted, etc.

4. Task-level planner

Following are the steps that the task-level planner has to take to accomplish the task. We only use command (1) to demonstrate the algorithm.

An example of assembly task

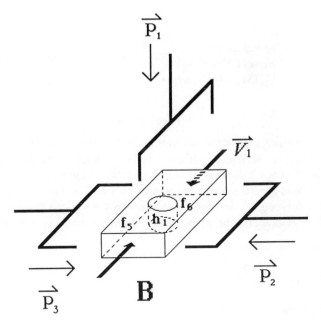

Fig. 4.8 Gripper approaching directions.

a. The spatial relationships would be:
   against (f_5, f_3)
   against (f_6, f_4)
   aligned (h_1, h_2)
   The equations could be obtained at the same time. Invoke the spatial relationship engine to resolve the variables so the final location of the *loose* object can be calculated.

b. As shown in Figure 4.8, if we just consider grasping the object by the center of each face, or edge, and choosing the direction such that the approaching vector of the gripper frame is pointing opposite to the face normal, there will be three ways to approach for grasping the *loose* object B. They are P_1, P_2, and P_3.

c. Since f_5 is a mating face, P_3 can be eliminated easily. The decision between P_1 and P_2 can be made by comparing the amount of translation and rotation it takes for these remaining approaches for grasping from the current locations to the grasping locations. This depends on the location of the robot and varies case by case.

d. After the grasping orientation has been decided, find the proper locations for the grasp and ungrasp of the gripper. Assume they are Q_1 and Q_2, which are actually 4×4 matrices. Notice that the orientation was determined relative to a surface/feature of the body being grasped which is known in the *world frame*.

e. Calculate the intermediate locations to avoid intrinsic collision, and assume that they are $M_1, M_2, \ldots, M_n$.

f. The internal mid-level command can be generated as:
place_onto([against $(B(f_5), C(f_3))$,
against $(B(f_6), C(f_4))$,
aligned $(B(h_1), C(h_2))$,
$\mathbf{Q}_1 [M_1, M_2, \ldots, M_n], \mathbf{Q}_2$)

5. Mid-level planner

The mid-level planner utilizes the information that comes along with the mid-level command and decomposes it into a corresponding general-level command sequence with numerical data. Figure 4.9 shows an example which has 3 intermediate locations.

The decomposed general robot-level command sequence will look like the following set, where the location $\mathbf{Q}'_1$ denotes the location on the top of $\mathbf{Q}_1$ with a default or user-defined offset, and D is the amount that the gripper should open or close.

move $\mathbf{Q}'_1$ (position component of $\mathbf{Q}'_1$)
align $\mathbf{Q}_1$ (orientation component of $\mathbf{Q}_1$)
approach $\mathbf{Q}_1$ (position component of $\mathbf{Q}_1$)
grasp (D)
depart $\mathbf{Q}'_1$ (position component of $\mathbf{Q}'_1$)

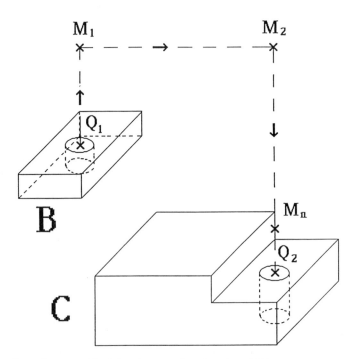

Fig. 4.9 Target locations (Q_1, Q_2) and intermediate locations $(M'_i s)$.

move M_1 (position component of M_1)

⋮

⋮

move M_n (position component of M_n)
align $\mathbf{Q}_2$ (orientation component of $\mathbf{Q}_2$)
approach $\mathbf{Q}_2$ (position component of $\mathbf{Q}_2$)
ungrasp (D)
depart M_n (position component of M_n)

This is just a simple assembly case that does not consider the working environment at all. In real assembly situations, many other complex factors have to be considered. These can include: dynamic world information from sensors, obstacle avoidance problems, etc. Also, the parts may be much more complicated. In the subsequent chapters, we show approaches for dealing with more complicated assembly situations.

4.8 PROGRAMMING ISSUES

Although the PROLOG language has a drawback in mathematical calculations, it is a powerful and convenient high-level programming language for reasoning. Since RALPH system largely involves object oriented reasoning, the version 1 of RALPH was programmed in POPLOG system which allows for PROLOG programming, and also provides major mathematical calculations and the capability to manipulate symbolic matrix operations by another language, POP11. The current version of RALPH is written in C and PROLOG.

One of the reasons that PROLOG is suitable for symbolic reasoning is its convenient way of storing information, *facts* or *predicates*. This is due to the flexible data structure, *list*, which is an *ordered sequence of elements* that can have any length. The planner program receives and updates world knowledge by *retracting* and *asserting* the *facts* out of/into the memory. This program could be treated as a control strategy that makes decisions among these facts. The C is excellent for the mathematical computations.

The heaviest load of reasoning happens in the geometric and function reasoning engine. Most of what the task-level planner does is a lot of information retrieving, updating, and calculations between vectors and matrices to prepare for the mid-level planner to decompose the task command. Unlike other ruled-based systems or theorem-proving systems, there are no production rules and rules-matching needed in this process, since the ultimate goal of the task planner is to generate a sequence of output (general manipulator-level commands) from the given input (task-level command). A hierarchical program structure with the *declarative* knowledge only [Nil80] is what is employed.

The program structure is quite straightforward, and modularized. It first acquires reasoning results from the geometric and function reasoning engine, and after the operator has been determined, the program flow switches to the corresponding operator group for further planning. From here on, the program concentrates on the current operator and the planning process varies from operator to operator. Of course this requires a well-defined function of each operator.

In order to ensure program efficiency and prevent memory overflow, the planner is programmed in a tree structure which spreads in width instead of depth, to prevent too many recursions.

4.9 DISCUSSION

The main goal of this chapter has been to provide a scheme which will enable the programming of robots by task specification. This saves the trouble of providing feature/surface information at the time of task specification. But this approach increases the difficulty of clarifying the ambiguity of a task whenever there is an unresolvable situation. Such ambiguous cases require detailed planning to resolve the ambiguity problem, otherwise, specify the mating feature/surface information of the interacting objects in the assembly task.

During reasoning, the task planner is dependent upon the geometric and function reasoning engine to a considerable degree. Without correct lists of mating and grasping sets, the planner will never be able to select the optimal motions. In order to reason about the mating faces efficiently, the classification of primitives plays an important role in the whole background work of the RALPH system. This part has been studied with useful results [NK88].

When RALPH is applied to the real working environment, one of the first practical problems to be encountered will be the position error between the world model in the CAD system and the real environment. According to Lozano-Perez[LP83a] using sensors is the best solution for the uncertainty. But sensing requires general I/O mechanisms for acquiring sensory data, and needs versatile control mechanisms for using sensory data to determine robot motions. In order to possess this capability, the current RALPH provides the flexibility of specifying predetermined sensory requirements along with the command. Besides sensor-based motion strategies which must be addressed, the obstacle avoidance system is another important feature that a useful system should possess and which RALPH now possesses.

CHAPTER 5

Sensors and representation

Sensors permit robots to interact with their environment in an *adaptive* and *intelligent* manner. There are numerous transducers available which measure the most important physical variable of interest to manufacturing.

5.1 BACKGROUND

For a robot programming system to be a truly dexterous and intelligent system, it must be provided with sensor capabilities. Because an automatic robot programmer is system independent, its sensor capabilities have to be able to accommodate any set of sensors which may be found in a particular robotic environment. By properly representing sensors to the automatic robot planner, it is possible for the planner to explicitly understand what can be perceived and is being perceived by the sensors about the environment, thus increasing its capabilities in planning the appropriate robot motions to perform a task. The objective of this chapter is to provide a set of principles by which this sensor functioning and information can be represented to an automatic robot programming system. The various events and interactions in a robot's manufacturing environment will be described.

In order for a proper representation to be designed, it is necessary to understand the various aspects of processing sensor information for robots, including the sensors themselves. This is especially important in attempting to process sensor information beyond the level to which it is traditionally processed; that is, to the level of interpretation. In order to do this, the type of information that a planner requires in existing systems must first be understood. In automatic robot programming, the sensor-based planning needs to occur off-line with *a priori* information and again during task execution with dynamic sensing information. Because there is an inherent hierarchical structure involved with processing raw data into abstract information, layers of sensor information can be described with the assumption that the highest level information is to be used by the robotic planner according to application.

Information from various sensors needs to be incorporated to give a more complete description of a scene, object, or situation. Consider that humans must in everyday life be able to combine all the various types of information they receive. For robots, this involves a series of issues that includes sensor modeling, control and processing strategies in the multisensor system, integration of the information, levels of representation of information, and evaluation of uncertainty[MHL88]. The problem of information integration is a portion of sensor fusion. Sensor fusion is more completely described as the intelligent combining of disparate sensed information in order to optimize the range and accuracy of the resulting conclusions. Many methods of sensor fusion are described for specific systems, but are not generalized for a variety of sensors. Some of the sensors used in robotics are briefly discussed below before we embark on their representation.

5.1.1 Tactile sensors

Usually, if objects which are to be manipulated are delicate or possess positional problems which cannot be addressed using vision techniques, then a sense of touch must be used to understand the attributes of the part. There are two main deficiencies in the vision systems which make tactile sensing attractive in some applications. These deficiencies are: low accuracy and inability to see hidden objects. Because of inaccuracy, parts may be positioned outside an acceptable range of tolerance. When this happens, the precise position must be sensed using some adaptive approach and this can be accomplished by a sense of touch. In a similar way, the fact that vision systems are unable to see the rear of the objects, allowing for at most $2\frac{1}{2}$D vision makes it impossible to know how to manipulate in the rear of the objects. Therefore, the use of tactile sensors becomes crucial here.

Sophisticated sensors try to emulate the human sense of touch. The human physiological sense of touch has two distinct aspects: the *cutaneous sense*, which denotes the ability to perceive textual patterns encountered by the skin surface; and the *kinesthetic sense*, which denotes the human ability to detect forces and moments [Sny85].

The tactile sensor typically possesses the capability to detect the following [Sny85]:

1. presence of objects,
2. part shape, location, orientation,
3. contact area, pressure and pressure distribution,
4. force magnitude, location, and direction, and
5. moments magnitude, plane, and direction.

A tactile sensor may work according to different principles. Typical sensors have the following structure:

1. A touch surface consisting of a matrix of pressure pins, variable resistors, capacitors or optical devices.
2. A transduction medium, which converts local forces or moments to electrical impulses,
3. A multiplexer which connects the elements of the tone matrix with a measuring and amplification circuit.
4. A control/interface to bring the measured data to a computer for evaluation.

There are essentially two types of tactile sensors: touch and force sensors. The touch sensors are generally used to detect the presence or absence of an object without regard to the contacting force. Included in this category are limit switches, microswitches, etc. The simpler devices are frequently used in the development of a robot's interlock devices. The force sensors measure local touch forces. In the case of a matrix sensor, the computer evaluates the touch pattern with the help of similar algorithms as they are used for machine vision.

5.1.2 Force sensors

These sensors are normally used to measure the robotic system's forces as it performs various operations. Forces which the robot uses in manipulation and assembly are usually of great importance. Force sensing can be achieved by using a *force-sensing wrist* [GWNO86] which consists of a special load-cell mounted between the gripper and the wrist. Another method is to measure the torque exerted at each joint and this can be done by measuring motor current at each of the joint motors.

5.1.3 Proximity sensors

Proximity sensors are used for detecting the properties of the surface of an object without touching it. The typical strategy is to detect the presence or absence of a surface. Proximity sensors are used in collision detection situations and in detecting approach situations for the arm. There are various types of proximity sensors. They include:

1. optical proximity detectors which project light beams and use reflectivity of any encountered surfaces to detect the presence of an object,
2. optical range sensor which projects a calibrated light source onto object surfaces and then measures the reflected intensity,
3. triangulation proximity sensors which project light at an angle to the object and then by using the triangle created by the light projection, the object and the detectors, the distance of the object can be detected.

5.1.4 Ultrasonic ranging

The ultrasonic transducer is another sensor which is used to sense the distance of an object. The transducer emits a pulse of high-frequency sound and then listens to the echo [Sny85]. Since the speed of sound is known, the time between emission and hearing of the echo can be measured to provide the distance.

5.1.5 Infrared

An infrared sensor can be used to measure the presence or absence of an object between the fingers of the gripper. When an object is located between the sensors, the infrared light emitted by the transmitter will be reflected to the receivers of the transmitting arrangement. In the case where the object surface is parallel to the transmitting arrangement, both receivers will detect an equal amount of reflected light. If the object deviates from parallel alignment, with respect to the gripper, surface signals of different intensity will be received. It will become possible for the robot to re-position its hand to accommodate this skewness. The infrared sensor is also used to sense the proximity of an object (workpiece) by evaluating the intensity of the two reflected signals.

5.1.6 Vision

The vision system is usually used for viewing the work area of the robot to provide information about the environment, in real time. It can also be used for augmenting work area information to make precise decisions for the robotic system. Finally, it can be used to confirm the existence of selected entities in the workspace. This last application usually amounts to pattern matching of new vision information against existing information.

Modern vision systems possess mechanisms which enable them to acquire models of objects from the user. These models can be CAD based representations of the objects to provide a richer understanding or they can be as simple as stored pictures of parts which are then extracted to be matched against the new information.

Machine vision is concerned with the sensing of vision data and its consequent interpretation using the computer. A typical vision system consists of the camera and digitizing hardware, a digital computer, and hardware and software necessary to process the vision information as well as to link the various devices. Vision is normally achieved by sensing through pictures and digitizing of the image data. The image is then processed and analyzed to refine it and infer the geometric entities involved which are then matched against existing duplication data.

The processing and analyses of images are intended to reduce the data for interpretation. Clearly, imprecise data can lead to high inaccuracies in data which in turn leads to inaccuracies in object position and orientation. An image is normally thresholded to generate a binary image. The binary image will then undergo various feature measurements to further reduce the data representation and produce more precise data. The data reduction can, in fact, result in data which is of the order of one thousandth the size of the original data.

Image characteristics can be obtained as features of objects. Features can be inferred from the geometric elements which are obtained from approximation of the original image in a pattern. Feature descriptors are normally programmed as software and a variety of geometric entities could combine to yield a class of features.

In 2D image processing, features can include grey scale (maximum, average, or minimum), area, perimeter length, diameter, minimum enclosing rectangle, etc. Other indirect features include, center of gravity, which can be calculated as follows. Let x and y designate a pixel location of the xy-plane. The center of gravity can be calculated for n pixels as:

$$CG_x = \frac{1}{n} \sum_x x \tag{5.1}$$

$$CG_y = \frac{1}{n} \sum_y y \tag{5.2}$$

Other basic indirect features include, eccentricity which is a measure of "elongation;" thinness — a measure of how thin an object is; number of holes, moments; aspect ratio — the ratio of length to width of an enclosing rectangle, etc.

In 3D feature recognition for vision, CAD models of the objects are used. Stereovision is an attempt to obtain a 3D image of an object. 3D images can be obtained by using two cameras or by using structured lighting and optical triangulation techniques to obtain a stereoscopic view of the scene. Such images normally apply to the types of algorithms used in feature extraction for CAD data in understanding the 3D data.

In image processing and analysis, the vision system is normally trained by having it store a series of images which can then be recalled for pattern matching when the system is in real operation.

The ability of a vision system to recognize objects with precision is dependent on a number of factors. The main ones include:

1. the power of the camera, such as field of view, focal length, resolution, geometric distortion etc.
2. the robustness of the associated software,
3. the power of the associated computational device.

5.2 INTERNAL AND EXTERNAL SENSORS

Another way to classify sensors is whether they are internal or external to the robotic device. This new categorization comes from the work of the author in RALPH [BNA88] and the recent work of E. Lin on sensor planning for RALPH [Lin91] upon which the rest of the presentation in this chapter is based. Internal sensors are commonly described as the variety of encoders in robot joints used to measure the current position and velocity of that particular joint or link. External sensors are the sensors which come into contact with, or are activated as a result of, an interaction with the environment outside the robot. Force and torque sensors, which are located in the joints of robots, are considered external sensors in traditional classification schemes.

Because sensors are devices which are inherently inaccurate to varying degrees, it is necessary to consider their reliability. For accurate representation and appropriate decision making concerning information from multiple sensors, it is necessary to provide some modeling of reliability.

5.2.1 Internal sensors

Internal sensors can be either digital or analog. Digital measurement of position and velocity is generally done through the use of encoders [Sny85]. For signals from sensors to be read by a microprocessor, it is necessary that all analog signals be converted to digital. Digital encoders exist in incremental and absolute forms, and can have either a light detector or a magnetic detector, and come in linear as well as rotary types [WCBW86].

A common analog internal sensor is the potentiometer, which is based on a resistance that varies according to the position of the "wiper." As the position of the joint changes, the position of the wiper also changes. The voltage associated with the resistance is measured, and can be used for both linear and rotary joints.

Linear Variable Differential Transformers (LVDTs) are used to convert linear motion into analog signals, providing an output voltage which is proportional to how much the transformer core moves with respect to the windings [BBM82]. Similar devices for rotary joints are resolvers, which also work on the inductance principle.

5.2.2 External sensors

External sensors allow robots to perceive the world in which they operate. Without them, robots would require all objects to be exactly located, the environment to be carefully monitored for no surprises, their joint movements to be extremely accurate, and the programming to be infallible. These requirements are obviously too restrictive. We can characterize external sensors in terms of contact and non-contact.

There is a wide variety of non-contact sensors, from binary vision system to infrared range sensors. Although this type of data is cumbersome, it does provide more complete information about a situation with one data set than most other sensors do. A great deal of research has been conducted on reducing the vast amounts of data into a form which is comprehensible by a robotic controller; the data is segmented, grouped, and interpreted [BC82, GL88].

Non-contact sensors provide information about the scene, resulting in intensity maps of various types. Vision produces maps of light intensity; infrared range sensors detect varying heat intensity; sonar range sensors measure intensity in the form of distance of the object from itself. Another type of non-contact sensor is the proximity sensor. Most proximity sensors detect the presence of objects through changes in inductance [FGL87].

Contact sensors include two types: those which come into contact with the robot's world; and those which are activated as a result of a contact. The sensors in the first category are referred to as tactile or touch sensors. The second category includes all those sensors in the joints of the robot to measure the force and/or torque which is being experienced at that location. Although the force/torque sensors are usually located at the wrist of the robot, they may be placed at any other joint.

Force and torque sensors are based on some type of deflection, which is measured using one of the many methods available. Deflections which are proportionally large compared to the force experienced indicate a more sensitive force sensor [BBM82]. The most common method of measuring deflection is the use of strain gauges. Figure 5.1 depicts a wrist sensor which is capable of measuring force and torque with its strain gauges.

Although tactile sensors can be arrays of single sensors, arrays of electrodes have been coupled with compliant material to produce artificial skins [FGL87]. These are

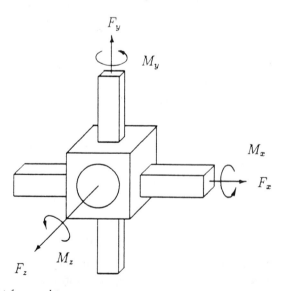

Fig. 5.1 A wrist force and torque sensor.

capable of making continuous measurements rather than the discrete data available simple tactile arrays. All tactile sensors can be considered as arrays of values.

5.3 SENSOR FUSION

In examining robotic sensors, it can be seen that different sensors have different strengths and weaknesses. By combining the information from different types of sensors, it is possible to obtain a wide range of information about the same object, scene, or situation. Information can also be combined when the same type of sensor is used at different locations. This may or may not require moving the sensing device to obtain the different information. With data coming from many sources, however, it is possible to actually have *too* much information to sift through. Therefore, it is necessary to be able to selectively incorporate sets of sensor data for maximum knowledge appropriation. Sensor fusion provides redundancy and increased reliability in a set of sensors. With redundant information, it is possible to verify information from individual sensors, judge when information from a sensor may be false, and compensate for weaknesses in particular sensors. It is also possible for sensor failure to be handled more gracefully, as the failure of one particular sensor does not necessarily mean that a particular piece of information cannot be obtained; it is possible that another sensor or set of sensors can provide the same information.

Most of the research in the area of sensor fusion has been concerned with optimizing the sensing efforts of particular sensor combinations. These combinations are generally designed so that the sensors compensate for each other's weaknesses with their own strengths. Flynn [Fly88] has combined two sensors which both provide a dense set of data, requiring that data smoothing be used to eliminate "bad" data points, as defined by the different useful ranges of the sensors. A different approach by Allen [All88] uses the dense information from a vision system to form hypotheses to be substantiated by the sparse information from a tactile exploration. Others, such as Wang and Aggarwal [WA88], actively reconstruct the 3D image of an object by taking multiple views of it with a vision system.

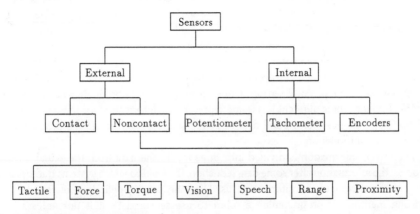

Fig. 5.2 Hardware categorization of sensors.

During processing of sensor information, the important issues are speed and reliability. To facilitate the reasoning during task execution, a reasonable approach is to take advantage of *a priori* information. Kak *et al.* [KVC⁺] modify CAD models of objects into intermediate models called sensor-tuned representations, which are the type of information a range sensor will produce. This is a representation which can be derived both off-line from CAD models and on-line from range data. Eggleston and Kohl [EK88] also make use of prior knowledge to improve decision making during processing. Their three-tiered architecture supports the integration of various sensor sources using target and clutter models. Stansfield [Sta88] presents work in describing objects with a generic representation to be used in recognition.

Strategies for obtaining information are presented by Durrant-Whyte [DW88], who considers sensors as team members working for a common goal, and Henderson *et al.* [HWMH88], who further develop the work previously done by Henderson and Shilcrat [HS84] in logical sensor specifications and Henderson *et al.* [HFH84] in a multisensor kernel system. These two approaches consider sensor fusion from a generalized view which is not dependent on actual sensor hardware.

Information from one source can be supported or refuted by information from another source. The combination of these pieces of information, however, is not simply the averaging of the multiple data sets, for data points which are very "off" should not be considered.

5.4 SENSOR ARCHITECTURE

The nature of an automatic robot programming system such as RALPH[BNA88] precludes the possibility of planning the use of particular sensors at higher levels. Sensor usage at the task level is not determined by the programmer, but is reasoned by RALPH according to the lower-level plans which are developed by the *planner*.

An architecture which supports the type of information flow needed is based on the task decomposition hierarchy as illustrated in RALPH. The highest level of the hierarchy should contain the type of information that a robotic planner needs; the sensors at this level can be referred to as virtual sensors, as they do not exist in a physical sense. A system which needs to accept signals from a variety of sensors cannot be dependent on the physical sensors themselves, but must be able to work with the information from these sensors without being burdened by the specific details of each device [HHB85]. A generalized method of sensor fusion cannot be accomplished without this separation. This presents the enormous problem of creating abstract information for virtual sensors using raw data from physical sensors. Considering this, it is appropriate to divide sensor information into three different levels and represent the information for these three levels [EK88]. This requires that a good deal of planning occurs during off-line programming using *a priori* information.

The sensor information is divided into three levels: the data from the actual sensing process; the information after some data reduction, the level to which systems traditionally process the data; and the information used for the actual application of the data. In existing methods of programming a robot to use sensor data, it is not necessary to interpret the raw data for application purposes, because the robot has been programmed

Sensor architecture

with the particular function of the sensor in mind. The data merely have to be conditioned to remove noise and to be processed with the appropriate software. In RALPH, the generalized sensor plan needs to be transportable from system to system, which requires that there be a great deal of interpretation in order for the information to be useful for a particular application.

Sensor architecture in RALPH is depicted in Figure 5.3. It is divided into three distinct levels of information, as described above. The highest level, the general sensor level, is the least dependent on physical sensors and can be referred to as the applications level. In the middle is the generic sensor level, where *types* of sensors are considered. On the lowest level, the specific sensor level, specific information about a physical sensor is considered.

The planning, which occurs top-down, converts general sensor plans into more specific ones until a plan which is feasible to the particular system being used exists. The bottom-up processing conditions the input signal, extracts information from the data in the form of tokens, and fuses the different sources of information, if applicable, to produce an interpretation of the data.

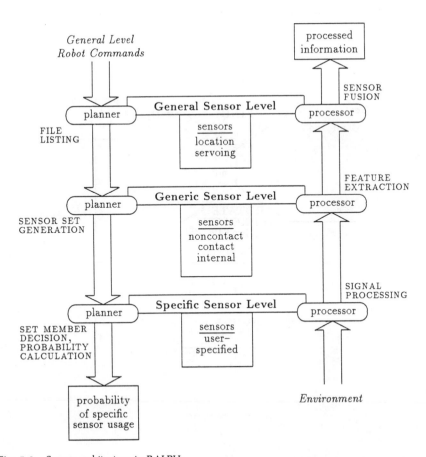

Fig. 5.3 Sensor architecture in RALPH.

5.4.1 General sensor level

On the applications level, there are only two types of information which a robotic assembly system may need: data for the location of objects and data for servoing purposes. These generalized categories allow RALPH to be able to find the information through whatever sensors are available to a particular system. The generalized information works with the general robot-level commands in the *planner*, as was described in Chapter 4.

The locations of objects can be either a specific location needed for robot interaction with the object or a general view of the locations of all objects which can be perceived by sensors. In robotic assembly, servoing functions are those which involve dynamically using sensor information in order to plan further actions while a gripper is interacting

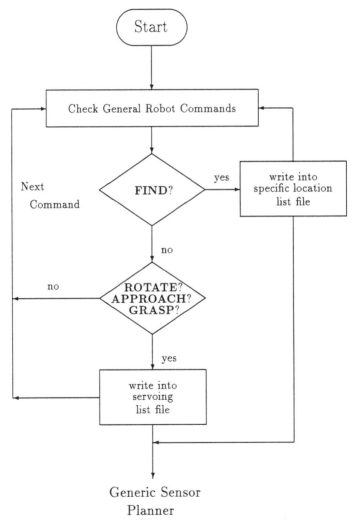

Fig. 5.4 General flowchart of the general sensor planning.

with an object. During these times, the movements between sensor data are very small, and tend to slow down robot operation.

Figure 5.4 shows the general flow of planning at this level. The input to this level of sensor planning are the general robot-level commands. As the commands are read one by one, the key commands – *FIND, ROTATE, APPROACH,* and *GRASP* – are sought. In RALPH, when a *FIND* appears, information is written into a file containing a list of specific object locations to be found. When any of the other three commands appear, information is written into a file containing a listing of servoing information needed.

Other than these specific subgoals, locations of objects in the robot's environment must be known at all times, for a robot's dynamic understanding of all objects in its physical world allows it to grasp specified objects, assemble objects together, and avoid collisions with objects which are obstacles. This is not the same as finding the specific location of an object, but provides a more general picture of the scene, with a variety of sensors working together to provide this information. A complete picture of the objects in the world must be complemented by a knowledge of the forces and torques experienced by the robot due to its interactions with the world, particularly during servoing, when further robot action is dependent on the information. The sensors which provide servoing information, usually force and tactile sensors, are processed fairly quickly. However, because a great amount of information is needed for a relatively small robot movement, it is not efficient for all information of this type to be fully processed at all times. The information can be "masked" during the times when it is not needed. The hierarchical structure which exists in processing as well as planning allows processing to stop before the level of interpretation. By planning for the times during task execution when servoing is needed, the information can be fully processed only at the appropriate moments.

5.4.2 Generic sensor level

On the intermediate level of sensor information, the sensors are divided into three types: *contact, non-contact,* and *internal* sensors. It must be noted, however, that these are not necessarily the same divisions as before.

At this level of sensor planning, the modeling of sensor operation is begun. For the two types of information described above in the general sensor level, there are some types of sensors whose usage is preferable over other types. This also applies to combining the efforts of various sensor types. To differentiate between the choice of sensor type to be used, the use of a particular type or set of types is prioritized. Preferences may be made according to the reliability provided by the choice. They may also be made according to the type or types which provide the quickest information processing. Because these two criteria tend to contradict each other in strength for a single type of sensor, the priorities assigned need to be the result of more than a straightforward consideration. Factors such as actual sensor and positional availability are considered in the specific sensor planning and may change the priorities assigned at this level. Figure 5.5 shows the general flow of generic-level planning, reading the specific location list file and the servoing list file in order to create a generic-level sensor plan.

Priorities can be assigned through the use of conditional probabilities: the probability of using data from a particular sensor or set of sensors, given that a certain type of general sensor level information is required. This is the main idea behind sensor planning

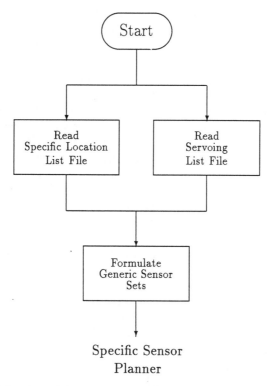

Fig. 5.5 General flowchart of the generic sensor planning.

at the generic level. Given the knowledge of the information required, as obtained through the general level of planning, this probability can be assigned. This method of planning uses knowledge of how sensors work and in what situations they perform best, defining the more desired sensor or set of sensors to be used. At the generic level, however, only generic sets are created, not taking into consideration the actual sensors available and their abilities. Once the sets are planned specifically, the result is utilized during the general level processor, sensor fusion.

5.4.2.1 Non-contact sensors

The *non-contact* sensor types are the largest and most varied group at this level. They include all the varieties of vision, range, and proximity sensors, and are grouped together because the representation of their information is very similar and can be manipulated in identical ways during sensor fusion. These sensors can provide a wide variety of information: the presence of an object in a binary fashion; a two-dimensional picture of the scene in an array; the distances from the sensor to objects within its range. Table 5.1 shows some of the varieties of non-contact sensors available.

The location of *non-contact* sensors can vary a great deal as well. Cameras may be simply attached to a fixed structure, or they may be attached to a portion of the robot and move as the robot moves. Many range sensors may have to be moved about to

Table 5.1 Table of varieties of non-contact sensors

Non-contact sensors	
Vision	Binary
	Gray scale
Proximity	Inductive
	Hall effect sensors
	Capacitive
	Eddy currents
	Ultrasonics
	Optical
	Whiskers
Range	Triangulation
	Structured lighting
Time-of-Flight	Pulsed laser
	Continuous beam laser
	Ultrasonic
	MMW

work properly, as in the case with structured lighting and triangulation. The data for these ranging methods result from consideration of detector locations. Proximity sensors are not particularly useful unless they are located near the gripper of the robot, because these sensors are generally used in order for the robot to recognize when the gripper is nearing a target.

5.4.2.2 Contact sensors

Contact sensors include all the various types of tactile sensors which may exist, from simple microswitches to complicated tactile arrays. The sensors which are activated as a result of a contact, but are not actually physically contacted, for example wrist force sensors, are *not* included in this type. Various types of contact sensors are shown in Table 5.2.

Although most force sensors are located within a joint, a single force sensor located at the fingertip of a gripper may also exist. This particular type of force sensor is considered a contact sensor rather than an internal sensor, because it will come into direct contact with the outside world. Force sensors are the only sensors which may be included in one category or the other.

Table 5.2 Table of varieties of contact sensors

Contact sensors	
Tactile	Strain gauges
	Resistive arrays
	Capacitive arrays
	Magnetoelastic
	Piezoelectric
	Electrooptical
Force	Microswitch
	Continuous

5.4.2.3 Internal sensors

Force sensors, torque sensors, encoders, and all other sensors located on the robot which never come into contact with the world outside the robot are included in the *internal* types. It is their locations which determine their classification into this category. These sensors are capable of measuring joint angles, joint forces, joint torque, and joint velocities. The joints can be gripper joints or those that possess degrees of freedom.

5.4.3 Specific sensor level

It is at this level that the specific sensors in the user's system are included, and specific sets of sensors and probability assignments are made. All of the sensors at this level are specific examples of the sensor types found at the generic level. Working with this level is dependent on being able to access specific details about the sensor being used in the system. Specific sensor information such as the useful range of the sensor, the type of data which is collected by the sensor, and the user classification of the sensor into one of the sensor types described in general sensor planning needs to be modeled by the user in order for planning the use of specific sensors to occur.

The resulting sets of sensors and their probabilities are used during sensor fusion to facilitate the fusing process. Without some plan of *how* and *what* information is to be fused, the process is very confusing and inefficient. The use of goals and hypotheses, along with the sensor sets, provides the processing with a direction. The sensor set with the highest probability value is used first during the processing. If, for some unforeseen reason, the set is not available or is known to be giving false information, the set with the next highest priority is used. This can go on until there are no sets left.

5.5 REPRESENTATION

Representation of sensors is needed for two main purposes, sensor planning and sensor processing. Planning occurs *a priori* before the execution of the task, and processing occurs while actual information about the environment is being obtained. The processed information is used for further dynamic planning by the RALPH robotic *planner*. Because of the *a priori* planning, some of the information may be used for verification of expected results.

The representation of sensors must reflect the way in which they function as well as the information they provide. It is this sensor modeling which allows sensor planning to occur, for it is not possible to predict the behavior of sensors without a descriptive mathematical model.

This section is divided into two large sections with a summary: planning, during which the operation and behavior of sensors is considered; and processing, during which the information from sensors is represented to the *planner*. In the planning section, sensor models and their usage are described. The processing section works backwards, describing how raw data are transformed into representations of information which can be used by the planner. The results of the sensor modeling in the planning stages are used to aid in the fusion of sensor data.

5.5.1 Planning

Similar to the planning of robot joint movements, sensor usage must be planned in successive levels. From the level of the general robot commands, the planning is not complete without plans for sensor usage. In accordance with the type of robot commands at the general level, the sensor plans at this level are also independent of the hardware being used. The general sensor plans are further processed at generic and specific sensor levels using heuristics until a plan specific to the sensors in the system is produced. The tasks which are eventually set to sensors are the result of the planning stage.

5.5.1.1 General sensor planner

The types of sensor information required by the robotic *planner* can be divided into two categories, location and servoing. One of the goals of the *general sensor planner* is to create lists of objects to be found. With these lists it is possible to reduce the amount of time necessary to recognize objects, for *a priori* expectations require more of a verification process rather than a complete recognition. This is true even when identifying several objects of a known group; the effort to match sensed objects with one in the group is much less time consuming than an effort to match them with all objects in the CAD database. Another goal is to create a list of the commands which require sensor information for servoing purposes.

Specific location list

The general level robot *FIND* command requests that the location of a particular object be found. There are three different types of request for finding the location of an object, depending on how much information is known before task execution. Obviously, in the interest of saving time and effort, it is best to use as narrow a search as possible to find a particular object. The first situation is a prediction of the precise location of the object being sought, referred to as a *location search*. With this information, finding the location of the object is merely a verification or relatively small adjustment of the prediction. A more common situation is that a part is known to be in a particular space or area, which can subsequently be searched first. This is called a *fine search*, a limited yet not completely knowledgeable search. The third situation, a *global search*, is used when there is no prior knowledge of where an object exists in the workspace, in which case a great deal of searching is needed to find a particular object.

For each situation, the parameters of the *FIND* command are as follows:

FIND([search mode], obj#, [location parameter])

The *search mode* parameter is one of the types of searches described above: g for global search; f for fine search; or l for location search. The second parameter, *obj#*, is the number given to the CAD model of the object to be found. All objects are given a number in the CAD database. The *location parameter* depends on the type of search to be performed: a global search always has a NULL character, because there is no particular portion of the workspace which particularly needs to be searched; a fine search describes a volume of the workspace in which *obj#* may be found; and a location search indicates the estimated location at which *obj#* may be found.

For a global search, all sensors may be called upon to contribute information. A fine search is more limited, and only those sensors whose useful range includes the volume specified in the *location parameter* are used. A similar concept is applied for a location search. In both of the more limiting cases, a *search sphere* is used to describe the volume of space to be searched for an object. It is expected that the object lies within the boundaries of that sphere. To determine whether or not a search falls within a sensor's useful range, the intersection of the two volumes, $search\_sphere \cap sensor\_range$, is calculated. Methods similar to those used for collision avoidance (Chapter 7) can be used.

An estimated location is represented with a 4 × 4 homogeneous transformation matrix which describes the position and orientation of the body-attached object coordinate frame (OCF) introduced in Chapter 2. With the 4 × 4 matrix, it is possible to perform linear translations as well as rotations with one matrix operation. Because the OCF of an object is permanently affixed, the matrix indicates the location of the object with respect to the fixed world frame, regardless of what transformations are performed on the object. Grasping an object, then, is a matter of assembling the gripper with the object using knowledge of the object location. From the object location, it is possible to reason about the required gripper location. Using inverse kinematics, knowledge of the gripper location is used to calculate the joint angles needed for the robot to acquire the appropriate gripper location.

The format of the specific location list file is similar to the format of the *FIND* command:

seq#_obj#_mode_parameter

where *seq#* refers to the sequence number of the object to be found. In later sensor planning, this is used to reference the particular listing in the file. Figure 5.6 shows an example of a specific location list file. Consider the first row of the file,

l1 4 f 500 250 550 100

'l1' indicates that the command line is the first listing of a specific location. The number '4' identifies the object number of the part to be found, and the 'f' shows that the search mode is a *fine search*. The next four numbers are parameters of the search area; the first three are the $x, y,$ and z locations of the center of the *search sphere*, and the last number is the radius of the sphere.

```
l1 4 f 500 250 550 100
l2 15 1 0 125 500 50
l3 2 g NULL
l4 10 f 10 50 250 75
l5 5 f 500 250 550 100
l6 6 f 500 250 550 100
```

Fig. 5.6 Example of a specific location list file.

Related to the *FIND* command are the various gross movement commands *MOVE*, *LMOVE*, and *CMOVE*. While *FIND* requires specific knowledge about a specific object, the *MOVE* commands require general knowledge about object locations, which is crucial in planning paths around those objects considered to be obstacles. Although *MOVE* commands seem to require more information, processing information for the *FIND* command is actually more intensive. Rather than giving merely the boundaries of where potential obstacles exist, recognition and specific location determination are necessary. Because a robot is generally making or planning moves when not actively seeking the location of an object or performing a fine motion, the outlining of obstacles is an ongoing process. This is similar to the awareness that human beings have of their surroundings even when they are not focusing on a particular object or task.

Another issue of concern in the *MOVE* commands is the maintenance of forces, torques, and orientation. Changes in force and torque readings due to orientation changes may not be of concern unless slippage is a problem. Because a *MOVE* statement is not a particular request for information, but requires monitoring of the environment, readings due to gravity forces must be interpreted as such and are not cause for alarm if within a reasonable range. The concern here is that an unexpected change in force or torque may imply a collision or other problem.

Servoing list

The *ROTATE, APPROACH,* and *GRASP* general robot-level commands require sensor information for servoing. Each of these commands can be associated with a *characteristic parameter*: *ROTATE* should indicate the maximum torque which can be applied in the motion; *APPROACH* needs to have a maximum force; and *GRASP* is equipped with the appropriate grasping parameter. These specific values are needed at the *specific sensor level* in order to predict whether or not the sensors available are capable of providing the information. The format of the servoing list file is as follows:

seq#_action_characteristic parameter

where *seq#* is the number referring to the order in which the actions occur; *action* is one of rotate, approach, or grasp; and *characteristic parameter* is as described above.

Figure 5.7 shows a sample servoing list file. The first line,

s1 a 7 30

```
s1 a 7 30
s2 a 0 50
s3 g 12 grasp parameter
s4 r 12 100
s5 a 12 20
s6 r 12 100
```

Fig. 5.7 Example of a servoing list file.

begins with 's1,' indicating that this is the first servoing listing, and is followed by 'a,' showing that the action is an *approach*. Because the type of action is an *approach*, the following number is a maximum force. The fourth line of the file,

s4 r 12 100,

indicates a rotating listing, with the ending number being the maximum torque allowed.

Other than the times when specific servoing information is required, an overall awareness of forces and torques being experienced by the robot must exist. For this reason, a map of forces and torques is constantly updated.

5.5.1.2 Generic sensor planner

As briefly discussed in earlier sections, planning at the intermediate level is based on the probability of using a set of sensors when a given piece of information is needed. The *generic sensor planner* provides only generic groups of sensors as candidates to be the chosen sources of information. The sets need to be specifically determined by the *specific sensor planner*.

5.6 PROBABILITY OF SENSOR USAGE

The probability of using a generic sensor or a set of generic sensors given that a particular type of general sensor is needed is $P\{g_i | \mathcal{G}\}$ for $i = 1, 2, \ldots, n$. The sensor most likely to be used is g_i if

$$P\{g_i | \mathcal{G}\} = \max [P\{g_j | \mathcal{G}\}] \text{ for } i = 1, 2, \ldots, n \text{ and } j = 1, 2, \ldots, n$$

g_i and g_j refer to a particular generic sensor or set of generic sensors, n is the number of sensor sets to be considered, and $\mathcal{G}$ refers to the general sensor which is required. Recall that a general sensor is either a *location* or *servoing* sensor. The conditional probability is derived from the following equation:

$$P\{g_i | \mathcal{G}\} = \frac{P\{g_i \cap \mathcal{G}\}}{P\{\mathcal{G}\}}$$

Initially, the probability of using a particular generic set g_i over another set g_j is preset for a given general sensor $\mathcal{G}$. These probabilities are later adjusted for penalties due to negative effects and rewards due to positive effects. For instance, if an object is located at the edge of the useful range of a sensor, the probability of using that set is reduced. If the vision sensor being used has 256 levels of gray, the probability of using that set is increased due to the sensor being more powerful than the average. Negative and positive effects are determined at the next level of sensor planning according to the specific members of the sensor sets.

The set of all non-contact sensors can be used to locate objects and to map out the environment. If these sensors are located correctly, these are the most reliable sources of information for locating objects. At other times, it is necessary to use active tactile exploration to provide the information. When this is the case, information from contact and internal sensors needs to be combined. Active use of non-contact sensors is also

Table 5.3 Table of ranked generic sensor combinations

General sensors		$\mathscr{L}$	$\mathscr{S}$
Generic sensor combinations	Rank		
	1	N	IC
	2	NIC	I
	3	IC	NC
	4	NC	N
	5	C	C
	6	NI	NI
	7	I	NIC

N: Set of all Non-contact Sensors
I: Set of all Internal Sensors
C: Set of all Contact Sensors
$\mathscr{L}$: Location Sensor
$\mathscr{S}$: Servoing Sensor

possible in those cases when using one of the ranging techniques, that yields only one point measurement at a time such as a triangulation setup. If a camera can move, it is also possible to acquire two viewpoints of an object or scene to more fully interpret it.

The numerical values of the conditional probabilities of using different combinations for certain information have not yet been thoroughly researched. There are, however, sensor usages which are preferred over others for particular applications. Their values are therefore ranked to indicate the order between them.

Let N signify a non-contact sensor, C signify a contact sensor, and I signify an internal sensor. Each sensor type can be used alone or in conjunction with one or two other sensor types. This results in seven different possible groups. These are N, I, C, NI, NC, IC, and NIC. A single letter represents the entire set of those sensors, as there is not yet more specific information about the sensors in the RALPH user's system. Table 5.3 shows the priorities which are assigned the different generic sensor combinations. Accordingly, the probability ranking of these groups for the general sensors *location* and *servoing* is:

$$P\{N|\mathscr{L}\} > P\{NIC|\mathscr{L}\} > P\{IC|\mathscr{L}\} > $$
$$P\{NC|\mathscr{L}\} > P\{C|\mathscr{L}\} > P\{NI|\mathscr{L}\} > P\{I|\mathscr{L}\}$$

and

$$P\{IC|\mathscr{S}\} > P\{I|\mathscr{S}\} > P\{NC|\mathscr{S}\} > $$
$$P\{N|\mathscr{S}\} > P\{C|\mathscr{S}\} > P\{NI|\mathscr{S}\} > P\{NIC|\mathscr{S}\}$$

where $\mathscr{L}$ represents a *location* general sensor and $\mathscr{S}$ represents a *servoing* general sensor.

5.6.1 Generic plan

Generic sensor sets are created as the result of planning at this level and are written into a file as shown in Figure 5.8. The file is a listing composed of the specific location file and the servoing file created by the general level sensor planner. An 'I' at the

```
l1 1 f 500 250 5 100 P1[N,NIC,IC,NC,C,NI,I]
s1 a 20 P2[IC,I,NC,N,C,NI,NIC]
s2 a [grasp parameter] P3[IC,I,NC,N,C,NI,NIC]
l2 11 f 0 100 5 100 P4[N,NIC,IC,NC,C,NI,I]
s3 a 20 P5[IC,I,NC,N,C,NI,NIC]
l3 2 f 500 250 5 100 P6[N,NIC,IC,NC,C,NI,I]
s4 a 20 P7[IC,I,NC,N,C,NI,NIC]
s5 g [grasp parameter] P8[IC,I,NC,N,C,NI,NIC]
l4 11 f 0 100 5 100 P9[N,NIC,IC,NC,C,NI,I]
s6 a 20 P10[IC,I,NC,N,C,NI,NIC]
```

Fig. 5.8 Example of a generic plan.

beginning of the line indicates a specific location search, and an 's' indicates a servoing information requirement. The set of probabilities at the end of each line is ranked in the order that has been predetermined for this level and given the appropriate probability values. The first line of Figure 5.8,

l1 1 f 500 250 5 100 P1[N,NIC,IC,NC,C,NI,I],

is a generic-level sensor plan to find the location of an object, 'l1' is the location sequence number; '1' is the object number; 'f' is the search mode; the next four numbers are the parameters of the search sphere; and the brackets surrounding the ranked sensor sets with 'P1' indicating that it is the first series of sensor set probabilities.

The different probabilities of sensor usage for the two applications are put into data structures based on those available in the C Programming Language. For further information on these data structures, the reader may refer to [Lin91].

5.6.1.1 Specific sensor planner

Planning of the specific sensor level actually considers the sensors which are available. Using the user's specific information described in earlier sections, it is possible to plan sensor usage specifically. The sets of generic sensors from the generic level planning are processed with the following procedure for sensor planning at this level:

- Adjust set members for availability considerations.
- Set probability of empty sets equal to zero.
- Adjust set members for positional considerations.
- Adjust probability values to reflect positional considerations.
- Adjust probability values to reflect other considerations.

First, actual availability of sensors is checked; every sensor which is eligible to be a member of a set is specifically declared as a member of that set. The sets which become empty as a result of the availability check are assigned a probability value of zero; it is impossible to assign a non-zero value to the probability of using a non-existent sensor.

Next, positional considerations are taken. It is at this point that the *search mode* becomes important, because the location of the workspace to be searched helps to determine which sensors are in the proper position to provide information. A feasible set of sensors can be determined by considering whether the location of the search area is within range of a sensor. The sets are thus pared down to be specific to the sensors which are positionally available.

Similar to the positional consideration is the determination of whether or not the required range of servoing values can be measured by the sensors available. If the maximum value of torque, for instance, is beyond the range of the torque sensor, then the probability of using that set of sensors is lowered accordingly. This can be done for all the sensor sets, after which a new ranking may appear. Penalties are appropriately assigned to reflect these problems.

Other considerations are designed to further adjust the probability of using a set of sensors according to the members of that set. These include the following:

- The probability of using a set is decreased if there is only one member in the set.
- Within a set, the sensors are organized in a hierarchy, so that in the event of sensors providing conflicting data, one sensor's data automatically has priority over another's data.
- Reduce probability of a $\{N\}$ location set if the part has features which may not be detected with non-contact only.
- Discount the use of a sensor in a set if it is presently occupied in another task.

5.6.2 Sensor range

As mentioned earlier, the intersection of a sensor's useful range and the search range must be calculated to determine whether a sensor is positionally able to provide information. The exact location provided by a *location search* is merely a point in space unless the object itself is considered. It is necessary to know the boundaries of the object in order to determine whether its entire body is within the range of a sensor. However, it is cumbersome and not entirely useful to know the location of every corner and edge of an object. In fact, it would be most useful to replace all objects, regardless of their specific shapes, with a uniform geometric primitive.

For ease in mathematical calculations, an object is replaced by the smallest sphere which will completely surround it. The use of a sphere also provides the maximum volume of space to be searched with one primitive without waste. Another primitive such as a parallelpiped is limited in its use, because it is not easily be used to describe many objects adequately. A sphere also allows for some "play" in the object's location description, making it possible to search the largest limited space described. The concept of enveloping objects has been described in Chapter 2.

This method of replacing objects with a primitive is a common technique employed in collision avoidance [Cam89] and is appropriate for finding whether two volumes of workspace intersect. The sphere's coordinate frame is attached in the center of the sphere, as shown in Figure 5.9. In a similar manner, a *fine search* is represented by a sphere

Fig. 5.9 Search sphere surrounding object.

which contains the volume of space to be searched for an object. A *global search* does not consider just a portion of the robot's workspace, and therefore does not require intersections of space to be found; all sensors of a set remain as candidates for a feasible set.

Obviously, the useful range of a contact sensor is dependent on the joint positions of the robot. This requires some control over the robot in order to move the sensor(s) into position. Once deliberate movement is started, the general level sensor becomes a servoing subset of the location sensor, and the information is represented accordingly.

Non-contact sensors, on the other hand, usually need only to have their locations noted in order for their intersections with search volumes to be found. The shape of the useful region of a non-contact sensor is estimated to be a cone with its vertex at the origin (Figure 5.10):

$$x^2 + y^2 = a^2 z^2$$

where a is the cone shape factor, determining how much the cone fans out. This can also be written in cylindrical coordinates as:

$$r = az$$

with

$$r^2 = x^2 + y^2$$

The extent of the cone in the z direction, to be referred to as z_{max}, is the maximum distance which the sensor can measure.

Non-contact sensors may be attached to the robot so that their locations are dependent upon the robot's joint angles. In these situations, the location of the sensor can be found with a matrix multiplication similar to the following:

$$^{world}T_{sensor} = {}^{world}T_{robot\_base} \cdot {}^{robot\_base}T_{joint1} \cdot {}^{joint1}T_{joint2} \cdot {}^{joint2}T_{sensor}$$

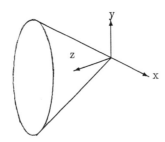

Fig. 5.10 Cone representing non-contact sensor range.

Probability of sensor usage

In this case, the relationship between *sensor* and *joint2* is known, and is described by $^{joint2}T_{sensor}$. This type of calculation is also useful for finding the location of contact sensors.

The first step in finding the intersection of the *search_sphere* with the *sensor_range* is to eliminate those sensors of the set whose intersection with the search volumes is null. If *search_sphere* ∩ *sensor_range* = $\{\emptyset\}$ for a particular *search_sphere*, the sensor associated with the *sensor_range* no longer needs to be considered as a member of the location search set. The test for $\{\emptyset\}$ occurs as follows:

- Calculate the transformation matrix $^{sensor}T_{search}$ from the known locations of the sensor and the search area in the world.

$$^{sensor}T_{search} = {}^{world}T_{sensor}^{-1} \cdot {}^{world}T_{search}$$

- Determine the location of the sphere in sensor coordinates

$$\begin{bmatrix} x'_{sphere\_center} \\ y'_{sphere\_center} \\ z'_{sphere\_center} \\ 1 \end{bmatrix} = \begin{bmatrix} {}^{sensor}T_{search} \end{bmatrix} \begin{bmatrix} x_{sphere\_center} \\ y_{sphere\_center} \\ z_{sphere\_center} \\ 1 \end{bmatrix}$$

- Rule 1:
  if $(z'_{sphere\_center} + r_{sphere}) < 0$, then *search_sphere* ∩ *sensor_range* = $\{\emptyset\}$ (Figure 5.11).
- Rule 2:
  if $(z_{cone\ max} < (z'_{sphere\_center} - r_{sphere}))$, then *search_sphere* ∩ *sensor_range* = $\{\emptyset\}$ (Figure 5.12).

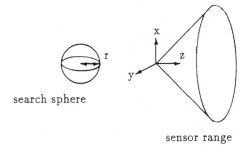

Fig. 5.11 No intersection: rule 1.

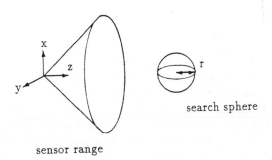

Fig. 5.12 No intersection: rule 2.

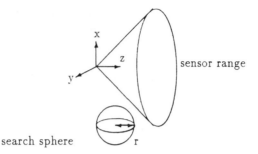

Fig. 5.13 No intersection: rule 3.

- Rule 3:

  if $az_{cone} = (\sqrt{(x'_{sphere\_center})^2 + (y'_{sphere\_center})^2} + r_{sphere})$,
  then $search\_sphere \cap sensor\_range = \{\emptyset\}$ (Figure 5.13).

If the operation *search_sphere* ∩ *sensor_range* does *not* yield the null set, then the volume of the intersection is considered. For the sphere to be completely within the cone, *both* of the following properties must be true:

- $\sqrt{(x'_{sphere\_center})^2 + y'_{sphere\_center})^2} + r_{sphere} < az'_{sphere\_center}$
- $(z'_{search\_center} + r_{sphere}) \leq z_{max}$

If either of these properties is violated, then a portion of the object is out of range of the sensor, implying that the use of only this particular sensor is not sufficient to completely identify the object. The position of a resulting volume which is equal to that of the search sphere is checked for its proximity to the maximum edge of the *search_sphere*, for this may imply that the data is not as reliable as that from the middle of the range.

5.6.3 Set considerations

Sensor sets must be evaluated from the viewpoint of the entire set. After sensor range considerations have been taken, the members of a set will not change any further. The value of the probability of using a sensor set is the only change to be made at this point. The following rules apply:

- If there is only one member in a set, multiply the probability of using that set by the penalty value p_{single}. For example, consider a situation where either a sensor set of $\{N\}$ sensors or a sensor set of $\{NIC\}$ sensors are available to find the location of an object. If the $\{N\}$ set contains a single vision sensor, while the $\{NIC\}$ set contains tactile and internal sensors as well, the use of the $\{N\}$ set may not be as reliable as the use of the other, although the probability of using a $\{N\}$ set is initially higher than that of the $\{NIC\}$ set. The value p_{single} is determined empirically by experimenting with different values. It is the same value regardless of the situation.
- If a location set of $\{N\}$ sensors has only vision members and is being used to find an object which is not a primitive, but has features, multiply the probability of using that set by the penalty value $p_{features}$. The use of merely vision sensors does

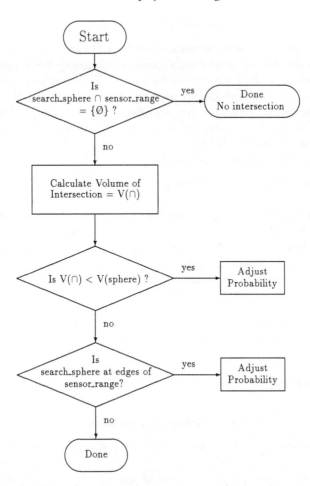

Fig. 5.14 Search sphere and sensor range intersection procedure.

not guarantee the recognition of features; a camera may be angled in such a way that the one distinguishing feature of the object cannot be seen. Again, this value is determined empirically and stays constant.

- If multiple sensors of a set are not located such that different perspectives of an object, scene, or situation can be obtained, multiply the probability of using that set by the penalty value p_{view}. The value of the penalty is dependent on the combined locations of the sensors, their measuring capabilities, and the geometry of the object itself; it is the only penalty of this group which changes for each situation. An object which is a primitive shape, for instance, does not require as many views as an object with features. A spherical object with no features requires the least number of views for recognition.

The above set considerations are multiplied together to produce the combined effect:

$$P\{g_i|\mathcal{G}\}(adjusted) = p_{single} \cdot p_{features} \cdot p_{view} \cdot P\{g_i|\mathcal{G}\}$$

where p_{single}, $p_{features}$, and p_{view} are all less than or equal to 1.0.

5.6.4 Individual sensor properties

Consideration of the individual properties of sensors is used to adjust the overall probability assigned to a sensor set and to establish a voting priority among the members of a set. There are two phases to these adjustments: the first is adjustment of the probability value according to the sensor manufacturer's reliability data; and the second is adjustment according to particularly positive sensor properties. In the second type of adjustments, only rewards are made, strengthening the use of particularly good sensors, but not weakening the use of average sensors.

Sensors which are rewarded are as follows:

- A continuous beam laser, for its capability to provide more information than a pulse beam.
- Use of sonar and infrared range sensors from the same orientation to provide three-dimensional information.
- A vision system with gray level according to how many levels of gray are available.
- A proximity sensor if the range between its minimum and maximum values is small compared to the characteristic length of the object being sensed.
- Linear Variable Differential Transformers (LVDTs) for their reliability and speedy response.
- Microswitches, contact and internal, for their reliability and simplicity.
- Capacitance-based sensors for their high resolution.
- Optical non-contact sensors for their dependability.

Because all of the above considerations are rewards, the values are greater than 1.0.

5.6.5 Gripper considerations

An essential sensor-related component on a robot is its gripper. Grippers vary from those which are very simple and specialized to those which are quite complex and versatile [Owe85]. There is a great variety of sensors which may be attached to a gripper, as shown in Figure 5.15. It is impossible, however, for all the sensors shown to be attached to a single gripper. The amount of information from such a variety would

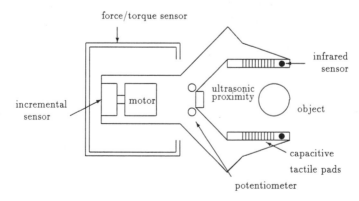

Fig. 5.15 A gripper equipped with a variety of sensors.

be overwhelming and time-consuming to process, therefore reducing the usefulness of each individual sensor.

Grippers are usually equipped with a set of sensors useful for their particular application. A vacuum suction gripper, for example, can be equipped with only a microswitch monitoring the up and down motion of the gripper. For the applications of this type of gripper, it is not necessary to have more information. A two-jaw gripper, however, is different in its needs, because it is generally used to perform tasks which require more feedback information, such as grasping object sides which are not parallel to the surface of the gripper.

5.7 PROCESSING

During actual task execution, processing of sensor information occurs. The processing begins with raw sensor data and ends with the interpreted information needed by the planner in a manner similar to the successive planning described above. At each stage of the processing, a certain level of information is provided by all the sensors. The final sensor sets and their probabilities which were determined during *specific sensor planning* control the paths which are taken to interpret the sensed data. The use of this type of goal-driven processing reduces the effort and time needed to process the data properly[Gar87]. It is assumed that traditional signal processing occurs before the data enters the RALPH sensor system, with the signals having had low-level processes such as noise conditioning already performed on them.

5.7.1 Specific level processor

In order for all data sets to be handled during processing, the data should be formatted according to the generic sensor type of which the particular sensor is a representative. Each sensor is provided with a sensor coordinate frame as well as an object coordinate

Table 5.4 Table of coordinate frames for generic sensor types

Generic sensor type	Sensor coordinates frames
Non-contact	y, x, z axes
Contact	y, x axes (grid)
Internal	z, x, y axes

frame, as shown in Table 5.4. *Non-contact sensors* have their coordinate axes attached to the center of the detector end, which points at the scene of interest. *Contact sensors* are provided with coordinate axes at one corner of the tactile grid. If the sensor detects only a point force, the origin of the axes is at that point. For all *internal sensors*, the coordinate axes are attached in the center of the sensor.

Uniform output formats

Recall that *non-contact sensors* consist of vision, range, and proximity sensors. The data sets obtained by *non-contact sensors* may be the result of their one sampling or multiple samplings over time. Multiple samplings may be needed for those sensors which are capable of providing only a single datapoint with each sampling.

Vision and range sensors are all capable of providing an $[n \times m]$ array of values measuring light intensity, distance, or some other value, as shown in Figures 5.16, 5.17,

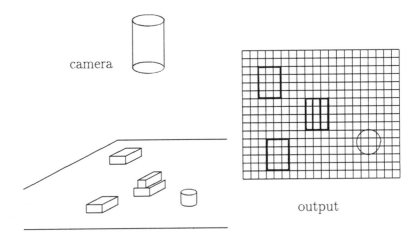

Fig. 5.16 Vision sensor.

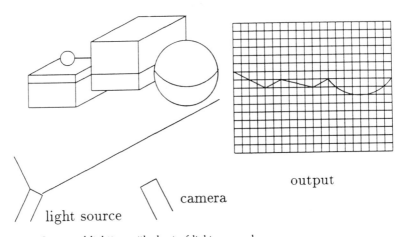

Fig. 5.17 Structured lighting with sheet of light approach.

Processing

5.18, and 5.19. The data from these sensors, then, are converted to these uniform formats by the *specific level processor*. Single sampling points may be put into a [1 × 1] array or into larger arrays to be filled over time. A proximity sensor is not as sophisticated as the other *non-contact sensors*, providing only binary information, shown in Figure 5.20, of whether or not an object is in its sensing range.

The outputs from *contact sensors* are like most of those from *non-contact sensors*, easy to format as a two-dimensional array of values (Figure 5.21). These identical formats are an advantage during higher level processing, because many of the same token extraction software can be used for different sensor types.

Internal sensors provide information in very simple formats that do not need to be altered at this level. The data are simply passed on to higher levels of processing.

5.7.2 Generic level processor

The objective of the *generic level processor* is to extract tokens of information from the data sets from each sensor. These tokens are chosen to contribute to information for the two general level sensors. Tokens are pieces of information from a data set not yet

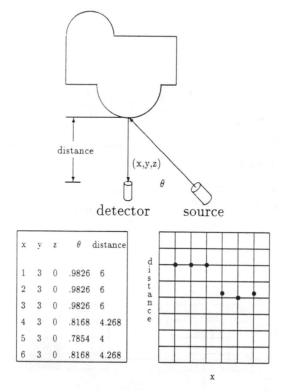

Fig. 5.18 Triangulation ranging technique.

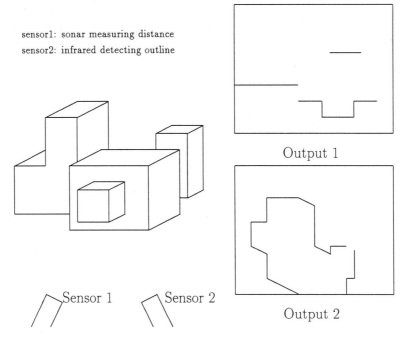

Fig. 5.19 Time-of-flight devices with different forms of output.

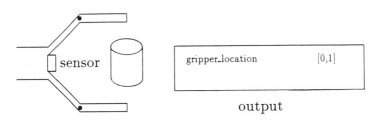

Fig. 5.20 Proximity sensor.

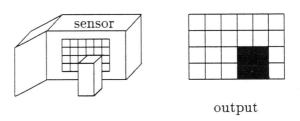

Fig. 5.21 Tactile pad on a gripper.

interpreted enough so that a robotic motion planner can use the information. For *location* and *servoing* general level sensors, the appropriate tokens are extracted.

Extraction of location entities

The location of objects relies on the system's ability to recognize objects. The easiest tokens of information to use in object recognition are geometric tokens. Almost every sensor can provide some sort of geometric information. The main sources of this information are, of course, non-contact sensors, which provide a great deal of geometric information with one data set. However, sensors such as tactile sensors are capable of providing geometric information as well as information about the robot's physical interaction with the world.

Geometric entities are defined as follows:

- Line segment – a straight line segment with a beginning and ending point.
- Arc segment – a curved line segment which is identified by a center location, radius, beginning and ending angles, and the direction of the arc.
- Vertex – the point at which two or more lines or edges meet.
- Edge – a line which is created by two planes.
- Void – a space within a set of faces which contains no substance.

All extracted tokens are referenced in the individual sensor coordinate frames.

Arrays from non-contact sensors are analyzed by first considering each pixel, its value, and its relationship to surrounding pixels. Although the word "pixel" will be used in this discussion, this analysis is not limited to vision systems alone. Except for the outermost rows of pixels in the array, each pixel is surrounded by eight other pixel (Figure 5.22). A pixel can be considered connected to other pixels surrounding it which are similar in color, intensity, or distance. "Blobs" of pixels are identified and analyzed for octal chains.

Octal chains are composed of pixels which, in any one direction of the chain, are connected to another pixel in only one of the eight possible directions. Figure 5.23 shows two chains. The chain on the left is not an octal chain, because it contains an extra pixel, which is marked by a cross. The chain on the right, however, is a true octal chain.

The octal chains are used to identify the outlines around blobs of different intensities. Line segments are extracted from these octal chains. Segments extracted from one octal chain do not necessarily belong to one object, although that is the case more often than not. They are stored as a group of segments at this level. This grouping aids in object recognition in general level processing. Tactile arrays as well as non-contact arrays can be analyzed in this manner. For each sensor which creates output in the form of an array of values, the octal chain representations are analyzed for geometric entities.

Fig. 5.22 Pixel neighbors.

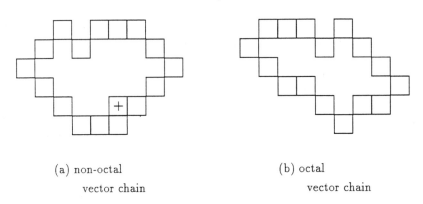

(a) non-octal
vector chain

(b) octal
vector chain

Fig. 5.23 Pixel chains.

Straight line segments are found through analyzing the changes in direction between the pixels. Arcs, curved line segments, are found by combining small straight line segments which form portions of a many-sided regular polygon. Vertices are identified as the point at which at least two line segment endpoints exist. Although an arc cannot make a true vertex with a straight line segment or another arc, it is useful to know that the two geometric entities meet at that point.

True edges can be found only with contact sensors at this processing level. Although most line segments found through non-contact sensors are eventually interpreted to be edges, it is possible that etched letters or other non-edge line segments will be detected. The absolute knowledge that a segment represents the division between object and space can be obtained from a contact sensor. Similarly, voids can be known with high certainty from the use of a contact sensor. Voids are important components of features used during object recognition; it is possible that the mere presence of a particular feature containing a void can uniquely identify a part.

Other tokens that can also be extracted at this level are the nongeometric attributes which can contribute to object recognition. Although they are not described here, they should eventually be a part of the recognition system:

- Color patches – the color of an area, including gray level. The boundaries of the color are identified.
- Surface roughness – related to the texture of an object, it is measured by the average height of the roughness.
- Texture – wavelength of the surface roughness, indicating the density of the roughness.
- Reflectivity – the extent to which a surface acts like a mirror.

Extraction of servoing entities

Unlike finding the location of objects, finding servoing information does not require a great deal of interpretation on the part of the sensor system. That job falls to the robotic *planner*. The information needed for *approach, grasp,* and *rotate* commands are derived from understanding how the robot is coming into contact with the world. Therefore, contact entities are found for this purpose. Contact entities are those tokens which are associated with the robot coming into contact with other objects, as follows:

- Stress patch – the area over which contact between two surfaces is being experienced. This information generally comes from contact sensors, although a non-contact sensor could possibly provide the information.
- Force entity – internal forces being experienced by the joints of the robot.
- Torque entity – internal torque being experienced by the joints of the robot.

5.7.3 General level processor

The *general level processor* combines the different sets of data and interprets the information in order to come up with the types of conclusions which have been planned for. The sensor data sets are combined according to the set with the highest probability value determined during the planning stages. Information for a *location sensor* request is obtained by integrating the various location entities found for each individual sensor. Likewise, information for a *servoing sensor* request is obtained by integrating the various servoing entities found for each sensor. In addition to the specific pieces of information, general maps of the world are created. The general locations of all objects and the contact stresses experienced by all the robot joints are mapped out at all times. This allows the robot to be able to be "aware" of the aspects of the world which are not planned for, including emergency types of situations.

Servoing information

Providing servoing information generally does not require actual reasoning about the information. Generally these are forces from different sources which combine to determine whether or not multiple sensors are associated with the same contact point or area.

Stress patches are accumulated into a map which describes the changing patterns of contact over time and space. In a similar manner to the constant updating of a world map of objects for collision avoidance, the map of contact stresses constitutes a global awareness of the forces and torques being experienced at all times. The maintenance of expected forces is essential in being able to recognize when an unexpected problem has occurred. Areas of expected change, of course, occur during the moments when the necessary servoing information is focused upon.

Recognition of related servoing entities is important in being able to combine information from the different sources of a sensor set. Once different entities are known to be related, forces can be resolved into single vectors. The information is represented as a description of the forces and torques experienced at the wrist, and whether or not the gripper is in contact with another object.

Wrist interactions are described by a force vector and torque vectors about x-, y-, and z-directions, as depicted in the wrist sensor in Figure 5.1:

$$\boldsymbol{F} = F_x \cdot \hat{\boldsymbol{i}} + F_y \cdot \hat{\boldsymbol{j}} + F_z \cdot \hat{\boldsymbol{k}}$$

$$\tau_x, \tau_y, \tau_z$$

When contact sensors on the robot are activated, the information is described by the location of the interaction, the array of values showing the variation throughout the contact, and the joint involved:

$$([3 \times 4] \text{ location matrix}) \quad ([n \times m] \text{ array of values}) \; joint_n$$

Each wrist interaction can possibly be associated with a fingertip interaction. When this is the case, the association is indicated in the representation.

The map of all contacts is represented as a series of patches,

$$\{S_1, S_2, S_3, \ldots, S_n\}$$

for n patches of contact. Each patch is described by an array of values and the location of the patch. If the patch is perceived by a non-contact sensor, the array values are binary throughout the patch, merely indicating whether or not a contact point exists.

Location determination

Location determination of objects is also divided into two levels. The more global information is the constant updating of where objects exist, identifying where they are roughly located with primitives replacing the exact objects. A map of the compiled primitives is used by the *planner* for path planning around the obstacles. Precise object location is needed for robot interaction and is required only at the times when the information is planned for. The appropriate sensor set is used during object recognition, according to the probability value found during planning.

The fusion of different views of the same scenes mainly involves recognizing where matching lines and endpoints occur. Once datapoints from different sets can be identified to be matched, it is possible to interpret the resulting three-dimensional geometric entities which belong to particular primitive families. Using the boundaries of the primitives, the appropriate coordinate frame can be attached and the parameters of the shape identified.

The following outlines the process for determining the primitive envelope of a polyhedral object (approaches for determining generalized primitives can be found in [Kan91]).

- For all the surfaces, if vertices of the object exist on only one side of the half space created by the surface, then the surface is a member of the envelope. For instance, consider Figure 5.24. While faces 1 and 2 (f_1 and f_2) are obviously members of the envelope, $f_3, f_4,$ and f_5 obviously are not. f_6 and f_7 also are not members of the envelope.
- Once the set of surfaces is determined, the surfaces must form a complete closed object. If not, one or more surfaces must be added to the set to close the object. Figure 5.25 shows the planes extracted from Figure 5.24. Because neither f_6 nor f_7 were members of the primitive, the primitive envelope formed by the planes is not closed; the right side of the figure contains no plane. A plane is therefore added to complete the primitive.

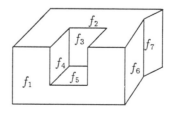

Fig. 5.24 Original polyhedral.

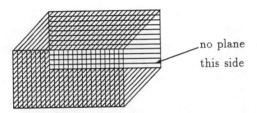

Fig. 5.25 Primitive envelope planes.

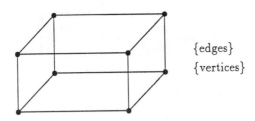

Fig. 5.26 Vertices and edges of primitive envelope.

- From the set of surfaces, intersections of the surface planes produce edge and vertex sets of the primitive envelope. Figure 5.26 shows the edges and vertices of the resultant primitive.

Using information to identify particular objects is made easier through the use of the sensor sets created during planning. The tokens from the highest ranked set of specific sensors should provide the information required. However, if the data is found to be inadequate, the next highest set of sensors may have to be used. With the use of another sensor set, it is possible that a very different approach to the problem is now necessary. Specifically, further acquisition of information is necessary if the switch is from all non-contact sensors to active exploration with a contact sensor.

A robot programmer such as RALPH should be equipped with a graphical simulator that executes a set of robot commands on the workstation screen ahead of the execution of the commands by the actual robot. Because the simulator is the dynamic world model for RALPH, it must work closely with the sensor system in order to create an accurate world model. While it is evident that the dynamic world model is dependent on sensor information, it may not be evident that the sensor system uses the simulator as a tool in object recognition.

The sets of tokens extracted by the *generic level processor* can be predicted to some extent by the simulator. By knowing the precise location of a sensor, the simulator can project the image of an object onto the vision viewing plane and predict which geometric entities from that object can be extracted by that sensor. These entities become a hypothesis as to the set of tokens to be found. Naturally, the actual sensor data will contain more information, a good portion of which is not useful during the recognition process. If the tokens extracted from the actual sensor match the simulated tokens closely enough, supporting the hypothesis created by the simulator, it is not necessary to go through a complete recognition process, saving valuable time during task execution. If the tokens are not familiar, or if not all the sensors of a set provide the predicted information, then the complete recognition process must occur. Whether the hypothesis is accepted or rejected, the actual data are used by the simulator to update its world model.

The location of an object, then, is identified by the following homogeneous 4×4 matrix, which indicates the position and orientation of the attached object coordinate frame.

$$\begin{bmatrix} 3 \times 3 \text{ rotation matrix} & 3 \times 1 \text{ position vector} \\ 1 \times 4 \text{ perspective transformation} & 1 \times 1 \text{ scaling} \end{bmatrix}$$

The maps of potential obstacles are presented to the *planner* as a set of primitives. In cases where the obstacle avoidance software is based on rectangular blocks, the occupied space is presented as a set of rectangular prisms:

$$\{p_1, p_2, p_3, \ldots, p_n\}$$

for n primitives, where each primitive is described by the location of its object coordinate frame, its length, width, and height.

5.7.4 Summary

The sensor representation scheme discussed in this chapter is dependent on the sensor architecture presented in section 5.4. Each of the three levels has a unique representation during planning of sensor usage and during processing of sensor information. Using the *general level robot commands* from the robotic *planner*, the *general level sensor planner* creates the listings of the general level sensors, *LOCATION* and *SERVOING* sensors, needed to provide information.

These lists are then processed by the *generic level sensor planner*, with sets of *generic level sensors* being created for each *general level sensor* listed. The sets of *generic level sensors* suggest the combination of different sensors to provide one piece of information, as in sensor fusion. By providing each *general level sensor* with a group of sensor sets, alternative *generic level sensor* usage is also planned for.

At this point in the planning, the specific information about the sensors in the system is considered. This information comes from the user models created before-hand, shown as the upper-left portion of Figure 5.27. Notice how "sensor file" is routed as an input to the *specific sensor planner* in the planning block. The *specific level sensor planner* considers the physical availability of each *specific level sensor* – whether members of particular *generic level sensors* exist, whether a sensor's position renders it able to provide the proper information, and whether the sensor is currently occupied in another task. The sets themselves are then looked at as a whole, with consideration of whether the combination of sensors suggest that the usage is not "ideal". The further the set is from "ideal," the more the set's probability of usage is penalized. Finally, the individual sensors are considered for each set. Some specific sensors of one generic type work better than others. The existence of strong members in a set causes the set to have its probability of sensor usage rewarded. During the adjustment of probabilities, it is possible that the ranking between the different sensor sets may change due to particularly strong situations in lower-ranked sets or particularly weak situations in higher-ranked sets.

Sensor data are processed by the *specific sensor processor* into formats dependent on the sensor's *generic sensor* type. The *generic sensor processor* analyzes individual sensors' data in order to extract tokens of location and servoing entities. The *general sensor processor* combines the different entities according to the sensor sets created as a result of the planning stage. In Figure 5.27, the specific sensor sets are routed from the lower-right portion of the diagram to the processing portion in the lower-left part of the diagram. If the set with the highest probability value is discovered to be an infeasible set due to

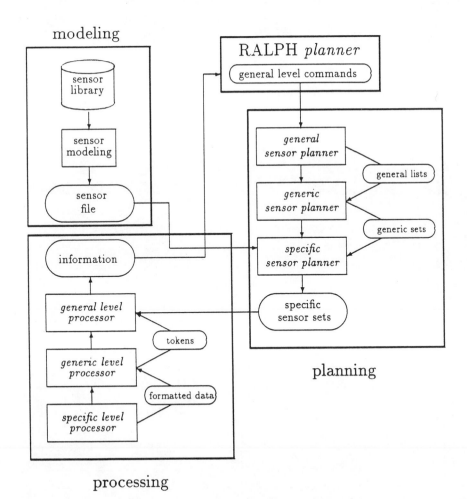

Fig. 5.27 Interaction of sensor representation modules and RALPH planner.

a sensor malfunction or other problem during processing, the next highest ranked set is used. In cases where no set produces results after the first consideration, the sets are used in an iterative manner in order to ensure that all the possibilities have not been completely exhausted; the situation may have changed in such a way that higher ranked sets produce results when given a second chance. The iterations cease after say, three complete loops. In this way, a broken sensor does not necessarily halt task execution.

CHAPTER 6

World modeling and task specification

In Chapter 1, the basic elements of the world modeling were introduced. The issue of world modeling and task specification can be addressed together under the same framework because of the developments in CAD techniques and spatial relationships. In this chapter, we will present the concepts of world modeling and task specification.

6.1 WORLD MODELING

World modeling is the task of creating a representation of a robot's operating world including the robot itself with all the characteristics including constraints of that environment. The world modeling task can be divided into the following:

1. Geometric description of all objects and machines in the task environment.
2. Physical description of all objects, e.g. mass inertia;
3. Kinematic description of all linkages;
4. Description of manipulator characteristics, e.g. joint limits, acceleration bounds, and sensor characteristics.

6.1.1 Geometric description

In world modeling, the geometric description constitutes the major task. The static data that describe the robot and its world in entirety can be obtained from the CAD database. This approach is more convenient and accurate than other methods such as using the vision system to obtain either static or dynamic world information. Even in cases where the vision is used, there must be pre-existing models which can be used to validate the vision information for recognition.

One of the main advantages of using the CAD data is that the object (product) description is available, usually in a complete 3D model. If the CAD data are in boundary form then the shape of the product which is to be manipulated by the robot is easily computable. The CAD data which describe the world are usually not as detailed as the product description. The goal in representing the world objects such as jigs and fixtures and machines is to provide enough information to deal with them as obstacles in the workspace and also to ensure that critical interactions are possible and accurate. It is therefore obvious that the model of the world objects which are not directly part of the task description need not have a detailed model. It is possible to merely represent

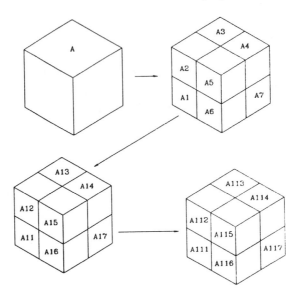

Fig. 6.1 The octree decomposition principle.

the objects which form obstacles as primitive solid shapes. Even in cases where portions of an object are part of the task, the geometric description need not be completely detailed.

Because of the availability of a spatial relationships engine, it is now possible to specify the relationships between world objects very easily. We will discuss this further in section 6.2. World modeling then reduces to providing a geometric description of objects and using spatial relationships to state the locations and orientations with each other.

Creation of the geometric description can be approached in several ways. One method which has shown some success is the *octree* encoding scheme [RSF89]. The octree scheme is a hierarchical representation of finite 3D space. It is similar in concept to both the spatial occupancy enumerations and cell decomposition approaches; and is a natural extension of the quad-tree concept which is used for 2D shape representation.

Figure 6.1 shows how the octree concept works. Imagine that the world can be enclosed by a cube A. It is possible to decompose this world into 8 equal parts, $A_1 \ldots A_8$. Each of these A_i sub-objects can be divided into 8 equal parts $A_{i1} \ldots A_{i8}$, etc. It can be seen that soon each object in the cubic workspace can be represented by unions of these tiny cubes. The level of subdivisions is usually dependent on the computational ability of the computing device and also on whether the object requires detailed modeling. Figure 6.2 is the graphical representation of the octree decomposition.

With the octree approach, it is possible to represent the robot and the objects in particular octants or a combination of octants.

World modeling

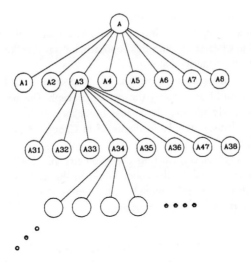

Fig. 6.2 Graphical representation of the octree concept.

6.1.2 Parametric world modeler

We present an approach which combines the advantages of the octree, spatial relationships and parametric specifications.

We imagine that any robot can be decomposed into its many physical components for the purpose of geometric and kinematic modeling. This means that a PUMA robot can be decomposed as shown in Figure 6.3. A database containing all the possible components can be established. Other machines can also be similarly decomposed. Each component carries with it the joint coordinate frame and a map to other components which it may combine with. A graph of inheritance can be created to show the possible product structure for the robots.

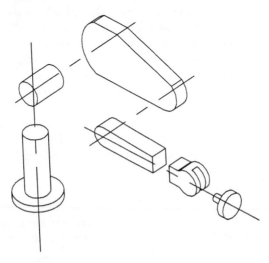

Fig. 6.3 A decomposed robot model.

In producing a linkage, the joint coordinates of a component will be required to articulate with another component using pre-established spatial relationships for mechanisms (as described in Chapter 3). With this approach, linkages can be easily created with all the needed parameters specified at joints.

Having created the linkages, it is now possible to position each linkage system in a world which can be partitioned using the octree concept. The partitionability of the world allows us to easily identify the spatial occupancy of the world entities. Thus computation of the obstacle avoidance discussed in Chapter 7 becomes connected to the modeling scheme.

There are other ways of constructing the geometry of objects in the world. For example it is possible to think of the robot world in the form of a sphere. It is also possible to deal with more realistic models of the world. But it is obvious that such a model would be computationally less tractable.

6.1.3 Physical description

The physical characteristics of the object in a robot world play a very crucial role in robot planning. How fast or how much force to apply to objects are determined through these characteristics. Stability of components can be determined through the physical characteristics. Problems in assembly such as jamming can be modeled or determined through knowledge of coefficients of friction between the assembly components. Of course the physical characteristics of the robot are used in robot dynamics. We will illustrate the importance of physical characteristics by discussing more about stability in section 6.3.

6.1.4 Kinematics of linkages

Kinematics deals with the geometry and time dependent aspects of motion without considering the forces involved in the motion. Through a kinematic model, the spatial behavior of the linkages can be designed. Parameters such as position, orientation, displacement, velocity and acceleration and time can be computed. The relationships of components including that of the robot can be covered by kinematics. We have discussed kinematic and spatial relationship concepts in detail in Chapter 3.

We will assume that $^{i-1}\mathbf{A}_i$ is the matrix which describes the position and orientation of link i with respect to the $i-1$ link.

From the $^{i-1}\mathbf{A}_i$ matrices, the homogeneous matrix $^0\mathbf{T}_i$ which specifies the relationship of the ith coordinate frame to the base frame can be obtained as a chain product of successive coordinate transformation matrices of $^{i-1}\mathbf{A}_i$ as follows:

$$^0\mathbf{T}_i = {^0\mathbf{A}_1^1} \mathbf{A}_2^2 \mathbf{A}_3 \ldots {^{i-1}\mathbf{A}_i} = \prod_{j=1}^{i} {^{j-1}\mathbf{A}_j}, \quad \text{for} \quad i = 1, 2, \ldots, n$$

From this matrix, a five degree of freedom manipulator can yield a transformation matrix $\mathbf{T} = {^0\mathbf{A}_5}$ that specifies the position and orientation of the endpoint of the manipulator relative to the base coordinate frame.

6.1.5 Description of the robot characteristics

To establish a more realistic model of the world, some bounds must be set on the varieties of robot characteristics. Characteristics such as joint limits must be set to ensure a manipulator is not expected to rotate 360° when it is not capable of such a rotation or a continuous one-directional rotation about the base when such a rotation is not feasible. Bounds must be set on other characteristics such as velocity, acceleration, etc.

Although we discussed the issue of sensor modeling and characteristics in Chapter 5, it is here that these models play a role in world modeling.

6.1.6 Complexity of the world model

Modeling of the robot is clearly the most demanding aspect of world modeling. This is because the other aspects of the world such as objects in the world and the assembly objects are expected to be modeled as part of the product design and facilities design tasks. Without the models of other objects of the world this way world modeling will be a potentially intractable task.

6.2 TASK SPECIFICATION

We introduced the problem of task specification in Chapter 1, namely that it is the business of specifying the sequence of configuration states from the initial state to the final state of the task.

We picked symbolic spatial relationships as the viable approach to doing the task specification and developed this concept in detail in Chapters 3, and 4.

If tasks are specified directly by symbolic spatial relationships, then some problems can be created by this approach. One major problem is that geometric and kinematic models of an operation's final state may not always be complete specifications of the desired operation. For example, how much to tighten a bolt is not specified through this approach since it does not deal with the actual operations.

A more viable approach is to describe a task to include parameters of the operations needed to realize the various configuration states. Thus, we can specify a sequence of operations instead of the configurations. The description of tasks in this approach must be object-oriented (task dependent) rather than robot. For example, the torque needed to tighten a bolt can be specified relative to the bolt and not the robot.

Another advantage of using operations specifications is that special relationships between objects are typically implied in operations specifications. The spatial relationships in the operation specification not only state the configuration, but also show the physical relationships between the interacting objects that should be achieved by the operation. In specifying that a bolt *fits* a nut, there should result a compliant motion that moves until the contact is actually detected, rather than a motion to the configuration which is expected to occur.

With this approach, we show an example of an operation-based task specification using Figure 6.4 as follows:

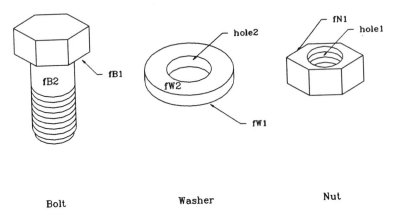

Fig. 6.4 An assembly task.

INSERT BOLT INTO WASHER SO
  (BOLT FITS WASHER) AND FACE (f_{B1})
  AGAINST FACE (f_{w1}).
SCREW NUT ON BOLT SO (HOLE 1 ALIGNED
  HOLE 2) TO (TORQUE = T).

During the task planning process, the task statements are decomposed. Then the symbolic spatial relationships among objects in the SO clauses are transformed to equations on the configuration parameters of the objects in the model. These equations are then simplified to determined the legal range of the configurations of world objects. We will discuss this determination in Chapter 7. The symbolic form of the relationships can be used during program synthesis (Chapter 11).

6.3 ASSEMBLY STABILITY MODEL

For automated assembly, disassembly, fixture, and grasp planning, it is necessary to determine the sequence of steps in which the required goal will be achieved. However, this sequence of steps cannot be arbitrarily chosen. It is necessary to satisfy certain constraints such as geometric and manufacturing resource constraints. Moreover, it is important that the components of the system retain their configuration when subjected to the assembly forces and torques, and frictional and gravitational forces during actual manufacturing.

This section deals with reasoning quantitatively and qualitatively about the stability of a subassembly[VNss]. The subassembly stability is analyzed quantitatively by deriving the dynamic differential equations of motion of the subassembly and by perturbing it with the assembly forces and torques. If any of the components of the subassembly tends to move away from its intended location, the subassembly is considered unstable. As friction plays an important role in maintaining stable equilibrium of the components, the model incorporates friction. In addition to analyzing the stability by the numerical simulation technique, the stability of the system is also determined by studying the effective mass, inertia, stiffness, and damping. The subassembly stability is examined qualitatively by the use of velocity, force, stiffness, damping, inertia, and non-linear forces ellipsoids. Furthermore, to employ the stability analysis procedure in the design stage, a stability index has been derived based on the variations in the velocities, forces, inertia, stiffness, and damping in the system. The stability index enables comparison of stabilities of various subassemblies and choosing the best assembly design.

The stability reasoner is part of the general planner (e.g. RALPH). It has been implicitly assumed in the past [LPB85, LP87] that parts and subassemblies will retain their configuration all through the manufacturing process. For instance when a one-handed robot picks up a part and places it in some location, it is assumed that the configuration of the part will not change until the robot moves it. In cases where the parts "looked unstable" during design, either (i) fixturing was taken as the default without even considering changes in the part or assembly design that would make the system stable and eliminate unnecessary fixturing, or (ii) changes were made to the design to make the subassembly stable, without giving a thought to what minimal changes in the design will bring in stability. This kind of a blind-folded approach to part and assembly design results in non-optimal designs that in turn increase the production costs.

The stability problem has been ignored for the following reasons. A mechanical assembly normally consists of several components with a variety of interconnecting joints, and therefore, it is extremely difficult to develop a model that will determine the stability of the entire system. In addition, there is the action of friction that complicates the modeling process. Part configurations that appear to be unstable may in fact be stable because of friction between the components. Further, the kinematic relationships between the components constantly change while the assembly is in progress and this requires constant updating of the model. Finally, there are computational difficulties that are to be surmounted in order to make stability analysis realistic. Nevertheless, the stability problem has important connotations in automation.

An assembly process is basically a sequence of steps in which the parts are to be placed on a subassembly to achieve the final goal [HdMS86]. However, this sequence cannot be arbitrarily chosen. It is necessary to satisfy certain constraints such as geometric and manufacturing resource constraints to realize the assembly. In addition, it is also important to consider the stability constraints. At every instant of the manufacturing process, the components of the subassembly must be stable. Otherwise, further manufacturing operations cannot be performed on it. This fact is illustrated in Figure 6.5 for two cases: one, a gear box assembly where the stability of the previously positioned gears on shafts dictates whether assembly shall be completed successfully; and the second, a pin and washer assembly, where the intermediate stability of the washers is critical for achievement of the goal.

Analyzing for stability is also helpful in determining the limiting values of the assembly forces and torques. If excessive forces or torques are applied while mating a certain component with a partial assembly (e.g. snap-fitting), then that can result in instability of the subassembly. This is illustrated in Figure 6.6. If forces exceeding the stability margin must be used in the manufacturing process, then fixtures that would take care of the excess forces may have to be employed. A stability analysis model and analysis thus makes both the design and manufacturing processes optimal.

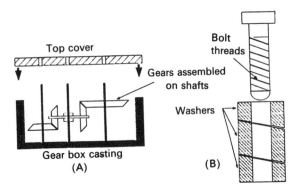

Fig. 6.5 Assemblies where intermediate stability is critical: (A) Gear box (B) Insertion of a bolt into washers.

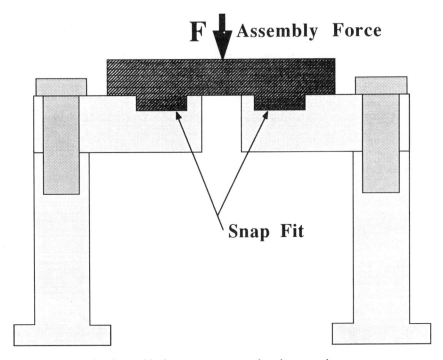

Fig. 6.6 Magnitude of assembly force/torque is critical in this example.

To summarize, a subassembly stability analysis is advantageous for the following reasons:
- It is possible to develop better designs of parts and assemblies knowing their stability. The designer can suitably alter the mass distribution or interfacial friction between components in order to prevent toppling of parts or assemblies during production and also during use.

- Fixturing is a very time consuming operation in manufacturing. The cost of manufacturing increases tremendously as more fixtures are used. The purpose of fixtures is to keep the parts in stable condition during production. If the parts and subassemblies are by themselves stable, there is no need to fixture them. Also, for flexibility and easy access of the tools and robot end-effector, it is preferred to have less bulky fixture elements, if fixtures are indispensable to achieve stability. This implies that one has to decide on the rigidity and type of fixtures. Therefore, a stability analysis helps in optimal fixture selection.
- In situations where fixturing is needed, a stability analysis is useful in determining if a component will retain its configuration while it is being fixtured. Also, when assembling the fixture itself, the "flexible" fixture elements must be stable while they are being fastened in place by the robot. Another situation is when a robot has to re-grasp a part after placing it in a particular configuration. It is again necessary to know the stability of the part in this case.
- A stability analysis is indirectly useful in determining the tolerances [Req83] for the parts and subassemblies during the design phase. This is because tolerances dictate the degree to which the components of the subassembly retain their relative locations or are stable.

6.3.1 Quantitative approach to analyzing stability

For automated assembly, it is importat to know the degree to which a partially completed assembly is stable. While an assembly or each of its components need not be equally stable in all task directions, it must be sufficiently stable or possess enough structural impedance in the directions in which assembly forces or torques are applied. No doubt, the subassembly can be made stable by fixturing. However, fixturing is expensive and it is therefore necessary to consider methods that would eliminate excessive fixturing operations.

We begin by developing a dynamic model for a simple two-block assembly, which actually represents a typical assembly task. We will then evaluate the stability of the system under different conditions, such as when the mass distribution is changed, when fixtures are used, etc.

6.3.2 Description of the two-block system

The two-block system is comprised of two rigid bodies, B_i ($i = 1, 2$), as shown in Figure 6.7 with B_2 supported on B_1. Dextral sets of unit vectors $\langle x_i, y_i, z_i \rangle$ are attached to B_i and are aligned with the principal axes of inertia of B_i. P and $\bar{P}$ are respectively points on B_1 and B_2 at which the two bodies contact each other when there is relative motion between them. B_i has its mass-center at B_{i*} and has mass m_i. The principal moments of inertia of B_i are I_{ix}, I_{iy}, and I_{iz}.

B_1 can translate along x_1 and z_1, and rotate about y_1. B_2 can translate along x_2, and rotate about y_2 and z_2. The system has in total 6 degrees of freedom (DOF). Corresponding to each DOF, generalized coordinates q_i and generalized speeds u_i ($i = 1, 2, 3$) are defined for B_1. Similarly q_i and u_i ($i = 4, 5, 6$) are defined for B_2. B_2 is acted upon by an assembly force $\langle \tilde{F}_x, \tilde{F}_y, \tilde{F}_z \rangle$ at B_{2*} in N, and an assembly torque $\langle \tilde{T}_x, \tilde{T}_y, \tilde{T}_z \rangle$ in N, where N is the

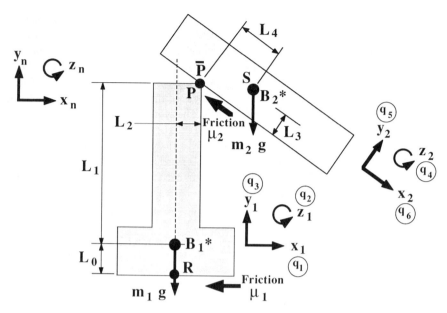

Fig. 6.7 Analysis of a simple two-block assembly for stability.

Newtonian reference frame. The assembly force and torque acting on B_2 are assumed to be functions of time. Frictional forces act on B_1 at R and P, and on B_2 at $\bar{P}$. In addition, there are frictional torques acting on B_1 and B_2. The system is also acted upon by gravity, g. Table 6.1 gives a typical set of values for the parameters of the two-block system.

6.3.3 Equations of motion of the two-block system

The equations of motion for the two-block system are derived using Kane's Method [KL85]. This method reduces an n-DOF system to n 1-DOF systems by the use of partial

Table 6.1 Typical values for the parameters used in the dynamic simulation of the two-block system

Generalized coordinates	q_1	q_2	q_3	q_4	q_5	q_6
m, deg	0.0	0.0	0.0	0.0	0.0	0.0
Generalized speeds	u_1	u_2	u_3	u_4	u_5	u_6
m/sec, rad/s	0.0	0.0	0.0	0.0	0.0	0.0
Dimensions	L_0	L_1	L_2	L_3	L_4	
$\times 10^{-2}$ m	5.0	10.0	2.0	2.0	20.0	
Masses	m_1	m_2				
$\times 10^1$ kg	3.0	2.0				
Inertia scalars	I_{1x}	I_{1y}	I_{1z}	I_{2x}	I_{2y}	I_{2z}
$\times 10^{-3}$ kg $-$ m^2	60.25	8.0	60.25	5.33	419.33	419.33
Frictional Coefficients	μ_1	μ_2				
$\times 10^{-2}$	5.0	5.0				
Acceleration due to gravity m/s^2	g 9.81					

velocities. Detailed derivations are included in Appendix A. The auxiliary variables used in the derivations are described in Appendix B.

If $u_1, \ldots, u_n$ are the generalized speeds of a simple holonomic system, S possessing n DOF, then n quantities $F_1, \ldots, F_n$, called the holonomic generalized active forces for S in N, and n quantities $F_1^*, \ldots, F_n^*$, called the holonomic generalized inertia forces for S in N may be defined. Then, the dynamical differential equations of motion of S in N are written as,

$$F_r + F_r^* = 0 \quad (r = 1, \ldots, n) \tag{6.1}$$

If B_i ($i = 1, \ldots, l$) are rigid bodies composing S, and a set of contact forces acts on B_i, the resultant forces and torques acting on each B_i can be expressed as ${}^N\mathbf{R}^{B_i/Q_i}$ and ${}^N\mathbf{T}^{B_i}$ respectively, where the line of action of ${}^N\mathbf{R}^{B_i/Q_i}$ passes through a point Q_i on B_i. Then,

$$F_r = \sum_{i=1}^{l} ({}^N\boldsymbol{\omega}_r^{B_i} \cdot {}^N\mathbf{T}^{B_i} + {}^N\mathbf{v}_r^{Q_i} \cdot {}^N\mathbf{R}^{B_i/Q_i}) \quad (r = 1, \ldots, n) \tag{6.2}$$

Here, ${}^N\boldsymbol{\omega}_r^{B_i}$ and ${}^N\mathbf{v}_r^{Q_i}$ are respectively the rth partial angular velocity (component of the angular velocity in the direction of the rth generalized velocity) of B_i in N, and the rth partial velocity of Q_i in N. In addition, if ${}^N\mathbf{T}^{*B_i}$ is the inertia torque for B_i in N, and ${}^N\mathbf{R}^{*B_i}$ is the inertia force of B_i in N, then,

$$F_r^* = \sum_{i=1}^{l} ({}^N\boldsymbol{\omega}_r^{B_i} \cdot {}^N\mathbf{T}^{*B_i} + {}^N\mathbf{v}_r^{B_{i^*}} \cdot {}^N\mathbf{R}^{*B_i}) \quad (r = 1, \ldots, n) \tag{6.3}$$

With reference to equations (6.2) and (6.3), ${}^N\boldsymbol{\omega}_r^{B_i}$, ${}^N\mathbf{v}_r^{B_{i^*}}$, ${}^N\mathbf{v}_r^{Q_i}$, ${}^N\mathbf{T}^{*B_i}$, and ${}^N\mathbf{R}^{*B_i}$ are described by the following relationships:

$$^N\boldsymbol{\omega}^{B_i} = \sum_{r=1}^{n} {}^N\boldsymbol{\omega}_r^{B_i} u_r + {}^N\boldsymbol{\omega}_t^{B_i} \quad (i = 1, \ldots, l) \tag{6.4}$$

$$^N\mathbf{v}^{B_{i^*}} = \sum_{r=1}^{n} {}^N\mathbf{v}_r^{B_{i^*}} u_r + {}^N\mathbf{v}_t^{B_{i^*}} \quad (i = 1, \ldots, l) \tag{6.5}$$

$$^N\mathbf{v}^{Q_i} = \sum_{r=1}^{n} {}^N\mathbf{v}_r^{Q_i} u_r + {}^N\mathbf{v}_t^{Q_i} \quad (i = 1, \ldots, l) \tag{6.6}$$

$$^N\mathbf{T}^{*B_i} = - {}^N\boldsymbol{\alpha}^{B_i} \cdot {}^N\mathbf{I}^{B_i/B_{i^*}} - {}^N\boldsymbol{\omega}^{B_i} \times {}^N\mathbf{I}^{B_i/B_{i^*}} \cdot {}^N\boldsymbol{\omega}^{B_i} \quad (i = 1, \ldots, l) \tag{6.7}$$

$$^N\mathbf{R}^{*B_i} = - m_i {}^N\mathbf{a}^{B_{i^*}} \quad (i = 1, \ldots, l) \tag{6.8}$$

In the above equations, ${}^N\boldsymbol{\omega}_r^{B_i}$, ${}^N\mathbf{v}_r^{B_{i^*}}$, ${}^N\mathbf{v}_r^{Q_i}$ ($r = 1, \ldots, n$), ${}^N\boldsymbol{\omega}_t^{B_i}$, ${}^N\mathbf{v}_t^{B_{i^*}}$, and ${}^N\mathbf{v}_t^{Q_i}$ are functions of $q_1, \ldots, q_n$, and time, t; ${}^N\boldsymbol{\alpha}^{B_i}$ is the angular acceleration of B_i in N; ${}^N\boldsymbol{\omega}^{B_i}$ is the angular velocity of B_i in N; ${}^N\mathbf{I}^{B_i/B_{i^*}}$ is the central inertia dyadic of B_i; and ${}^N\mathbf{a}^{B_{i^*}}$ is the acceleration of the mass center, B_{i^*}, of B_i in N.

Before the partial velocities, the inertia torque, and the inertia force can be computed for use in equations (6.2) and (6.3), it is first necessary to determine the velocities and accelerations of the bodies comprising S. This is accomplished by the repeated use of the following equations:

- The angular velocity ${}^N\boldsymbol{\omega}^{B_i}$ of a rigid body B_i in a reference frame N can be expressed in a form involving n auxiliary frames with the help of the addition theorem for angular velocities as,

$$^N\boldsymbol{\omega}^{B_i} = {}^N\boldsymbol{\omega}^{A_1} + {}^{A_1}\boldsymbol{\omega}^{A_2} + \cdots + {}^{A_{n-1}}\boldsymbol{\omega}^{A_n} + {}^{A_n}\boldsymbol{\omega}^{B_i} \qquad (6.9)$$

- The angular acceleration $^N\boldsymbol{\alpha}^{B_i}$ of a rigid body B_i in a reference frame N is defined as the first derivative in N of the angular velocity of B_i in N, so that,

$$^N\boldsymbol{\alpha}^{B_i} = {}^N\frac{d}{dt}{}^N\boldsymbol{\omega}^{B_i} \qquad (6.10)$$

- If P and Q are two points fixed on a rigid body B_i having an angular velocity $^N\boldsymbol{\omega}^{B_i}$ in N, then the velocity $^N\mathbf{v}^P$ of P in N, and the velocity $^N\mathbf{v}^Q$ of Q in N are related to each other as,

$$^N\mathbf{v}^P = {}^N\mathbf{v}^Q + {}^N\boldsymbol{\omega}^{B_i} \times \mathbf{r}_{qp} \qquad (6.11)$$

where $\mathbf{r}_{qp}$ is the position vector from Q to P.

- If a point P is moving on a rigid body B_i while B_i is moving in a reference frame N, the velocity $^N\mathbf{v}^P$ of P in N is related to the velocity $^{B_i}\mathbf{v}^P$ of P in B_i as,

$$^N\mathbf{v}^P = {}^N\mathbf{v}^{\bar{B}_i} + {}^{B_i}\mathbf{v}^P \qquad (6.12)$$

where $^N\mathbf{v}^{\bar{B}_i}$ denotes the velocity in N of the point $\bar{B}_i$ of B_i that coincides with P at the instant under consideration.

- This acceleration of P in N is,

$$^N\mathbf{a}^P = {}^N\frac{d}{dt}{}^N\mathbf{v}^P \qquad (6.13)$$

Finally from equations (6.2) and (6.3), equation (6.1) is obtained.

The equations of motion in equation (6.1) can be expressed in a form involving the generalized coordinates ($\boldsymbol{\theta}$), speeds ($\dot{\boldsymbol{\theta}}$), and accelerations ($\ddot{\boldsymbol{\theta}}$) of the n joints of the assembly as,

$$M(\boldsymbol{\theta})\ddot{\boldsymbol{\theta}} + B(\boldsymbol{\theta})[\dot{\boldsymbol{\theta}}, \dot{\boldsymbol{\theta}}] + C(\boldsymbol{\theta})[\dot{\boldsymbol{\theta}}^2] + G(\boldsymbol{\theta}) + \boldsymbol{\tau}_e = \boldsymbol{\tau} \qquad (6.14)$$

where $\boldsymbol{\theta} = (q_1 \cdots q_n)^T$ are the generalized coordinates; $\dot{\boldsymbol{\theta}} = (u_1 \cdots u_n)^T$; $\ddot{\boldsymbol{\theta}} = (\dot{u}_1 \cdots \dot{u}_n)^T$; $M(\boldsymbol{\theta})$ is the $n \times n$ mass-matrix; $B(\boldsymbol{\theta})$ is the $n \times n(n-1)/2$ matrix of Coriolis coefficients; $C(\boldsymbol{\theta})$ is the $n \times n$ matrix of centrifugal terms; $G(\boldsymbol{\theta})$ is the $n \times 1$ vector of gravity terms; $\boldsymbol{\tau}_e$ is the $n \times 1$ vector of torques due to non rigid-body effects; $\boldsymbol{\tau}$ is the $n \times 1$ vector of joint actuator torques; $[\dot{\boldsymbol{\theta}}, \dot{\boldsymbol{\theta}}]$ is an $n(n-1)/2 \times 1$ vector of joint velocity products; and $[\dot{\boldsymbol{\theta}}^2]$ is an $n \times 1$ vector of squares of joint velocities. Equation (6.14) is the configuration-space representation of the equations of motion. For purposes of numerical simulation, it is convenient to express equation (6.14) as:

$$\ddot{\boldsymbol{\theta}} = M(\boldsymbol{\theta})^{-1}(\boldsymbol{\tau} - \boldsymbol{\tau}_e - B(\boldsymbol{\theta})[\dot{\boldsymbol{\theta}}, \dot{\boldsymbol{\theta}}] - C(\boldsymbol{\theta})[\dot{\boldsymbol{\theta}}^2] - G(\boldsymbol{\theta})) \qquad (6.15)$$

The motivation for expressing the equations of motion in the above form is that in the case of dynamic systems with three or more DOF, the dynamic differential equations of motion of the system can be complex. It will then not be possible to find an exact solution for the configuration of the system at different time periods using the closed form equation in equation (6.14). To overcome this problem, the equations of motion are expressed as in equation (6.14). Then, using a numerical integrator, the configuration and velocities of the bodies composing the system are determined.

Before we can rely on the results of the simulation, it is necessary to verify if the equations of motion have been derived correctly. It is known that a dynamical system has to conserve its energy when there is no gain or loss of energy. This fact is exploited in verifying the equations of motion of the two-block system.

The kinetic energy K of S in N can be expressed as,

$$K = K_\omega + K_v = K_0 + K_1 + K_2 = \tfrac{1}{2}\sum_{i=1}^{l} {}^N\omega^{B_i} \cdot {}^N\mathbf{I}^{B_i/B_{i^*}} \cdot {}^N\omega^{B_i} + \tfrac{1}{2}\sum_{i=1}^{l} m_i({}^N\mathbf{v}^{B_{i^*}})^2 \quad (6.16)$$

where K_ω is the rotational kinetic energy of S in N, and K_v is the translational kinetic energy of S in N; K_j is a function of $q_1, \ldots, q_n, u_1, \ldots, u_n$, and t is homogeneous, and is of degree j ($j = 0, 1, 2$) in $u_1, \ldots, u_n$; and l refers to the number of bodies composing S.

Given generalized active forces $F_r (r = 1, \ldots, n)$, all of which can be regarded as functions of $q_1, \ldots, q_n$, and t, but not of $u_1, \ldots, u_n$, a potential energy V for S in N can be found, if V exists. S has a potential energy only if the following equations are all satisfied [KL85]:

$$\frac{\partial F_r}{\partial q_s} = \frac{\partial F_s}{\partial q_r} \quad (r,s = 1, \ldots, n) \quad (6.17)$$

When S possesses a potential energy V in N, so that, $F_r = -\dfrac{\partial V}{\partial q_r}$ $(r = 1, \ldots, n)$ then,

$$V = \int_{\alpha_1}^{q_1} \frac{\partial}{\partial q_1} V(\zeta, \alpha_2, \ldots, \alpha_n; t) d\zeta + \int_{\alpha_2}^{q_2} \frac{\partial}{\partial q_2} V(q_1, \zeta, \alpha_3, \ldots, \alpha_n; t) d\zeta$$

$$+ \cdots + \int_{\alpha_n}^{q_n} \frac{\partial}{\partial q_n} V(q_1, q_2, \ldots, q_{n-1}, \zeta; t) d\zeta + C \quad (6.18)$$

where $\alpha_1, \ldots, \alpha_n$ and C are functions of t. Usually, $\alpha_1, \ldots, \alpha_n$ are set equal to zero to facilitate determination of V in equation (6.18).

If V does exist, one can construct an integral of the equations of motion of S in N that can be expressed as,

$$H \triangleq V + K_2 - K_0 = C \quad (6.19)$$

where H is the Hamiltonian of S in N, and C is a constant.

6.3.4 Simulation of the two-block assembly

Routines for simulation of the motion of the two-block system can be written and the system is excited with the assembly torques and forces. The response of the system varies with its structural properties, such as the inertia, mass, damping, and stiffness of the joints. In addition, the response is also dependent on the geometric characteristics, such as the location of the mass centers of the bodies.

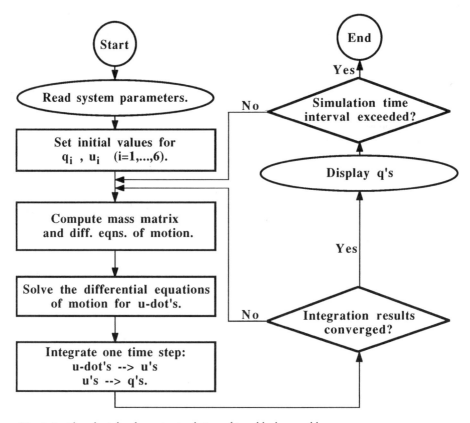

Fig. 6.8 Flowchart for dynamic simulation of two-block assembly.

As the initial configuration and velocity of the system are known, the initial acceleration is computed from equation (6.15). Then by numerical integration, the velocity and configuration at the following time step are determined. Next, the acceleration at the current time step is computed, and the new velocity and configuration vectors are determined. This process is continued one step after another until the end of the simulation time interval. The flowchart in Figure 6.8 illustrates this procedure.

The simulation is useful in analyzing the response of the system for given structural and geometric properties, and assembly conditions. It is also possible to design a new system with impedance characteristics that will yield the required response. In the either case, the magnitude and transient characteristics of the assembly forces/torques must be known beforehand.

Figure 6.9 shows that the two-block system conserves kinetic energy under zero gravity conditions, thus verifying the correctness of the equations of motion. When only gravitational forces act on the system, and there are no externally applied assembly forces, the rotational and linear motion of the system are as shown in Figures 6.10 and 6.11. The simulation results confirm the fact that the two-block system is unstable. The system can be stabilized in several ways, such as by varying the mass distribution, or

Assembly stability model

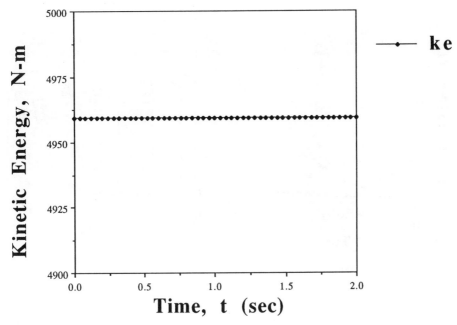

Fig. 6.9 Conservation of kinetic energy under zero gravity conditions verifies correctness of the equations of motion.

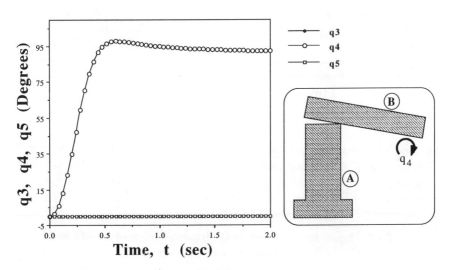

Fig. 6.10 Rotational motion of the two-block system.

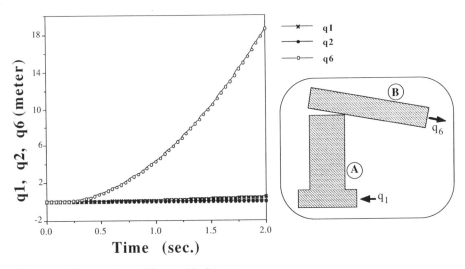

Fig. 6.11 Linear motion of the two-block system.

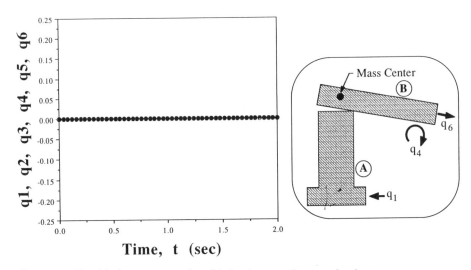

Fig. 6.12 Two-block system is made stable by changing the mass distribution.

by using a fixture element. This is clear from Figures 6.12 and 6.13. It is estimated that as much as 60% of the setup costs during manufacturing is accountable to fixturing. Obviously, since fixturing is a very costly process, it should be the last resort, and more emphasis must be laid on designing components and subassemblies for stability.

6.3.5 Equivalent parameters for transformation of a multi-body system to a single body

The dynamic behavior of a multi-DOF kinematic mechanism is complex due to the interaction between the translational and rotational motion of the joints, and also because

Assembly stability model 147

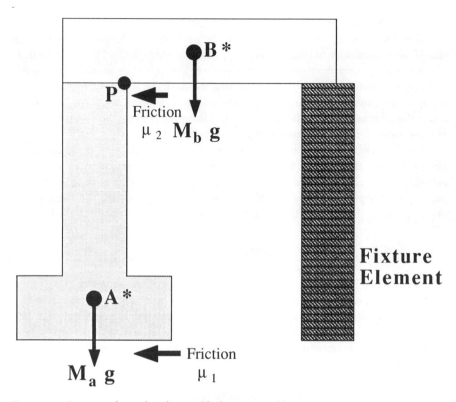

Fig. 6.13 Fixturing also makes the two-block system stable.

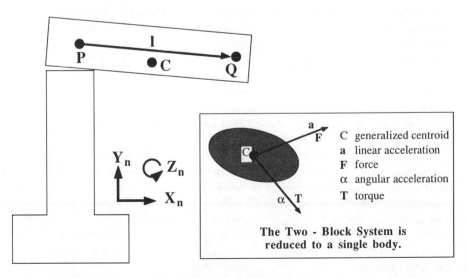

Fig. 6.14 Determination of virtual mass and effective inertia of a system of rigid bodies.

of non-linearities arising from the varying mass distributions of the components of the assembly. However, in order to create a stable assembly, the designer must have an idea of which parameter to tweak to design a stable assembly. Should the effective mass of the assembly in the direction in which the forces are applied be altered? Can a stable assembly be created by changing the stiffness and damping alone rather than altering the effective mass? If so, changing the stiffness and damping is much more easily accomplished than changing the mass. Before making the decision, however, the values of these parameters for the current design must be known.

Once the designer knows the effective mass, the limiting values of assembly forces can be decided. Effective stiffness will show by how much the critical values of the assembly forces can be exceeded and still have the system spring back to its original position on completion of the assembly. Effective damping will tell the designer the degree to which the system will permanently change its position once disturbed.

6.4 DESIGNING FOR STABILITY

Mechanical assemblies have complicated dynamical interactions among multiple joints, nonlinear effects such as Coriolis and centrifugal forces, and varying inertia depending on the configuration of the system. In order to achieve the required stability, it is necessary to eliminate the non-linearities by minimizing the variations in inertia from point to point on the assembly, and obtain a mass distribution that matches the assembly task requirements.

6.4.1 Singularity and stability

The transmission ratios can serve as fundamental performance criteria in the design of an assembly. When the velocity ratio becomes zero, no output velocity is generated at that instant for a finite input velocity. The mechanism is the said to be in a toggle or singular configuration. The force ratio, on the other hand, is the reciprocal of the velocity ratio. Hence, it is infinite in a toggle position. At a singular point, the velocity ratios are zeros, and the mechanism cannot change its orientation in the direction of the eigenvector corresponding to a zero eigenvalue.

In addition to the isotropy of the task space, the magnitudes of the generalized velocity ratios and the generalized force ratios must be within appropriate ranges. If the average of the generalized velocity ratios is excessively small, the magnitude of the output velocity will be too small. If the average of the generalized velocity ratios is excessively large, the magnitude of the output torque will be too small. This will result in poor load capacity of the mechanism.

The ultimate goal is to achieve isotropic stability with proper magnitudes for the transmission ratios so that the subassembly and components are equally stable in all directions and cannot be disturbed by the assembly forces and torques (within the range of isotropicity). Modifying the design by reducing the number of joints or varying the mass distribution, and by choosing the correct assembly directions will result in isotropic stability. The elimination of redundant joints brings down the net DOF of the assembly. This improves the load capacity of the assembly and makes it more attractive for industrial applications.

6.4.2 Inertial effects and non-linearities

Large variations in inertia can cause large non-linear forces. Large differences in the lengths of the major and the minor axes of the generalized inertia ellipsoid imply that excessive variations in inertia will result at the point of interest. Furthermore, when the generalized inertia ellipsoid changes appreciably both in shape and in configuration from point to point in the workspace, non-linearities occur to a greater extent. For greater stability of the assembly, the non-linear forces must be minimized.

The generalized moment of inertia should be made uniform in all parts of the workspace so that the above mentioned problems can be annulled. This may be accomplished by modifying the dimensions and hence the mass distribution of the assembly. When the generalized ellipsoid is a sphere at every point in the workspace, the mechanism is isotropic [YTA87] and there are no non-linearities. In addition, an isotropic system [TT88b] also helps the designer to forget about the stability issue as long as the assembly and other forces are within the limiting value of isotropy (Figure 6.15).

6.4.3 Generalized centroid

Further to reducing the non-linear effects to achieve greater stability, it is essential to optimize the location of the centroid of the mechanism so as to obtain a mass distribution well suited to the task at hand. Since an assembly is a multi-body system, this centroid is referred to as the generalized centroid [AO87].

The inverse-inertia matrix with reference to a point on the assembly is given as:,

$$\bar{M}_x = \begin{pmatrix} \bar{M}_{x_{11}} + \bar{M}_{x_{12}} R^T + R\bar{M}_{x_{12}}^T + R\bar{M}_{x_{22}} R^T & \bar{M}_{x_{12}} + R\bar{M}_{x_{22}} \\ \bar{M}_{x_{12}}^T + \bar{M}_{x_{22}} R^T & \bar{M}_{x22} \end{pmatrix} \quad (6.20)$$

Under certain conditions, there exists a point on the assembly at which translation and rotation are decoupled so that a force causes a linear acceleration, and a moment results in an angular acceleration. This point is the generalized centroid. The effective inertia of the system is a maximum at the generalized centroid, and decreases monotonically as the distance from the generalized centroid increases (Figure 6.14).

At the generalized centroid, the off-diagonal terms of the matrix in (6.20) become

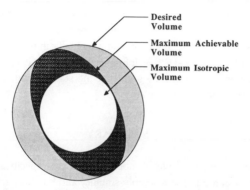

Fig. 6.15 Achieving isotropic stability in mechanical assembly design.

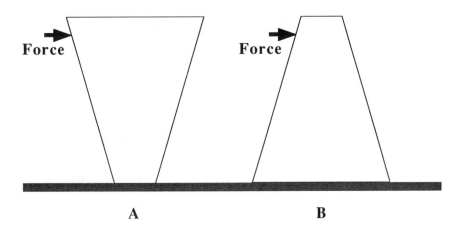

Component B is relatively more stable than component A

Fig. 6.16 Relative stability of the subassembly or its components.

zero. Therefore, no coupling between the translational and rotational inertias exists at that point. The necessary and sufficient conditions for the inverse inertia matrix to reduce to the block-diagonal form are that $\bar{M}_{x_{12}} + R\bar{M}_{x_{22}} = 0$ and $R^T = -R$. These two equations may be used to determine the location of the generalized centroid.

The concept of the generalized centroid may be gainfully employed in determining optimal configurations of the mechanism to obtain the desired mass characteristics at the point of assembly. It is desirable to have the line of action of the assembly forces orthogonal to the vector from the contact point to the generalized centroid. Moreover, the inertia ellipsoid should be oriented in such a manner that the assembly has maximum inertia along the line of action of the force.

6.5 RELATIVE STABILITY

So far, we have considered methods to determine whether a subassembly and its individual components are stable or not, and to create designs with enhanced stability. However, there are times when we may have several subassemblies and need to determine which one is the most stable. Therefore, it is necessary to consider the relative stability of the subassemblies. With reference to Figure 6.16, component B can be concluded to be more stable than component A. To make this inference, the stability index derived in the previous section must be used. It will be seen that the stability index for component B is much greater than that for component A.

6.6 SUMMARY

In this chapter, we have discussed world modeling and task specification. We showed in detail one of the critical models in assembly-stability and how to evaluate components to ensure stability.

CHAPTER 7

Gross motion planning and collision avoidance

7.1 INTRODUCTION

It has been shown that a more advanced approach to robot programming is task-level programming, where the operator does not specify a sequence of manipulator motions and functions, but the general task (independent of the specific robot and the workcell) is in natural language. Thus the problem of automatic robot programming can be formulated [Lau88] as follows: *Given a description of the initial and of the goal situations along with a complete model of the robot world, find a robot motion allowing to reach the goal situation without generating any collision between the robot (the arm and the payload) and the objects belonging to the robot workspace; moreover, the generated solution must verify various constraints (contacts, accuracy, velocity, robustness...) depending on the context of the motion to be executed.* For manipulator robots the motion planning problem can be broken up in three separate parts: gross-motion-, fine-motion- and grasp-planning, while for mobile robots it is usually divided into route- and docking-planning [Rem89]. The objective in this chapter is to present methods of solving gross motion problems in order to generate collision-free paths for general robot arms.

7.2 GROSS MOTION IN RALPH

The gross motion planner is triggered by the RALPH planner which provides information about:

- The robot:
  - kinematic classification of the robot using Denavit Hartenberg parameters;
  - geometric information about the base of the robot as well as about the joints and the end-effector (i.e. gripper);
  - joint ranges and if necessary;
  - geometric information about the payload.
- The environment of the robot:
  - geometric description of all objects in the workspace; and
  - location of all objects (including the robot) in the workspace.
- The present and desired location of the manipulator links.

Given this information, the gross motion planner has to find a collision-free path from the present to the desired position. This path should be found in an efficient way and must be optimal to certain specified conditions (i.e. shortest path or maximum clearance).

The output of the path planner is returned to the RALPH planner and must include the following:
- Existence of a feasible path.
- If such a path exists it must be specified unambiguously and in a way that allows further processing (i.e. trajectory planning).

7.3 ROBOT MOTION PLANNING PROBLEMS

The *classical mover's problem** is to find a path for an n-dimensional rigid polyhedral object P from an initial position p_i to a final position p_f without intersecting fixed polyhedral obstacles O_i. One example is to move a closet through a door or through a staircase.

This problem can be expanded to the *generalized mover's problem* where not a single object but a system of linked polyhedrons P_j must be moved. The following section will show in increasing complexity some special cases of this general problem.

7.3.1 Findspace and findpath problems

The nomenclature for the following two planning problems *findspace* and *findpath* were first used in [LP83b].

7.3.1.1 The findspace problem

The easiest (but not simple) problem is to place an object P in a planar Euclidian space ε^2 occupied with k obstacles O_i so that:

$$P \cap O = \emptyset, \quad \text{with} \quad O = \bigcup_{i=0}^{k} O_i \tag{7.1}$$

Typically (but not necessarily) P and O_i are polygons (see Figure 7.1).

We now can form a set of free placements $FP = FP(P, O)$ which includes all admissible placements of P. Of course this problem can be generalized for 3D space where P has three translational and three rotational degrees of freedom.

The findpath problem is related to the findspace problem, as it is concerned with moving an object P from one location p_i to another location p_f without colliding with any O_i (see Figure 7.2).

7.3.1.2 The findpath problem

Findpath for one object
This was first addressed by Shannon in the late 1940s when he built an electronic mouse moving in a checkerboard-like maze [Yap87]. Later the same problem was studied

---
*The terminology is due to Reif [Rei86].

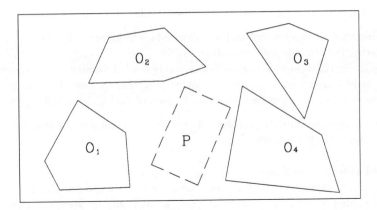

Fig. 7.1 The findspace problem in 2D.

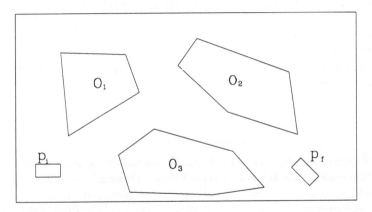

Fig. 7.2 The findpath problem in 2D.

when researchers tried to move a mobile robot with a nearly circular base (to avoid the rotation problem) in a plane. To simplify the problem often all relevant obstacles are projected onto a plane and a 2D object (the robot) must find a path between two positions, even though this method might not find all possible paths. Thus for a realistic robot model, rotation and height information must be included. Another example for the findpath problem is the routing of VLSI circuits, which can be seen as finding a path in a plane.

More generally the robot must not be limited to stay on the surface of a plane, but should be allowed six degrees of freedom (DOF), three for translation and three for rotation (for example, an airplane).

Motion for several connected bodies

This is more complicated than the motion of independent bodies, as the motion of one joint usually results in the displacement of another joint. For articulated robots this problem is hard to solve in Euclidian space, as the displacement is highly non-linear.

Motion coordination for several robots can be divided into two problems:

1. Coordination of independent mobile robots which must not only not collide with obstacles, but also not with each other [BLL89].
2. Coordination of several robot arms. The case of two robot manipulators with independent tasks is described in [OLP89]; in [MG91] a planner for several interacting arms is presented. Another problem is the coordination of the legs for a walking robot.

In the motion coordinating problem not only the path, but also the trajectory (time history along path) must be planned.

Asteroid avoidance problem
In all preceding cases the obstacles were assumed to be fixed and not allowed to move while the robot is in motion. This assumption is valid only for areas like work cells, but for mobile robots this is no longer the case. To complicate the problem the acceleration and velocity bounds of the robot have to be considered [Can87]. This problem is named after the problem of steering a spaceship through an asteroid field.

The shortest path problem
This is the simplest variant of an optimality criteria, as time history is not needed for the computation. It has to be differed between the shortest path problem in Euclidian space and configuration space (see below), as these do not correspond. Another possible optimality criteria for path planning is to maximize the minimal clearance (see section 7.3.4.3).

7.3.1.3 Compliant motion with uncertainty

In gross motion planning the world is usually assumed to be perfectly known. This assumption can be made as normally conservative approximations of the obstacles and the robot are used, and the robot is not supposed to come close to the obstacles. But for fine motion planning (for example inserting a peg in a hole) exact positioning is a must. When the world is not perfectly modeled (which in reality it will never be) sensors have to be used to achieve a task. Thus on-line planning must replace off-line planning. We refer the interested reader to [BHJ$^+$83, Chapter 5].

Before we present some solutions for the findpath problem, we first describe the currently most widely used tool in motion planning.

7.3.2 Configuration space

7.3.2.1 Definition

The positions of all points of an object (or of a set of objects) can be unambiguously specified by a set of independent parameters, which define the configuration space (C-space) of the object(s). The most widely used C-space parameters are the joint parameters.

Configuration space was first mentioned in [LPW79] and allows a unified treatment of different cases like a polyhedron in 3D space and a 6-DOF robot. First, the translational and rotational degrees of freedom are specified as each DOF represents one variable in

C-space. The C-space of an n-dimensional object P is denoted $C\text{-}space_P$ with $d = n + \binom{n}{2}$ dimensions (n parameters specifying the position of the reference point relative to the origin of the world coordinate system and $\binom{n}{2}$ parameters describing the orientation). For example, a polygon P in $\Re^2$ is defined by (x, y, θ) where (x, y) are the coordinates of a reference point in P and θ is the orientation of P (for a rotation about the reference point).

Of course not all configurations q in $C\text{-}space_P$ are feasible configurations for P, so the obstacles O_i must be mapped from the Euclidian space into $C\text{-}space_P$ where they form C-space obstacles CO_i. These regions are defined by:

$$CO_i = \{q \in C\text{-}space_P | P(q) \cap O_i \neq \emptyset\} \qquad (7.2)$$

where $P(q)$ is the subset of the Euclidian space occupied by P at configuration q [ZL89]. Thus a feasible path must be completely in the free configuration space C_{free} which can be found by:

$$C_{free} = C\text{-}space_p - \bigcup_{i=1}^{k} CO_i \qquad (7.3)$$

7.3.3 Computation of the configuration space obstacles

In this section it is shown how C-space obstacles can be created in the case of a planar polygon without rotation and for a manipulator. For the first case CO_i can be found by simply translating P around O_i in such a way that either one edge or one vertex of P is always in contact with a vertex or a edge of O_i (see Figure 7.3). This case is explained in depth in [LP83b].

Finding the exact CO for a manipulator is more difficult, as swept volumes must be calculated. Thus usually the C-space is discretized and tested only for the discrete values of q_j. This is explained in section 7.4.2. Figure 7.4 (adapted from [LP87]) shows the configuration space for a two-link manipulator.

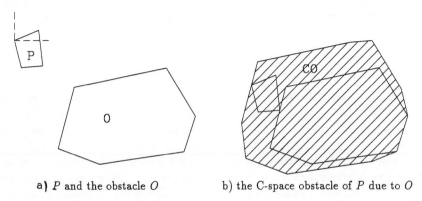

a) P and the obstacle O   b) the C-space obstacle of P due to O

Fig. 7.3 The C-space obstacle for a fixed orientation of P.

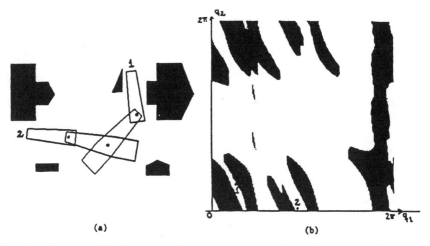

Fig. 7.4 (a) A two-link planar manipulator and obstacles; (b) two-dimensional C-space with approximated obstacles.

7.3.3.1 Advantages and disadvantages of C-space

The configurations space method provides an effective framework for robot motion planning problems like planning a safe path for mobile robots [LP83b], for general robot manipulators [LP87], for planning fine motion for assembly [Mas81], and grasp planning [LP80].

The advantages of a C-space description of the workspace are:

- The joint limits are easily defined, as the limits are usually straight lines in the C-space (not necessarily in the world space).
- In planning a trajectory, the maximum accelerations and speeds are more easily defined in C-space than in the Cartesian space, especially for non-Cartesian robots.
- The computed paths can be used directly for the robot, as no further computation (e.g. as for following straight lines) is necessary.
- The C-space defines the configuration unambiguously, as some robots can reach certain points in the workspace with different configurations.

These advantages of the C-space can be used in all of the techniques described in section 7.3.4. Some of the implemented algorithms are able to find a path in a very cluttered environment. But these advantages have a price:

- The configuration space must be calculated in advance, a process which, even with fast implementations, takes a relatively long time.
- To reduce the time to create the C-space, most algorithms use some kind of conservative approximation, thus some paths might not be found in the C-space search.
- C-spaces with more than three dimensions are hard to search, as the C-space obstacles cannot be described with simple geometrical shapes (only approximated, as straight lines in the workspace are complicated curves in C-space) and are usually stored in a multidimensional bitmap.

- Higher dimensional C-space uses a lot of memory, especially with a fine resolution.

To overcome some of these disadvantages, some researchers [Hoe87, AH] choose a different representation of the C-space. In the Cartesian C-space approach the configuration of a manipulator is not represented in joint angles, but in the position of a reference point (the point where the three axes of the hand intersect). This simplifies the calculation of the C-space obstacles and the search, but as the configuration of the robot is not unambiguously specified, different kinematic states of the robot must be searched.

7.3.4 Path planning algorithms

In this section the most important path planning techniques are reviewed and some successful implementations will be discussed.

7.3.4.1 Visibility graph

The idea is to connect the initial point p_i, all vertices of the configuration space obstacles, and the final point p_f of the motion with straight line segments which do not intersect any of the obstacles. The name visibility graph (V graph) is used as all connected points and vertices can "see" each other. Thus the path will touch some obstacles and is therefore very sensitive to inaccuracies (of either the model or the motion). This graph is searched by treating the vertices, p_i, and p_f as nodes and the line segments as arcs of the search graph (see Figure 7.5).

Even though this method guarantees the shortest path in a 2D configuration space [LPW79], it cannot be generalized to higher dimensional configuration spaces [LP80] and is mentioned here only for completeness.

7.3.4.2 Hypothesize and test

This method was the first implemented method to solve the automatic path generation problem [Pie68]. It can be implemented either in the robot's workspace or in C-space. A path is proposed which connects the initial position with the final position. In the next step this path is tested for collisions* with obstacles (either by sweeping the robot or by using a selected set of configurations along the path). If the path is not collision free the obstacles which caused the collisions are examined and a different path is proposed using applicable heuristics. Typically the first proposed path is divided into segments and only the segments in which collisions occurred are changed.

While it is easy to detect a collision, it can be extremely hard to propose a new path (especially in a cluttered environment), as a change often will result in a collision with a different obstacle.

Often the obstacles are approximated for example with a sphere to avoid dealing with concave obstacles, where a simple heuristic path planner might get stuck in (see Figure 7.6).

Another way is to use more sophisticated heuristics like in [Buc87] where a directed

---

*Sometimes the obstacles and the robot are mapped into cells and the intersection test is simplified to only checking if cells occupied by the robot are also occupied by obstacles.

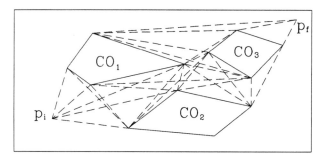

Fig. 7.5 Using the visibility graph to find the shortest path.

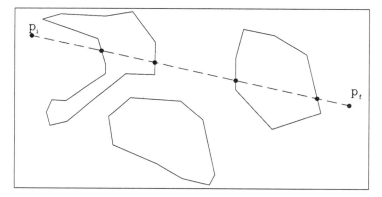

Fig. 7.6 Illustration of the hypothesize and test approach.

distance function Δ (which provides information also in the case of intersection) is used to find a path for a planar robot in a relatively cluttered environment. In contrast to this implementation in the workspace Glavina proposed in [Gla90] an algorithm in C-space which also has been implemented for a planar mobile robot. A point in C-space follows a straight line until a collision is detected, then a subgoal is created by a mixture of random and goal directed search.

In general these hypothesize and test algorithms are simple to implement and will find a path very fast in an uncluttered environment, but are inefficient for a more complicated workspace.

7.3.4.3 Voronoi diagram

A Voronoi diagram Vor(S) is defined as a subset of free space FP with each point simultaneously closest to two or more obstacles. The edges (resp. surfaces in a higher dimension) represent basically the maximum clearance between the obstacles and are thus intuitively the safest way to pass them. The initial point and the final point of the motion are connected to the nearest point in the Voronoi diagram with a straight line segment. Then these two line segments and the two points are linked to the diagram.

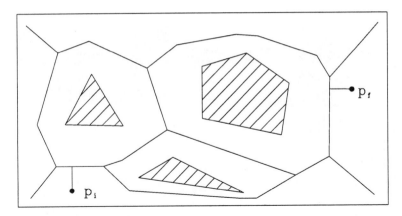

Fig. 7.7 Approximated Voronoi diagram in 2D.

One example for solving the findpath with rotation problem for a rectangle in a plane using a Voronoi diagram is given in [TS89], where the centroid of the rectangle is moved on the Voronoi diagram. The rotation is found by aligning the principal axis of the rectangle with the Voronoi diagram. The implementation seems to be fast, but is limited to a planar environment. While standard Voronoi (see Figure 7.7) diagrams are 1D subsets of a 2D space, a generalized Voronoi diagram had been introduced in [OSY87]. In that work an algorithm called *retraction* for moving a line segment in a planar space had been proposed. The Voronoi diagram surfaces found are retracted a second time so that only 1D curves have to be searched. This work also proves that if a feasible path for a disk exists in the workplane, then a collision-free path can be found using the Voronoi diagram.

7.3.4.4 Cell decomposition

Several researchers tried to map the space in a set of cells. These cells are labeled either full, empty, or mixed. A cell is full if it is completely within an obstacle, empty if the cell does not intersect an obstacle, and mixed otherwise. Usually an *adjacency graph* is created with nodes corresponding to empty cells and arcs between two nodes if these two nodes are adjacent*. In the next step the graph is searched using a heuristic search algorithm to limit the search. Usually the A^* algorithm or a modified version is used, see [BF82] for details.

As shown in [SS83] the cell decomposition is guaranteed to find a path if one exists. But in a very cluttered workspace the size of the cells might become very small in order to find a feasible path, which complicates the search.

Several different cell decomposition methods have been used. The simplest cell decomposition method is to create equisized cells [BX87, Hoe87]. But while it has the advantage of quick and easy cell creation and intuitive cell adjacency a huge number of cells is created which causes the search to be exhaustive. So better representations were searched for,

---

*Two cells a and b are adjacent if $(\bar{a} \cap b) \cup (a \cap \bar{b}) \neq \emptyset$ where $\bar{a}$ defines the complement of a.

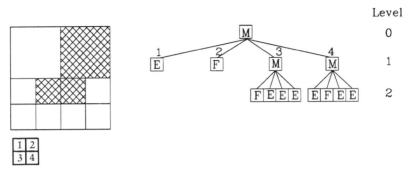

Fig. 7.8 Quad-tree structure.

which allow big cells in uniform space (either empty or full) and smaller cells only when needed (for example at the border of an obstacle). This limits the number of cells (and therefore the search graph) but complicates the adjacency graph. The most straightforward way is to use a *quad tree* [NNA90] (see Figure 7.8) or *octree* [Fav84] to represent the space.

This tree structure makes the cells easily representable but tends to produce a high number of cells as mixed cells are often divided into other mixed cells. So in [BLP83] and in [ZL89] mixed cells are cut in such a way that maximizes either the empty or full area. But the connectivity gets complicated (Figure 7.9).

These two methods also include mixed cells in the adjacency graph and refine mixed cells only if these cells are part of the currently searched graph.

One of the first fast algorithms for articulated robots which solves the path planning problem for pick and place operations was presented by Brooks in [Bro83a]. The free space is represented as a set of *generalized cones* [Bro83b], which are two-dimensional structures defined by the two facing edges of parallel obstacles. These cones are swept over a range of heights (which is defined by the surrounding obstacles) to form 3D *freeways*. The payload and the hand are moved in these freeways, while an approximated C-space is computed for the upper arm. The search graph is created using the overlap of two freeways as nodes. A freeway forms an arc between two nodes if it is large enough to allow a collision-free motion for the payload and the hand.

A different approach is taken by Lozano-Pérez in [LP87] where the C-space is stored in the form of a generalized octree. The legal ranges of configurations are stored recursively

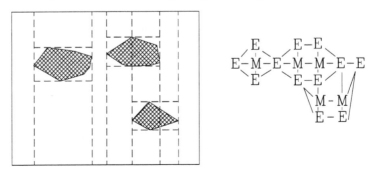

Fig. 7.9 Decomposition and connectivity graph for [BLP83].

in an n-level tree, where n is the number of joints. Thus the obstacles are described as a set of linear ranges, which are called *slice projections*. The free space is then represented as *regions* which are created by overlapping free ranges of adjacent slice projections. The nodes in the *region graph* are those regions and the arcs indicate regions with common boundary. To speed up the planning process, only the portion of the C-space is created which is bounded by the joint values of the initial and final configuration.

7.3.4.5 Potential field

In 1980 Khatib [Kha86] presented a new class of general robot motion planning algorithms. The idea is the following [Kha86]: *The manipulator moves in a field of forces. The position to be reached is an attractive pole for the end effector and obstacles are repulsive surfaces for the manipulator parts*. The algorithm creates a potential field with a minimum at the final position and superposes this field with positive fields created by obstacles. The resulting *artificial potential field* (see Figure 7.10) must be differentiable as the path is usually searched by following the flow of the negated gradient vector field generated by this potential field. The potential around the obstacles is usually inverse proportional to the distance to the obstacle. To simplify the distance calculation, the robot is usually approximated as a set of points.

The advantages of this method is that no expensive precomputation (as for C-space or Voronoi diagrams) is necessary, a smooth path is automatically created, and the joint torques/forces can be easily found as they are proportional to the gradient. As this approach uses only local information (which could be obtained by sensors) it is fast enough to run in real-time. However the problem with this strictly local method is that the planner might get stuck in a local minimum and would not reach the final configuration. To overcome this problem Rimon and Koditschek [RK88] transformed a *star world* to a sphere world in which the *navigation function* is a potential field without local minima. But the analytical description of the transformation is complicated and requires precomputation.

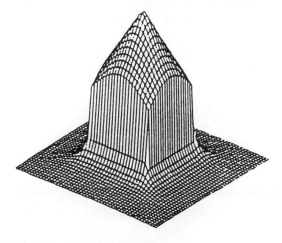

Fig. 7.10 Potential field of a 2-D square [War89].

Warren [WDM89, War89] creates the potential field in the configuration space and avoids local minima by combining overlapping C-space obstacles to one obstacle, as these could create a saddle point. Barraquand *et al.* showed in [BLL89] a simple and highly efficient algorithm to create a numerical potential field without local minima. But as the manipulator is approximated by several points which might compete to reach their goal position a stable position can occur before the final position is reached. In the same work several ideas have been presented to overcome this problem (like Brownian motion and creating a graph of local minima). This algorithm had been implemented for manipulators with up to 31 DOF.

Several approaches to gross motion planning have been presented. They can be classified as either local planning algorithms (hypothesize and test as well as potential field approaches) and global planners (which usually need a search graph). The advantage of the local planning algorithms is their speed, but they cannot guarantee the shortest path and often get trapped in a local minima. Global path planners otherwise will find a good path, but are computationally very expensive.

Most of the existing planners solve the problem only for a restricted class of robots. The first general approach was made in [SS83] using exact cell decomposition, but is not efficient. A much faster algorithm is given in [Can87]; this dissertation also discusses the time bounds of robot motion planning.

7.4 THE PATH PLANNING ALGORITHM

7.4.1 Outline of collision repelling algorithm

As shown in the previous sections, the C-space representation allows to treat all manipulator types uniformly. In [Can87] it was shown that the running time exponent must be at least linear to the number of degrees of freedoms. Having this is mind we decompose the manipulator in two parts, the arm and the hand including payload. This allows us to make a rougher approximation for the arm (which usually does not come very close to the obstacles). The C-space is created only for the first three joints and the hand is approximated with an envelope. Then the free C-space is retracted away from the obstacles

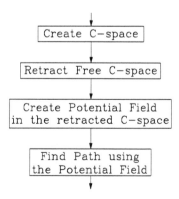

Fig. 7.11 Flowchart of the proposed collision repelling algorithm.

and the starting and ending configurations are linked to this retracted C-space. After finding the shortest path in the retracted C-space the hand is considered. While following the path for the arm the distance of the hand to the obstacles is computed and the hand is turned away from the closest obstacles. This algorithm is graphically represented in Figure 7.11 and called the *collision repelling method*.

It has to be noted that the transformation algorithm to find the C-space obstacles and also the C-space search is independent of the number of joints.

7.4.2 Creating the configuration space

In this section it will be shown how to transform the Euclidian-space obstacles to C-space obstacles for a serial robot manipulator. A serial manipulator consists of prismatic and/or revolute in a non-branching sequence. The C-space is recursively computed using an approach similar to [BX87, LP87]. As pointed out in the previous section, it is not necessary to create the C-space for all n joints. To economize this time-consuming task further, the joints are conservatively approximated by cylinders. Obstacles are assumed to be polyhedral and convex. This does not mean loss of generality, since non-polyhedral objects can be approximated by polyhedral objects and concave objects can either be enclosed in a convex hull or split up into several convex objects. Following the idea of [Fav84], the cylinders which approximate the joints are shrunk to line segments and the objects are grown by the radius of the cylinder and a factor to compensate for the inaccuracy of the discretized workspace. Thus for m approximated joints we have m sets of grown obstacles.

The following algorithm will recursively be applied to all links j starting with link 1 (the base) to link m [LP87] (see also Figure 7.12). In this implementation $m = 3$.

1. Ignore all links beyond link j and sweep link j around or along its axis (for revolute

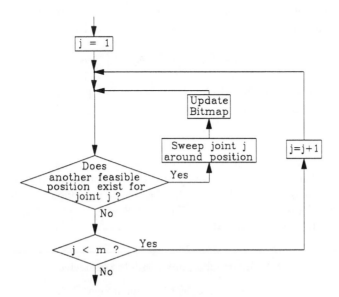

Fig. 7.12 How the C-space is created.

and prismatic joints respectively). The position of the joint is defined by the legal values of the preceding joints $q_1, \ldots, q_{j-1}$.
2. Find the legal values of joint j by intersecting at the swept volume with all obstacles iteratively.
3. Discretize the value range conservatively at the pre-specified resolution and update the bitmap.
4. If $j = m$ then stop, otherwise repeat steps 1–4 for link $j + 1$ for all legal values of link j.

This simple concept allows efficient implementation, but the bitmap representation of the C-space requires memory space exponential to the number of dimensionality. The individual steps are described in detail below.

7.4.2.1 Finding the position of the axis

Before the joint j can be swept, the coordinate frame of its axis j must be calculated. The most widely used method to represent the kinematics of robots is the Denavit-Hartenberg representation [DH55]. This method results in a 4 × 4 homogeneous transformation matrix $^{j-1}A_j$ representing the coordinate system of link j at joint j with respect to the coordinate system of the previous link $j-1$. This matrix can be described for standard kinematic structures by the four Denavit-Hartenberg parameters d_j, a_j, α_j, and θ_j.

$$^{j-1}A_j = \begin{pmatrix} \cos\theta_j & -\cos\alpha_j \sin\theta_j & \sin\alpha_j \sin\theta_j & a_j \cos\theta_j \\ \sin\theta_j & \cos\alpha_j \cos\theta_j & -\sin\alpha_j \cos\theta_j & a_j \sin\theta_j \\ 0 & \sin\alpha_j & \cos\alpha_j & d_j \\ 0 & 0 & 0 & 1 \end{pmatrix} \quad (7.4)$$

Using simple matrix multiplication the relation $^R T_j$ of the jth frame in respect to the base frame (with the number 0) can be calculated as follows:

$$^R T_j = {^0 T_j} = {^0 A_1} {^1 A_2} \cdots {^{j-1} A_j} = \prod_{i=1}^{j} {^{i-1} A_i} \quad (7.5)$$

The relation of the robot in respect to the world is $^W T_R$, thus the relation of the jth frame to the world is

$$^W T_j = {^W T_R} {^R T_j} \quad (7.6)$$

Knowing the position of the robot in the world and the current values of the preceding joints, the position of joint j can be easily computed. If one of the preceding joints i ($i = 1, \ldots, j-1$) intersects one of the obstacles, then this configuration is not feasible and the whole 1D C-shape slice is forbidden. Thus the following step is performed only for legal values of the $(j-1)$ C-space slice.

7.4.3 Intersecting link j with the obstacles

Having found the position of link j, the joint is swept around/along its axis. The swept area is iteratively checked for intersection with all obstacles. Each obstacle i determines

a) Rotation on Axis   c) Translation along Axis

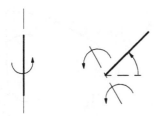

b) Rotation not on Axis   d) Translation not along Axis

Fig. 7.13 The four joint categories.

one forbidden range CO_{ji}. The complete set of forbidden ranges is defined by:

$$CO_j = \bigcup_{i=1}^{k} CO_{ji} \qquad (7.7)$$

To intersect the link with the obstacles it has to be differed between different joint types. As shown in [BX87] robot joints can be divided into four basic categories:

1. Rotation on axis (Figure 7.13(a)),
2. Rotation not on axis (Figure 7.13(b)),
3. Translation along axis (Figure 7.13(c)) and
4. Translation not along axis (Figure 7.13(d)).

Figure 7.13 illustrates the joint categories and Table 7.1 classifies the joints for the bodies of several robot types.

Table 7.1 The joint categories for four robot body types

Joint category	Puma	Stanford arm	SCARA	Cartesian
1	1	1	4	–
2	2, 3	2	1, 2	–
3	–	3	3	3
4	–	–	–	1, 2

The following subsections treat the intersection process for the four joint categories.

Rotation on axis
This is the simplest case because the prohibited interval of link j is either empty or covers the whole range of the joint limits. The infinite line defining the axis is intersected with all planes defining the obstacle. The intersection points are stored and checked if they are part of the obstacle. Note that for a convex obstacle there exist a maximum of two intersection points with a line. No intersection points and one intersection point are trivial cases, but for two intersection points some care has to be taken, as the line segment could be enclosed completely in the obstacle.

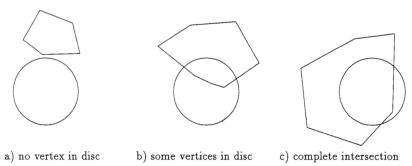

a) no vertex in disc   b) some vertices in disc   c) complete intersection

Fig. 7.14 The three cases when rotating perpendicular to the axis.

Rotation not on axis

As in all commercially available robots, the angle between the joint and the rotation axis is either zero or $\frac{\pi}{2}$, and to simplify the calculations, only the special case of the axis perpendicular to the joint is considered in this section. The axis of the cylinder which approximates the joint is swept 2π around the rotation axis. The swept area is a disc with radius equal to the length l_j of the joint and in a plane orthogonal to the axis. Now check if all vertices of the grown obstacle are on the same side of this plane. If so, then the joint does not intersect the obstacle, otherwise the connecting edges of the vertices on different sides of the plane are intersected with the plane and are stored together with the vertices on the plane. These stored points p_i are expressed in polar coordinates. The origin of the coordinate system O is the center of the disk (where the axis intersects the plane). The angles θ_i are measured relative to the line which is formed by the joint j when $q_j = 0$. R_i is the radial distance, i.e. the Cartesian distance between the point p_i and the origin of the polar coordinate system. It has to be differentiated between three cases (Figure 7.14).

If $\theta_a = \max_i \theta_i$ and $\theta_b = \min_i \theta_i$ (for the points p_a and p_b respectively) for an obstacle i and a joint j, then the C-space obstacle $CO_{ji} = [\theta_{COjmin}, \theta_{COjmax}]$ can be found by:

(a) if $(R_i > l_j \forall i) \wedge (\theta_a - \theta_b) < \pi$ then no vertex is inside the disc but care has to be taken that no edge intersects the disc. Defining the vector v_1 as the vector from p_a to p_b and v_2 as the vector normal to v_1:

$$v_1 = \begin{pmatrix} R_a \cos \theta_a - R_b \cos \theta_b \\ R_a \sin \theta_a - R_b \sin \theta_b \end{pmatrix} \quad v_2 = \begin{pmatrix} R_b \sin \theta_b - R_a \sin \theta_a \\ R_a \cos \theta_a - R_b \cos \theta_b \end{pmatrix} \quad (7.8)$$

Then all p_i can be expressed as $p_i = \lambda_i v_1 + \mu_i v_2$. The critical points p_l and p_m are defined as:

$$p_l = \{p_k | \mu_k = \min_i \mu_i \wedge \lambda_i \geq 0\} \quad p_m = \{p_k | \mu_k = \min_i \mu_i \wedge \lambda_i < 0\} \quad (7.9)$$

The shortest distance d of the line l_{lm} (defined by p_l and p_m) to the origin O can be found by projecting either one of these points on the line which is normal to l_{lm} and includes O. If $d > l_j$ then $CO_{ji} = \phi$, otherwise

$$CO_{ji} \in \left[\theta_m + \arccos\left(\frac{d}{R_m}\right) \pm \arccos\left(\frac{d}{l_j}\right) \right] \quad (7.10)$$

(b) if $\exists (p_i | R_i < l_j) \wedge (\theta_a - \theta_b) < \pi$
   For $\theta_{COj\max}{}^*$:
   Let

$$\theta_m = \max_i \{\theta_i | (R_i < l_j)\} \qquad p_m = \{p_k | (\theta_k = \theta_m) \wedge R_k = \min_i R_i\} \quad (7.11)$$

If $\theta_m = \theta_a$ then $\theta_{COj\max} = \theta_a$ else define the vectors v_1 and v_2:

$$v_1 = \begin{pmatrix} \cos\theta_m \\ \sin\theta_m \end{pmatrix} \qquad v_2 = \begin{pmatrix} -\sin\theta_m \\ \cos\theta_m \end{pmatrix} \quad (7.12)$$

The set P_l is defined as $P_l \in \{p_i | (\theta_i > \theta_m)\}$. Each p_l can be described as a linear combination of v_1 and v_2:

$$p_l = \lambda_l v_1 + \mu_l v_2 \quad \text{with} \quad \begin{aligned} \lambda_l &= R_l \cos(\theta_l - \theta_m) \\ \mu_l &= R_l \sin(\theta_l - \theta_m) \end{aligned} \quad (7.13)$$

Let

$$P_c = \{p_i | (\lambda_i \leq \lambda_k) \forall k \neq l, (p_k, p_i) \in P_l\} \quad (7.14)$$

$$P_l^* = \{p_i | (\mu_i < \mu_k) \forall (p_k, p_i) \in P_c\} \quad (7.15)$$

Intersecting the line l_{ml^*} (defined by the points p_m and p_{l^*}) with the disc results in:

$$\theta_{COj\max} = \theta_m + \arccos\left(\frac{d}{l_j}\right) - \arccos\left(\frac{d}{R_m}\right) \quad (7.16)$$

where d is the smallest distance of the line l_{ml^*} to O.
(c) if $(\theta_a - \theta_b) > \pi$ then $CO_{ji} \in [0, 2\pi]$.

Problems can arise when the axis is not close to one end of the link, as the rear end of the link is not considered in this algorithm. Of course the rear end could be checked separately, but this would nearly double the computation time for this kind of joint.

Translation along axis

If the joint translates along its axis, a similar strategy as in section 7.4.3 is applied. First the line defined by the axis is intersected with the obstacle and then the intersection points p_i (if they exist) are evaluated with respect to the line segment (defined by the joint swept over the full range of the joint limits). For efficiency the joint is shrunk to a point and the points p_i are moved by d_1 and d_2 (see Figure 7.15).

A single intersection point is therefore considered as two points. The intersection point p_i with the smaller u coordinate is displaced by d_1 in negative u direction and the other point is displaced by d_2 in positive u direction. The joint limits do not need to be transformed, as they are expressed directly in u.

Translation not along axis

The first part of this subroutine is similar to section 7.4.3 as the plane defined by joint j and the translation direction is compared to all vertices of the obstacle and if necessary the connecting edges of vertices on different sides of the plane are intersected with the

---

*The other end of the interval $\theta_{COj\min}$ can be found using a similar approach

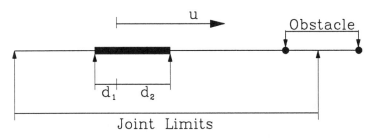

Fig. 7.15 Translation along axis.

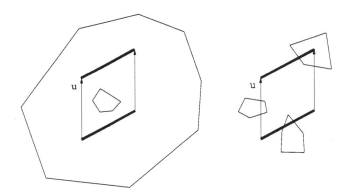

Fig. 7.16 Intersection of the hull of the obstacle with the swept joint.

plane. One way to calculate the forbidden interval is to create the convex hull of the intersection points p_i (see [PS85a] for full details of the algorithm QUICKHULL) which is intersected with the parallelogram of the swept joint. Figure 7.16 shows some cases of importance.

A faster way is to create two infinite lines l_1 and l_2 by infinitely sweeping the joint (see Figure 7.17). These two lines are defined by the translation vector $\boldsymbol{u}$ and the two end points of the joint. If there are points p_i on both sides of one or both of these lines then all points on one side are connected with all points on the other side. Intersecting these connection line segments with l_1 and l_2 results in the two sets of points I_1 and I_2 respectively. The points in set I_2 are now projected in joint direction n onto line l_1. Also all points in between the two lines are projected on l_1 (also using $\boldsymbol{n}$ as the projection vector). All projected points are added to I_1. The forbidden region is now defined by the points of I_1 with minimum and maximum u-values.

7.4.3.1 Updating the bitmap

After the location of the C-space obstacles is obtained, the 1D C-space slice for this joint is easy to calculate. The set C_j of allowed joint coordinates is:

$$C_j = J_j \wedge \neg CO_j \tag{7.17}$$

where J_j is the set of the joint limits for joint j and CO_j is the forbidden interval for joint

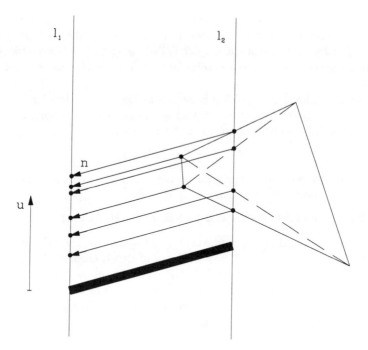

Fig. 7.17 Projecting points on l_1.

j due to the obstacles. Thereafter the C-space slice is discretized into finite steps of size ε_j, where ε_j is adjusted to the required precision. If the joint limits are $J_j = [q_{j\min}, q_{j\max}]$ then the number of steps is $k_j = (q_{j\max} - q_{j\min})/\varepsilon_j$. Thus after joint m is swept, an mD bitmap has been created with $\prod_{j=1}^{m} k_j$ cells.

7.4.4 Finding the path for the arm

After creating the C-space a feasible path must be found. While most previous approaches used some kind of cell decomposition (see section 7.3.4.4), in this research the free C-space is retracted using a numerical potential field. With the cell decomposition method either a huge number of cells must be searched or a different cell representation scheme must be adapted first to limit the search graph.

A numerical potential field had been chosen, as this method allows a fast path generation and does not need an analytical description of the obstacles. First a potential field is created in the dimensional free C-space C_{free} to find the retracted free C-space C_{free}^R. This ensures that the manipulator stays away from obstacles and limits the search space, as only C_{free}^R must be searched. Then the start- and end-point of the motion are linked to C_{free}^R, which is searched to find an optimal (i.e. shortest) path.

7.4.4.1 Creating the retracted free C-space

This method follows an idea presented by Barraquand *et al.* [BLL89], except that this implementation is in the C-space and not in the workspace as in the previous paper. Also the mD bitmap is not completely retracted to an $m-1D$ workspace skeleton. The C-space allows a better representation of the robot, as in the workspace the robot is represented as a set of *control points*, while the C-space is built using a 3D robot model. Also the problem of competition of several control points, which can lead to spurious local minima, is eliminated.

The k-neighborhood of a point p is defined as the set of points having maximal $k(k\in[1, m])$ coordinates differing one step s_j from the coordinates of point x. The set of k-neighborhood neighbors Y_k of point x can be described as:

$$Y_k(x) = \{y| \, \|y - x\|_k < 1\}$$

when the distance between two neighbors is normalized to 1. Thus in a 3D bitmap the 1-neighborhood neighbors of a point $x = (i, j, k)$ are:

$$Y_1(x) = \{(i-1, j, k), (i+1, j, k), (i, j-1, k), (i, j+1, k), (i, j, k-1), (i, j, k+1)\}$$

The distance of a pixel x in the bitmap of the free configuration space C_{free} to the closest obstacle (the joint limits are also considered part of CO) is called $d_1(x)$. It is calculated starting from the C-space obstacles CO_i. The calculation of $d_1(x)$ is terminated when $d_1(x) \geq \varrho$ or if two "wavefronts" originating from different obstacles meet. Points where two wavefronts meet and all remaining points are included in the list C_{free}^R. The effects of varying ϱ are described in section 7.4.4.3.

The retraction of the free C-space is done by the following algorithm (Figure 7.18):

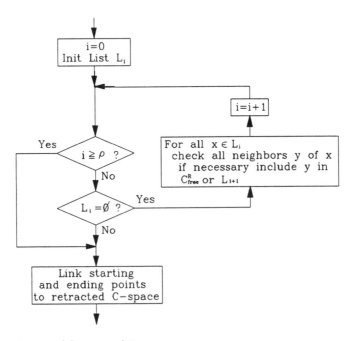

Fig. 7.18 Creation of the retracted C-space.

1. Initialization:

    If $\{x | x \in C_{free}\}$ then $d_1(x) = \infty$ (i.e. a large number M)
    If $\{x | x \in CO\}$ then $d_1(x) = 0$

2. Create list L_i with $i = 0$:

    $$L_0 = \{x | d_1(x) = 0 \wedge \exists(y | d_1(y) = M \forall y \in Y_1(x))\}$$
    $\forall x \in L_0$ do: set $O(x) = x$

3. Loop:
    While $L_i \neq \phi$ do:
    Set $L_{i+1} = \phi$
    $\forall x \in L_i$ do:
    $\forall y \in Y_i(x)$ do:
    If $d_1(y) = M$ then $d_1(y) = i + 1$, let $y \in L_{i+1}$, set $O(y) = O(x)$
    else if $\| O(x) - O(y) \|_1 > \Delta$ then let $y \in C^R_{free}$
    (typically $\Delta = m$ in mD C-space)

4. Set $i = i + 1$
    If $i < \varrho$ go to step 3.
5. $\forall \{x | d_1(x) = M\}$ let $x \in C^R_{free}$

This algorithm stops when all points x in the C-space are labeled or included in C^R_{free} (see Figure 7.19). $O(x)$ is the originating point of x, thus $d_1(x)$ is the L_1 distance between x and $O(x)$.

To link the starting point x_s and the final point x_f of the motion to C^R_{free}, use the following algorithm:

1. Set $x = x_s$ (or x_f respectively).
2. While $x \notin C^R_{free}$: Let $x \in C^R_{free}$, let $y_{max} = \{y^* | d_1(y^*) = \max d_1(y), y \in Y_1(x)\}$
3. Set $x = y_{max}$, go to step 2.

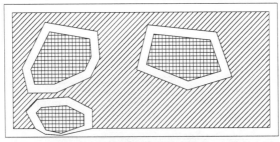

☷ C-space Obstacles

▨ Retracted Free C-space

Fig. 7.19 A retracted free C-space.

7.4.4.2 Creating the numerical potential field in C_{free}^R

While in the original algorithm the numerical potential is created in C_{free}, this algorithm limits the field to C_{free}^R. This speeds up the search, as the C_{free}^R is a subset of C_{free}, but the path will be longer, as the distance to the obstacles is maximized. The following algorithm is used to create the numerical potential field (Figure 7.20):

1. Initialization
   If $x_s = x_f$ then stop, as this is a trivial case.
   Set $d_2(x_s) = 0$ and $d_2(x) = M \forall x \in C_{free}^R, x \neq x_s$.
   Set $i = 0$ and $L_i = x_s$.
2. Loop:
   If $L_i = \phi$ then stop with error message.
   $\forall x \in L_i$ do:
   $\forall \{y | y \in Y_m(x) \land y \in C_{free}^R\}$ do:
   If $d_2(y) = M$ then $d_2(y) = i + 1$, let $y \in L_{i+1}$.

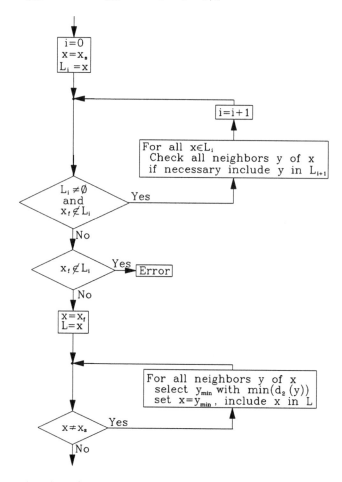

Fig. 7.20 Finding the path.

3. Set $i = i + 1$,
   if $x_f \notin L_i$ go to step 2.

If $L_i = \phi$ before $x_f \in L_i$ then this motion is not feasible as with this starting position and C-space resolution this final position cannot be reached. As soon as the distance $d_2(x_f)$ is calculated, the path $L_P = \{x_{p0}, x_{p1}, ..., x_{pn}\}$ (where $x_{p0} = x_s$, $x_{pn} = x_f$, and $n = d_2(x_f)$) can easily be found by backtracking x starting from x_f:

1. Set $x = x_f$, $i = d_2(x_f)$.
2. While $x \neq x_s$ do:
   $y_{min} = \{y^* | d_2(y^*) = \min d_2(y), y \in Y_m(x)\}$, set $x = y_{min}$,
   $x_{pi} = x$.
3. $i = i - 1$. Go to step 2.

7.4.4.3 Effects of varying ϱ

The characteristics of the path can be easily varied by adjusting ϱ. A small value of ϱ would allow the manipulator to come close to the obstacles, while a very large ϱ would enforce a path with maximum distance. If $\varrho = M$ then C_{free} is retracted to a $m-1$D surface. This surface is a general Voronoi diagram, as all points in C_{free}^R are at a maximum distance of at least two obstacles. In uncluttered environments this might lead to sub-optimal paths, as maximum clearance is not required. Setting $\varrho = 0$ would ensure the shortest path, but might not leave enough space for the end-effector, which will be considered in the next section.

A change in ϱ also changes the time-bounds of the algorithm. Small values of ϱ result in a fast creation of C_{free}^R, but as the retracted C-space is larger than for a large ϱ, the creation of the numerical potential field in C_{free}^R needs more time than for a small C_{free}^R. Generally a smaller ϱ will result in a faster search, as $d_1(x)$ must be calculated in the space $C_{free} \wedge \neg C_{free}^R$; $d_2(x)$ on the other hand must be calculated only in a part of C_{free}^R (until the final configuration is reached), so that usually only a part of the space will be enumerated. For uncluttered environments and small ϱ, a more time-efficient algorithm could be employed to take advantage of the large search-space. Thus the optimal value of ϱ is dependent on the number and size of obstacles as well as the used search-algorithm.

7.4.5 Moving the end-effector

Up to this point only the arm of the robot has been considered, but for complete motion planning the end-effector must also be taken into account. As the effect of a wrist movement is smaller then the effects of arm movement, the priority of the wrist can be considered lower than of the arm. Also the creation of a complete nD C-space would take too much computation time. The two proposed methods allow decoupled path planning for the arm and the hand.

7.4.5.1 Fixed arm configuration

The simplest and fastest method is to change the configuration of the hand only when close to the obstacles (fine motion planning) and keep the hand in a fixed configuration for transfer motions. The hand and the payload are approximated by an envelope H which is related to the mth arm-frame by the transformation matrix $^m\mathbf{T}_H$. This matrix is defined by the configuration of the joints $m+1,\ldots,n$. After the C-space for the first m joints has been created, the envelope is tested for intersection with obstacles at all valid configurations of the arm. If intersection is detected, then this configuration of the arm is marked as forbidden. Thus the C-space is valid not only for the arm but also for the hand at this specific hand configuration and a path can be found as shown in the previous section. The obvious disadvantage is the inflexibility of this method, as it restricts the robot to m-DOF instead of n-DOF. Subsequently the solution will most probably be sub-optimal and in a cluttered environment this method might not find a feasible path, even if one exists. Nevertheless, this method has been successfully implemented by several researchers [LP87, Bro83a] and performed well in uncluttered workspace.

7.4.5.2 Adjusting the hand configuration

In a cluttered environment a manipulator with a fixed wrist configuration may not be able to reach the goal. Thus the hand should not be fixed, allowing the hand to avoid the obstacles. In [BX87] the free C-space C_{free} is subdivided in two spaces E_2 and E_3. E_2 is the subspace in which the manipulator and the hand do not collide for all hand configurations, while in E_3 the hand can collide for some hand configurations. The subspace E_3 is refined by creating a sub-configuration-space for the joints $m+1,\ldots,n$. This sub-C-space is created only for the points of the path which are part of E_3. This algorithm will solve problems for a more cluttered workspace, but will not find a path in a complex situation, as backtracking of the path for the arm is impossible. Also in the worst case time-bounds are not much better than creating the full C-space for the whole path.

One way to build a planner with better time-bounds is to build a potential field only for the hand. While the arm is following the precomputed path, the distance of the envelope H (which encloses the end-effector and the payload) to the k closest obstacles is calculated. The following algorithm is executed for all joints $j = m+1,\ldots,n$:

1. Let q_j be the current joint angle of joint j and $q_j^* = q_j + \Delta q_j$ (Δq_j is a small positive step).
2. Let d_{ji} and d_{ji}^* be the smallest distance between H and the obstacle i in the workspace for q_j and q_j^* respectively.
3. If $d_{ji} = 0$ or $d_{ji}^* = 0$ then error, else

$$q_{j\text{new}} = q_j + \eta_d \sum_{i=1}^{k}\left(\frac{1}{d_{ji}} - \frac{1}{d_{ji}^*}\right) + \eta_q\left(\frac{1}{q_j - \underline{q}_j} - \frac{1}{q_j^* - \bar{q}_j}\right)$$

with $q_{j\text{new}}$ the new joint angle, $\underline{q}_j$ and $\bar{q}_j$ the lower and upper joint limits respectively. The gains η_q and η_d must be adjusted to the step size and the resolution of the C-space of the arm.

While this method will be faster then the sub-C-space method, it does not solve the problem of backtracking. Also the manipulator might get trapped in local minima, as this is a purely local method.

7.5 DISCUSSION

7.5.1 Evaluation of the proposed algorithm

The advantages of this algorithm are:

- It is modular, which makes it easy to implement and change. For example, the C-space could be created only for a specific robot type and would therefore be faster (although less general). Alternatively the path could be found using a search graph instead of a numerical potential field.
- The sole constraint on the manipulator is that it must be non-branching, thus this algorithm can theoretically be applied to all manipulators, with any number and kind of joints.
- The path for the arm will be found quickly when the degree of freedom is limited to be less than 4. This is achieved by decoupling the hand and the arm, also the use of approximated joints instead of the complete 3D model speeds up the process.
- Cluttered environments usually cause no problem, as the retracted free C-space repels the arm away from the obstacles. Especially in these environments the planner can find a path faster than a human robot programmer using an interactive robot programming system.

The shortcomings of the planner are:

- In very cluttered environments some legal paths might not be found due to the approximation of the arm and discretization of the C-space. In the worst case no path will be found.
- As with any recursive C-space algorithm (which are necessary for general manipulators), the worst case time-bounds are exponential to the number of degrees of freedom [SS83].
- The decoupling of hand and arm might cause problems if the addition of the hand requires a change in the path of the arm. If the two planning processes are independent, heuristics are needed to find the required alteration of the path of the arm.
- The planning algorithm for the wrist needs further improvements, as this research is centered on the path of the arm.
- The environment must be completely described in a world model, and it must be stationary while the manipulator is in motion.

7.5.2 Proposed improvements

Some of the disadvantages of the algorithm could be overcome by the following improvements:

- A more dynamic C-space model could be achieved by relating the obstacles i to the C-space obstacle CO_i [BX87]. If for example one obstacle had been moved, only a part of the C-space must be modified. The recreation of the complete C-space is therefore unnecessary.
- If the obstacles are not completely known, they could be grown by an additional distance to take into account this uncertainty.
- The cost structure of the manipulator movement could be reflected by selection

- of the bitmap resolution accordingly, such that the magnitude of a step is inversely proportional to the cost.
- When the final position is not unambiguously specified, all feasible final configurations could be linked to C_{free}^R. The search is then terminated when one of the final configurations is reached. This does not increase the search-time, but would speed up the process compared to a different search for all final configurations.
- The current algorithm restricts a revolute joint to stay within the joint limits, even when these are $[-\pi, +\pi]$. This could be improved by allowing a transition from $-\pi$ to $+\pi$.
- The planning algorithm could be speeded up by a hierarchical cell structure. The C-space could be created at the finest resolution and the search be performed in a coarser level [BLL89]. If the search is not successful, it will be repeated in a finer level. Also the C-space could be created in a rougher approximation and only the C-space obstacles could be refined if necessary.
- Usually it is sufficient to create only the portion of the C-space which is bound by the joint values of the initial and final configuration [LP87].

7.5.3 Trajectory planning

The output of most path planners (except some using the potential field approach [Kha86]) must be post-processed to fit the robot controller, which expects either a function of torque/force over time or a trajectory. This post-processing is done by a trajectory planner. While a path can be described as a function of the distance s along the path, a trajectory is defined as a function of time t. Both parameters can be normalized to 0 at the initial configuration and 1 at the final configuration. Velocity and acceleration can be found by the differentiation of distance to time. The constraints for a trajectory include joint limits, forbidden ranges caused by obstacles, maximum velocity and acceleration. As acceleration depends on joint forces/torques and on the inertia of the joints, trajectory planning is very closely related to control problems.

Usually trajectory planners use a feasible path as a base for the trajectory. However a path which is created by a path planner is usually not optimal for trajectories, especially if it consists of straight lines in the world space. Thus, better trajectory planners modify the initial trajectory to one that is better suited for the specific robot as in [SD89], where cubic splines are employed. The most widely used optimality criteria for trajectory planning is minimum time planning, but often minimum energy trajectories are more cost efficient, as minimum time trajectories increase wear on the robot. When a path is pre-specified, the minimum time trajectory can be found by switching between maximum acceleration and maximum deceleration for at least one joint. The switching point can then be found by evaluating the optimal joint speed [PJ87] and the following formula, which describes the dynamics of an n-DOF manipulator:

$$\tau = \mathbf{M}(\theta)\ddot{\theta} + \dot{\theta}^T C(\theta)\dot{\theta} + \mathbf{G}(\theta) \quad (7.18)$$

where θ, $\dot{\theta}$, and $\ddot{\theta}$ are the joint displacements, velocities and accelerations, respectively; τ is the n-dimensional vector of joint torques/forces; $\mathbf{M}$ is the $n \times n$ inertial matrix; C is an $n \times n \times n$ array of coefficients (representing the centrifugal and Coriolis forces); and the vector $\mathbf{G}$ takes the gravitational forces into account [SD89].

A trajectory planner could use the C-space created by the path planner to save precious computation time. If the trajectory planner is fast enough, it could be used on-line. This would allow avoiding unexpected obstacles by modifying the trajectory using sensor information.

7.6 SUMMARY

An efficient algorithm to find a good path for a robot manipulator arm in a cluttered environment has been presented. The hand and the arm are decoupled for higher efficiency. First the C-space is created for the first three links. The transformation algorithm is theoretically able to handle any kind and number of joints due to its recursive structure, but for more than three joints it is inefficient as the exponent of the time-bounds is linear to its number of degree of freedoms. To further speed up the transformation algorithm, the joints are approximated by line segments and the obstacles are grown accordingly.

To avoid the problems of constructing a search graph and performing a heuristic search, a numerical potential field is applied to a retracted free C-space. The retraction is accomplished by building a numerical potential field around the obstacles. In this retracted free C-space a second numerical potential field is created, which does not contain any local minima and in which the path is easily found. By varying the retraction parameter ϱ the path can be optimized to the two optimality criteria shortest path and maximum clearance. This modular path planner is relatively easy to implement and is capable of finding a path in a cluttered environment.

A concept for the end-effector had also been presented, which employs a simple potential field to force the manipulator to avoid the obstacles. However, this step needs further research as the decoupling of the manipulator in two parts creates some problems and a more sophisticated planning process will be necessary. The relationship of the amount of retraction ϱ and the computation time of the path should be investigated to find the optimal ϱ dependent on the particular workspace. In addition a trajectory planner should optimize the calculated path.

CHAPTER 8

Grasp planning

8.1 INTRODUCTION

Grasp planning is one of the three sub-planning topics in program synthesis for automatic robot programming. The grasp planner is a module that fits into the general planner and generates a prioritized set of feasible grasps. To illustrate the problem, the presentation in this chapter is limited to parallel-jaw grasp planning for assembly operations. The approach taken is to use the feature information and the assembly relationships of the object while finding the optimal grasp. Much of this presentation derives directly from the work of Dubey and Nnaji [ND] on grasp planning.

Task specification is examined to determine the attributes that affect grasp type and parameters. A framework that supports the consideration of a variety of grippers is needed. Possible grasping pairs are chosen by reasoning about the features and the information about the mating faces; in general, mating surfaces cannot be grasped. But some mating faces may have large non-mating areas that can be gripped. The geometrical relationships of the object with the other components in the assembly can be evaluated and constraints created to sort out the feasible face-pairs. The geometrical data of the object consist of faces, edges and vertices described with reference to the object coordinate frame. The sensory inputs, such as vision, giving the current position and orientation of the object are used to find the corresponding geometrical data in the world coordinate frame. Grasp points are postulated and then validated or rejected on the basis of this information. The approach direction for assembly also affects the choice of the grasp and helps in determining the suitable angle of the grasp. Finally, the stability of the validated grasp is determined. The three components of stability which are considered include the following: translational slippage, rotational torque that can be withstood without slippage, and the grasp's ability to withstand twisting by the object.

The action of grasping an object corresponds to the placement of the jaws of the gripper in contact with it. Grasp planning is concerned with the choice of these contacts: location, number of contacts, type, forces and torques involved. The grasp planner has to reason about the features of the object to be assembled and infer the set of face-pairs or face-edge pairs or face-vertex pairs that can be gripped in order to execute the assembly operation. Grasp planning effort is a conglomerate of the following:

- The problem is task-oriented – the choice of the grasp depends on the type of task that is to be executed. One of the aims of grasp planning as presented here is to outline an effective way of incorporating the task information, so that the grasp planner is able to reason out some characteristics of the optimum grasp.

- A generalized grasp representation that allows the grasp planner to handle a variety of gripper types is essential.
- Conceptual development of feature reasoning in grasping and its preliminary implementation is the third goal.
- Finally, this automatic grasp planner involves the organization and development of the various algorithms needed to compute the sets of object entities that constitute the feasible grasps and to choose the optimum grasp amongst these sets on the basis of stability.

8.2 BACKGROUND

One key issue in task-level programming is finding the optimum grasp configurations for the robot. The different aspects of this issue have been studied separately by various research groups. The ultimate goal of all the research has been to build a hand with a measure of versatile behavior comparable to a human hand. Several hands have been developed, ranging from the parallel jaw gripper to the four-fingered Utah/MIT dextrous hand [SJ86]. Each hand has demanded separate attention and the problem has been further multiplied by myriad considerations and approaches. Once a satisfactory model of the hand or gripper is attained, one has to get down to the details of choosing the specific grasp type and parameters. Integration of these steps is a third aspect of research in this area. The following sections categorize the history of the problem into these three parts.

8.2.1 General model of grasping

Multifingered robot hand has been the focus of much research in robotics [Fea86] [CH88] [HLB89] [GW90]. In an assembly task, a range of manipulation, placement and affixment abilities are called for. However, many simple assembly operations do not involve any manipulation and are best performed by parallel jaw grippers. The grasp analysis for parallel jaw grippers is the simplest amongst all hands and has been studied extensively.

8.2.1.1 Approaches to parallel jaw grasping

There are two classical approaches to grasping with such grippers. The first method has no pre-stored information about the object and obtains the relevant data from various sensors. Only the sensed portion of the objects can be used in grasping. Boissonnat's [Boi82] system finds a grasp on the silhouette from the sensor's point of view, where the fingers are parallel to the line of sight from the sensor. Bach *et al.* [JBP85] designed a system using light stripping techniques for 3D scene information to pick up the topmost object from the bin. The object is not recognized even though it is grasped. A large amount of on-line computation has to be performed in such a method.

The second method works with a model of the object and analyzes the entire model for choosing a suitable grasp. This method allows one to know the features of the object exactly and do a lot of the complicated geometrical computations off-line. But then, it assumes that the world is static and so the analysis has to be restricted to modeled environments, unlike the former approach. Laugier and his associates[Lau86] [DL82] have developed a filter based system to find one or several solutions for grasping objects modeled with polygons, cylinders and spheres as surfaces. But no method to find out the grasp points on the object is discussed.

Wolter et al. [JWW85] have worked on grasping objects with graspable surfaces restricted to polygons. Pairs of parallel surfaces which satisfied the geometric constraints required by the gripper e.g. parallelism, visibility etc., were broken up into discrete pairs of points. Approach directions were chosen perpendicular to the polygonal edges or bisecting the angle at each vertex. Most of the papers do not demonstrate grasps that include edges and vertices as possible entities that come in contact with the gripper face. Gatrell [Gat85] describes, in his thesis, how polygon-edge and polygon-point type grasps can be generated, besides polygon-polygon type, using CAD models of the objects.

The combination of partial geometric models and sensory information has been studied in recent times. Laugier and his associates [CLT90] have guided the grasp selection process by combining vision based information collected by a 2D camera and a 3D vision sensor with geometric models representing some information about the workspace and the object to be grasped. Three processing phases, one aimed at selecting a viewpoint avoiding occlusions, the other at modeling the local environment and the third at determining the grasp parameters, are developed.

8.2.2 Choosing grasps and grasp parameters

Numerous factors determine the effective grasp of an object. One of the earliest issues was stability of grasping. Jameson [Jam85] measured stability by the proximity of the line connecting two grasp points to the models' center of mass. Abel et al. presented equlibrium equations using forces at point contacts, coefficent of friction, weight and center of mass. Force analysis for different types of fingertip contacts is described thoroughly in [Cut85]. The Wolter group [JWW85] suggests another measure for stability—the torque at which slippage occurs divided by the distance from the center of the mass to the grasp point. This work has been extended by Gatrell [Gat85] to grasps on edges and vertices.

Although stability remains a basic requirement, other aspects specific to various tasks are important for choosing a grasp that suits the task. Researchers in robotics have tried to analyze what constitutes a good grasp [HA77] [LS88] [Lyo85]. Cutkosky cites nine analytic measures for describing a grasp: compliance, connectivity, force closure, form closure, grasp isotropy, internal forces, manipulability, resistance to slipping and stability. The connectivity between two particular bodies in a kinematic system is defined as the number of independent parameters necessary to specify completely their relative positions. Connectivity is used as a measure of how well the object can be manipulated in the grip. A form closure implies complete restraint of the body; it is a finite set of wrenches applied on the object, with the property that any other wrench acting on the object can be balanced by a positive combination of the original ones. A grasp on an object is force closure if and only if we can exert, through the set of contacts, arbitrary force and moment on this object [Nyu88].

Liu's group [HLB89] narrows down the set of task attributes to four members—stability, manipulability, torquability and radial rotatability. Some authors have defined grasping and manipulating forces for articulated multifingered hands and have presented an algorithm to decompose the fingertip force into these two forces. For a parallel jaw grip, manipulating force becomes zero and the force analysis is greatly simplified.

8.2.3 Building integrated systems

In spite of the distributed research in this area, there exists very little literature dealing with an overall framework for grasping and manipulation. The task-level planning system HANDEY [JLP90] attempts to deal in a general fashion with the interaction between the choice of grasp and the choice of paths to reach the grasps. Allen and his associates [All88] describe an integrated system for dextrous manipulation using a Utah-MIT hand that allows one to look at the higher levels of control in a variety of tasks. Lyons [Lyo85] presents a general framework that can be implemented in a distributed programming environment and suggests a simple set of grasps based on three basic contact configurations between the hand and the object—three non-collinear points, n line contacts and two planar contacts.

There is still no system that satisfactorily integrates the various aspects of the problem. For a range of task requirements, there exists a range of grippers, grasp types, approaches and optimum grasp parameters. The future grasp planners should encompass all the reasoning, selections and evaluations involved in the process of performing a task.

8.3 WORLD SPATIAL RELATIONSHIPS IN GRASPING

The description of an object without reference to its location in the evironment is incomplete. Each object has its own coordinate frame which is related by a transformation matrix to the world coordinate frame. The object descriptions are with reference to its body-attached frame. The spatial relationship of the fixed object to the world can be extracted from the CAD model of the workcell.

The relation between the robot end-effector and the world can be written as:

$$^{world}T_{end-eff} = {}^{world}T_{robot-base} \cdot {}^{robot-base}T_{link1} \cdot {}^{link1}T_{link2} \cdots {}^{linkn}T_{end-eff}$$

To obtain the dynamic information about the location of the movable objects which may initially be anywhere, sensors are used. The location of the sensors are known in the world frame and the object position and orientation are found with respect to the sensor. Camera and laser range finders are typical examples of such a sensor. Without this augmentation of the geometric model of the world with dynamic world information, the grasp planner will not be able to find out whether a certain entity on the object is actually accessible or not.

8.4 GRASPING CONCEPTS

The conceptual development of the grasp planner for RALPH is described in this section and used to illustrate the general concepts of the grasp problems. The planner has been conceived in a series of steps and each section below is devoted to one of these.

8.4.1 Task requirement

It is quite evident that the same object can be grasped in different ways for different purposes. An uncovered pen is held in a certain way while writing and in another way for placing it on the table and in yet another way while inserting it into its cap. The incorporation of task requirement is essential for a grasp planner that reasons out the best grip. The first step is thus in the method of capturing the key aspects of tasks that actually affect the grasping control [HLB89]. The mid-level command should be able to support the information the grasp planner will look for.

The domain of RALPH is the assembly of operations on typical mechanical products. Some of the typical operations that arise are:

- Placing an object on, against, or beside another.
- Inserting a peg into a hole. This is a generic version of analogous tasks like inserting a pin into the piston, inserting a bolt or screw through a hole etc.
- Grasping a wrench to tighten a nut onto a bolt.
- Grasping a screwdriver to turn a screw.
- Turning a nut onto a bolt.
- Fitting a part into a slot in another part.
- Slipping or fitting a part over another, e.g. putting a sleeve over a shaft.

A careful look at these sample operations highlights three important task aspects: the *action* (e.g. turning, inserting, placing onto etc.), the *object* or the part to be grasped, and the *context* or the second part along with the relationship adjective (e.g. onto the bolt, against a part etc.). An action like *turn* implies a torque requirement and *insert* implies precision. A *place onto* operation requires no maneuvre and can be accomplished by a simple parallel jaw gripper or even a suction gripper if the object is light enough. The main concern while holding both the wrench and the screwdriver is providing torque for rotation, besides stability. But the two rotations are clearly about two different axes – one being the long axis of the body and the other being an axis perpendicular to it.

8.4.1.1 Basic task attributes

Researchers have come up with a number of task attributes to define a good grasp. We present four major task attributes needed for a good grasp as presented in [HLB89].

Table 8.1 Task attributes for various tasks; on a scale of 0 to 2, where 0 means no requirement and 2 means strong requirement

Task	Task attributes			
	Stab.	Manip.	Torqu.	Rotat.
1. Placing object beside another	2	0	0	0
2. Inserting peg	2	2	0	0
3. Grasping wrench	2	0	2	0
4. Grasping screwdriver	2	0	0	2
5. Turning a nut	2	0	0	2
6. Fitting a part into a slot	2	2	0	0
7. Fitting a part over another	2	1	0	0

1. Stability, which measures the tendency of the object to return to equilibrium after being disturbed by some external forces. If the stability requirement increases, it means that the grasp should be able to encounter larger disturbances.
2. Manipulability, which is a measure of the ability of the fingers to impart motions to the object while maintaining contact at the contact points. Connectivity in the chosen approach orientation is maximized to get the highest manipulability.
3. Torquability (tangential rotatability), which measures the ability to rotate the long axis of symmetry with minimum force. The long axis has either a supporting point (e.g. a wrench) or a contact point (e.g. a hammer).
4. Radial rotatability, which measures the ability to rotate the object about its long axis or axis of symmetry, as in the case of a screwdriver or the turning of a nut.

Table 8.1 lists the task requirements in terms of these attributes for the sample tasks considered before. The requirement is measured on a scale of three units, where 0 means no requirement, 1 implies moderate and 2 means very strong requirement. The choice of three integers, though arbitrary, is reasonable as a rough but practical scale of judgement.

8.4.1.2 Suitable task description

The mid-level command should have a structure that can contain the information about the action, object and context. Once the format is fixed, a knowledge based approach has to be taken to derive the task attributes from the task description. The choice of the gripper type has to be made first and then the basic grasp type has to be inferred in case a multifingered hand is required.

Lyons [Lyo85] has suggested a set of three grasps as the basic set of grasps. The three useful set of contacts are shown in Figure 8.1 – contacts at three non-collinear points, n line contacts and two planar contacts. The first one can be chosen to completely restrain a grasped object. Addition of more contacts overconstrains the object. These redundant contacts may provide the grasp with better ability to withstand disturbances. The next two contact types therefore impose excess constraints on the object which

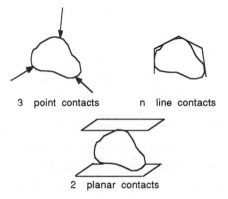

Fig. 8.1 Three basic contact configurations.

can be used as a measure of the rigidity of the grasp. The three grasps suggested are the following:

- An encompass grip where the fingers are used to envelope the object in a compliant fashion.
- A lateral grasp which uses the planar faces of the phalanges to produce a vice-like grip. The object size determines how many fingers are needed to oppose the thumb. This grasp can control the orientation of the long axis of the object by rotating it around the axis created by the thumb and fingers. Hence this grip can be used for precise manipulation of long objects.
- A precision grasp which has contact points at the fingertips and thus can be used for very precise manipulation.

These grasps have different degrees of stability and manipulability. A weighting system can be used to quantify each grasp mode's attributes on a uniform scale. The radial and tangential rotatability will aid the determination of the position of the grasp and the number of fingers used.

A set of heuristics is now employed to determine a good grasp posture and pose. Grasp posture is described in terms of the gripper type, grasp mode, number of fingers required (in the case of a multifingered hand) or distance between the fingers. The grasp pose includes the position of the contact points, lines, or planes and the orientation of the hand during approach. The development of a general set of heuristics for the overall grasp mode selection and a detailed set for the parallel jaw gripper has been attempted.

8.4.2 Feature reasoning for grasping

We employ the various feature reasoning principles developed in Chapter 2 to deal with geometric reasoning in grasping.

Developing a set of heuristics for grasp choice based on these features is crucial. A few examples of features and heuristics will illustrate how this is pursued.

- The first example in Figure 8.2 shows a body with beveled corners. The bevel faces A, B, C, and D are usually meant to remove sharp corners or to ease assembly of the part into another. These surfaces are usually not recommended for grasping and so a very low probability of choice will be attached to it.
- The screw shown in the second part in Figure 8.2 is best grasped at its head. The rest of the part is threaded and is not meant for grasping.
- The rib in the third part is the optimum choice if none of its constituent faces is a mating face. If one of them is a mating face, the entire feature can be rejected.

8.4.3 Geometric constraints in grasping

8.4.3.1 Parallelity and exterior grasp condition

The search starts for faces satisfying the first set of constraints – parallelity of the faces.

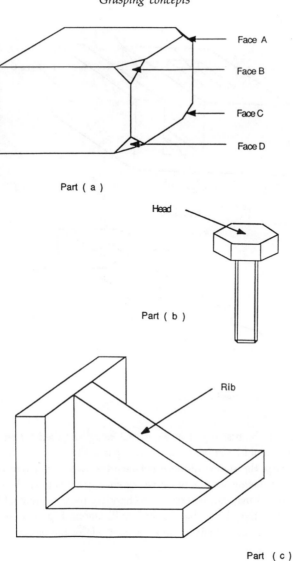

Fig. 8.2 Examples to demonstrate feature reasoning in grasping.

Definition 1.1

Two geometrical entities G1 and G2 are parallel if there exists at least two parallel planes P1 and P2 tangent respectively to G1 and G2.

If one of the members of the candidate pairs is a vertex, this condition is automatically satisfied. If both the members are faces, the dot product of their normals must be of unit magnitude. If it is a face-edge pair, the dot product of the face normal and the edge direction must be zero.

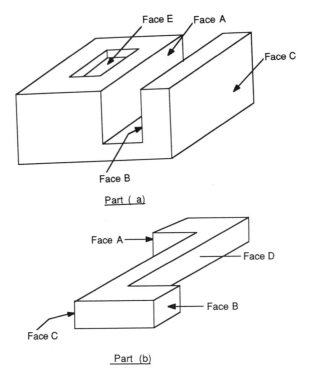

Fig. 8.3 Illustrated geometrical constraints.

Faces A and B in the first object in Figure 8.3 cannot be used for an exterior grasp for which matter should lie in between the two parallel faces. To detect the feasibility of an exterior grasp, the signed distance between the two faces along the normal has to be computed. If this distance is negative, exterior grasp is possible. Otherwise, the face-pair is rejected. It may not be possible to find face-pairs meeting all the conditions required – and then the choice has to be made amongst polygon-edge pairs and polygon-vertex pairs. For stability reasons, at least one of the pair members is always chosen to be a polygon.

8.4.3.2 Local accessibility

This analysis is aimed at eliminating those pairs which cannot provide contact with a jaw of the gripper. It is basically similar to local convexity analysis. The analytical characterization for each type of entity are presented in [PT88]. The filter can be visualized in two steps. The first one evaluates the local accessibility of each member of the grasp pair without considering whether geometry allows the two members to be contacted at the same time. Faces like E in Figure 8.3(a) are not locally accessible by themselves. The second step determines whether the geometry around the two entities of a grasp pair can lie between the two jaws of the gripper. Faces A and C of the object in Figure 8.3(a) have the wrong geometry and hence cannot be considered for grasping. The first filter uses the following definitions:

Grasping concepts

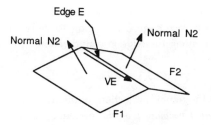

Fig. 8.4 Local accessibility of an edge.

Definition 2.1
An object O is locally accessible in the vicinity of edge E formed by the faces $F1$ and $F2$ if the inner product $[VE \cdot (N1 \times N2)] \leq 0$. $N1$ and $N2$ are the outward normals of $F1$ and $F2$ respectively. VE is the vector along the edge E in a direction such that the matter is on the right with respect to $F1$. (see Figure 8.4).

Definition 2.2
An object O is locally accessible in the vicinity of a planar face F if $\exists$ at least one edge E in its edge loop such that O is locally accessible in the vicinity of E.

Definition 2.3
An object O is locally accessible in the vicinity of a vertex V if $\exists$ at least one edge E ending at V such that O is locally accessible in the vicinity of E.

The second filter can be said to test for planar local accessibility – it checks whether contact between each member of a grasp pair and a particular plane (that of the corresponding jaw plane) is possible. The particular plane is represented by the normal N_P pointing into the interior of the gripper from the jaw. Since at least one of the members of the grasp pair is a face, the normal of this planar face corresponds to the N_P of the jaw plane for the other member. The analytical characterizations of this property for the different types of entities are presented below:

Definition 2.4
An object O is locally planar-accessible in the vicinity of the edge E belonging to faces $F1$ and $F2$ if

- O is locally accessible in the vicinity of edge E.
- $\exists$ two negative real numbers a and b such that $N_P = aN1 + bN2$ where $N1$ and $N2$ are the outward normals of $F1$ and $F2$ respectively.

Definition 2.5
An object is locally accessible in the vicinity of a planar face $F1$ if

- O is locally accessible in the vicinity of the face F.
- $N_F = -N_P$ where N_F is the outgoing normal of F.

Definition 2.6
An object is locally accessible in the vicinity of a vertex V if

- O is locally accessible in the vicinity of the vertex V
- $(V_E \cdot N_P)$ is negative for each edge E belonging to vertex V. V_E is the unit vector directed away from V along the edge in consideration.

8.4.3.3 Mutual visibility

The property that the orthographic projections of the two parallel faces should intersect, has been termed as *mutual visibility property* in [Lau86]. This property is illustrated in the second object in Figure 8.3. Faces A and B cannot be grasped because they are not mutually visible. Faces C and D possess this property and can therefore form a graspable pair.

Since the problem of computing the intersection of 3D surfaces is not trivial, the object model is rotated until the two parallel entities are parallel with the XY plane. Then the Z component can be ignored and the problem is reduced to two-dimensions. The algorithms for finding the projective intersection of a polygonal face with another, a face with an edge and a face with a vertex are described in the next chapter. If this intersection is non-null, it means that the two entities are mutually visible.

8.4.3.4 Finding the grasp point and approach direction

The next check is for the separation distance between the two entities. If it is more than the maximum separation of the gripper jaws, the pair is rejected. If a pair passes through this pruning stage, discrete grasp points are chosen and tested for approach directions placed at discrete angles.

The region of intersection of the orthographic projections of the two members of the grasp pair contain an infinite number of grasp points. Discrete points are generated at regularly spaced intervals to define a finite number of grasp points on the surface. For a face-face-pair, points are chosen over the entire intersected region in both X and Y directions. The spatial resolution can be user specified. For a face-edge pair, points are chosen at increments of the spatial resolution along the edge only. For a face-vertex pair, the orthographic projection of the vertex is the only point considered.

The gripper can actually grasp the object at a grasp point in a given approach direction if there is no obstacle in the way. The obstacle may be due to a certain part of the object itself or due to some part of the mating object that may intersect the gripper in its final configuration. The first case is studied in this thesis and can be extended to the second case, as suggested in the concluding chapter. The algorithm for finding the best approach direction for a grasp pair is described through the following steps.

1. Choose an initial grasp point and approach direction
Starting with the centroid of the region of intersection of the two chosen parallel faces as the first grasp point, an initial approach direction is chosen. This line of approach is called the θ line and the line perpendicular to it and passing through the grasp point is called the perp-line. The angular increment for the θ line is set at an arbitrary value which can be changed dynamically by making this increment a global variable.

2. Find the portion of the object between the two planes of the jaws
The edges, or edge elements lying between the proposed gripper-jaw planes are found. The faces lying completely within these two planes are then picked out. The faces that

have edges intersecting the planes are bounded by new edge segments connecting the intersection points of the edges and the jaw planes [JWW85].

3. Find the polygonal projections of the faces of this portion on the gripper plane
The line segments, when projected onto the gripper plane which is parallel to and equidistant from the jaws, define polygonal regions which the jaws have to avoid.

4. Choose the vertex of all the projections that occurs first during approach
The vertices of all the slice polygons and the face polygons are examined to choose the one which is at the farthest distance from the perp-line. This maximum distance is stored.

5. Find the point where the gripper actually hits the object during approach
A rectangle corresponding to the gripper jaw projection is created, using the maximum distance as length and the jaw width as breadth, through the grasp point with the longitudinal axis oriented along the θ line (Figure 8.5(a)). The rectangular box is intersected with the face polygons. The vertex of the various areas of intersection which is at the farthest distance from the perp-line is chosen. This vertex represents the point where the gripper base hits the object and prevents further approach of the gripper.

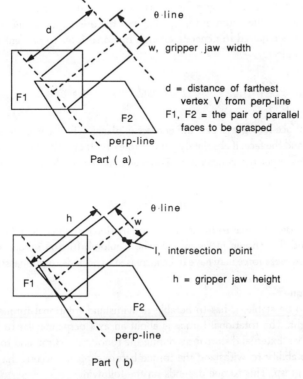

Fig. 8.5 Finding valid approach directions.

6. Check for intersection with a locally inaccessible edge

If the gripper box intersects a locally inaccessible edge of one of the parallel faces, the current approach direction is discarded and the search returns to step 1. Otherwise, it moves onto step 7.

7. Find the actual grasp area when approached in this direction

Another gripper box, this time corresponding to the actual size of the jaw, is created with the border passing through the vertex chosen in the previous step (Figure 8.5(b)). The intersection of this gripper box with the intersection region of the two grasp entities is found. The area of this intersection is found.

8. Find the best approach direction for the grasp point

Repeat steps 4 to 7 for the next θ line. The direction with the maximum area of intersection is chosen for the particular grasp point.

9. Evaluating the grasp points

The whole procedure is repeated for all the grasp points. Each grasp point with the chosen approach direction is rated according to the area of intersection. The more this area, the better the rating.

8.4.4 Grasp force and grasp evaluation

One more grasp parameter, the grip force, has to be computed. This force is normal to the jaw plane. Neither the width of the opening nor the orientation of the gripper affect it. The grip force is required to generate enough frictional force to prevent translational slippage of the object.

8.4.4.1 Analysis for a rigid gripper jaw surface

Normal force

In the case of a face-face grasp pair the contact is planar. The two surfaces in contact, the jaw surface and the face, if absolutely parallel to each other, will have contact pressure evenly distributed over the contact area. Force equilibrium results in the following:

$$F_N = \frac{W}{2(\mu \cos\theta + \sin\theta)} \tag{8.1}$$

where F_N is the force normal to the jaw, μ is the frictional coefficient, W is the weight of the object and θ is shown in Figure 8.6. The Coulomb friction model fits this case quite well. This squeeze force may have to be increased to provide rotational stability.

Rotational torque

For the grasp to be stable, it has to be able to withstand rotational torque caused by the object weight. The rotational torque is about an axis perpendicular to the contact face of the gripper. External disturbances during the robot motion can add to the torque. The greater the ability to withstand the torque before slippage occurs, the greater the stability of the grasp. This torque depends on the following:

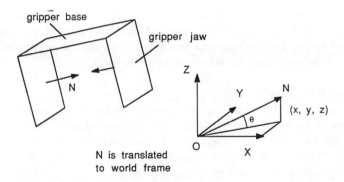

Fig. 8.6 Grasp force analysis for a parallel-jaw gripper with a non-compliant surface.

- pressure over the contact area between the gripper's jaw and the object's entity,
- shape of the contact area and
- coefficient of friction between the gripper's jaw and the object.

Since the pressure is evenly distributed, the torque can be said to be dependent upon the normal force F_N. The torque at which the grasp slips is derived by Wolter et al. [JWW85] for a polygon. The polygon is first decomposed into right-angled triangles each with an acute vertex on the center of rotation. Each side of the polygon has two triangles on it. The equation is:

$$\tau_{slip} = N\mu_s \frac{\sum_{i=1}^{\Delta} \text{sign}(i)\left(\frac{2}{3}a_i b_i c_i + \frac{1}{6}\ln\left(\frac{b_i + c_i}{a_i}\right)\right)}{\sum_{i=1}^{\Delta} \text{sign}(i)\frac{1}{2}a_i b_i} \qquad (8.2)$$

where $\text{sign}(i)$ is a function which is 1 when the ith triangle is positive, and -1 if the triangle is negative. To get the total torque that can be withstood, we have to take the sum of two values computed for each contact.

When the contact entity is an edge, the pressure is assumed to be evenly distributed over the edge. The axis of rotation is assumed to pass through the midpoint of the edge. Then the pressure P can be expressed as

$$P = \frac{F_N}{2R} \qquad (8.3)$$

where R is the length of the edge. The equation at which slip occurs over the line contact is (see [Gat85]):

$$\tau_{slip} = \frac{F_N \mu_s R}{2} \qquad (8.4)$$

For a vertex contact, the torque that can be withstood before slippage reduces to zero. This is because the contact area reduces to zero. And the grasp cannot take any external disturbances in the form of rotational torque. Hence we do not take vertex-vertex grasp pairs into consideration. In most real situations, the gripper jaws have a rubber pad to provide a soft fingered contact. This wives the contact at the vertex some ability to withstand torque. The maximum resisting ability of each of the acceptable grasps is stored as a factor for finding the overall rating of the grasp later on.

Twisting by the object
The object may twist about axes that are perpendicular to the gripper jaws. Due to a variety of forces and torques that the object may experience while the robot is in motion, the object may have a tendency to push the jaws apart and twist into a new orientation. Wolter et al. [JWW85] suggest a solution which involves computing the convex hulls of the 2D contact areas between the object and the gripper. The set of edges in this convex hull is called H. For each edge of H, the distance of the farthest vertex in the other hull (corresponding to the other contact area) is found. The minimum of these distances is chosen.

$$d_t = \min_{e_i \in H}[d_i] \quad (8.5)$$

The difference between this distance and the finger separation for the grasp gives the additional distance the jaws have to open to let the object rotate into a new orientation.

$$\Delta d = d_t - d_f \quad (8.6)$$

where d_f is the jaw separation. The greater the Δd, the lesser the tendency to twist and hence the better the grasp.

Overall rating
The rotational slippage and twisting tendencies are captured by a single rating function. A suggested form is:

$$k_1 \Delta d + k_2 \frac{\tau_{slip}}{\mu_s F_N} \quad (8.7)$$

where k_1 and k_2 are the weights assigned to the two criteria. The frictional coefficient and normal force are taken as constants over all grips and are therefore taken out of the formula for the rotational slippage measure.

Once the grasp with the best rating is chosen, the normal force for resisting translational slippage is checked to see whether it is enough to generate the torque needed to counteract the torque due to the weight of the object. The larger of the normal forces required by the two considerations is chosen. An estimate of the disturbance torques is made and a margin added to the force found. This provides the grasp force parameter.

8.4.5 Analysis for a soft contact

The Coulomb frictional model would not work for a deformable gripping surface. A rubber pad would provide a soft contact and a higher coefficient of friction. The gripper can then grip gently. For soft materials, the Coulomb model is replaced with a shearing model.

Real plane surfaces have micro-irregularities because of which only the peaks come in contact when two flat surfaces are brought into contact. The real area of contact, A_r, is the integrated area of all these little areas of contact. Thus it is smaller than the apparent area of contact A_a. The maximum shear force that the contact can sustain is given by the following formula [CP86]:

$$f_s \propto k A_r \tag{8.8}$$

where k is the shear strength of the material. For a non-elastic contact, A_r is directly proportional to F_N. Thus f_s is also proportional to F_N and can, therefore, be equated to the normal load through a proportional constant μ. For an elastic contact, A_r increases as F_N^n where n is smaller than 1 and can be experimentally found. Since A_r increases less rapidly than the normal load, the coefficient of friction decreases as the load increases. So, the following expression can be written:

$$f_s = k_1 k (F_N)^n \tag{8.9}$$

where k_1 and n can be experimentally found.

Figure 8.7 shows the jaw with a compliant contact surface like a rubber pad attached to it. The following analysis is based on the soft-finger model described in [CP86]. It is assumed that there is no rolling and that the compliant medium is elastic.

The grasping forces at the object surface can be expressed as integrals of the stresses over the contact area:

$$f_l = \int_A \tau_{nl} dA$$

$$f_m = \int_A \tau_{nm} dA$$

$$f_n = \int_A \sigma_{nn} dA$$

$$t_l = \int_A m \sigma_{nn} dA$$

$$t_m = \int_A l \sigma_{nn} dA$$

$$t_n = \int_A (\sqrt{l^2 + m^2}) \tau_{ml} dA$$

where f_l denotes the force along the l direction (see Figure 8.7), t_l denotes the torque about the l axis and so on. It is reasonable to assume that since the stresses τ_{ml}, σ_{mm} and σ_{ll} are zero at the outer surfaces, they are approximately zero throughout and hence, t_n is zero. τ_{nl} has to be equal to τ_{nm} for equilibrium. Thus f_m and f_l are equal and can be expressed as a function of F_N. Force equilibrium results in the following:

$$(F_N + k_1 k F_N^n) \sin \theta + k_1 k F_N^n \cos \theta = W \tag{8.10}$$

The equation can be solved to get F_N.

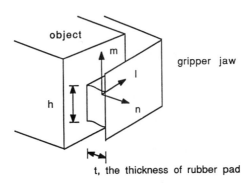

Fig. 8.7 Elastic gripper surface in contact with the object surface.

Torsional stiffness

The torsional rigidity of the pad gives us an idea of the resistance of the pad to rotational torque. Timoshenko and MacCullough [TM49] give an approximate formula for the maximum torsional stress in a rectangular section:

$$\tau_{max} = \frac{T}{ab_2}\left(3 + 1.8\frac{b}{a}\right) \quad (8.11)$$

where a and b are the length and breadth of the cross-sectional area respectively. b is the smaller of the two dimensions, T is the torque applied. The maximum shear stress can be found using the shear force given in equation (8.9) as f_s/A, where A is the cross-sectional area of the pad. Thus equation (8.11) gives us a measure of the torque at which slippage is initiated.

The above analysis is valid only for rectangular areas of contact. It can then be used to determine whether a chosen grasp can withstand the torque due to the weight of the object and has a reasonable margin for disturbance torques.

Table 8.2 Task attributes for mid-level commands

Mid-Level Commands	M	T	R	Grasp Type	Gripper Type
place(onto)	0	0	0	lateral	parallel jaw
place(over)	0	0	0	lateral	parallel jaw
place(besides)	0	0	0	lateral	parallel jaw
drop(into)	1	0	0	precision	3 finger
drop_insert(into)	1	0	0	precision	3 finger
push_insert(into)	2	0	0	precision	2 finger
screwR(into)	1	0	0	precision	3 finger
screwL(into)	1	0	0	precision	3 finger
flip	0	0	0	lateral	parallel jaw
turn1	0	0	1	lateral	parallel jaw
turn2	0	2	0	encompass	dextrous
hold	0	0	0	lateral	parallel jaw
hold(against)	0	0	0	lateral	parallel jaw
remove	0	0	0	lateral	parallel jaw
pull (tool)	0	1	0		

8.5 DESIGN AND IMPLEMENTATION

In this section, we describe the design of a knowledge based grasp planning system to incorporate the three important aspects mentioned in the previous section — task requirement, feature reasoning and geometric analysis. The planner has three modules, one for each of these aspects. The basic requirements for building such a planner have been implemented and are discussed in this section.

8.5.1 Task requirement

The initial model of the task is derived from the mid-level commands in RALPH. They come along with the geometrical description and feature information of the object, spatial relations and process information from PAM (section 3.1.1.1) and sensor information giving the current location of the object. The first level of grasp type selection is made on the task attributes broadly associated with each of these commands in the knowledge base of the planner. Table 8.2 illustrates the task requirement for all the mid-level commands included in the most recent version. The gripper and grasp types are suggested by the requirement pattern. Stability of the object is a primary specification and is required for all the jobs. So it does not affect the selection process and has not been included in the table.

In the beginning, the likelihood of a particular gripper or grasp type to be chosen is the same for all the cases. This fact can be captured by assigning equal specific strength values to all choices. After the first categorization, the supports for the chosen gripper and grasp types increase by a certain prechosen value. The planner can then go ahead and reason about the features and analyze the geometric constraints. The final set of possible grasps are evaluated according to some criteria and the feedback can be utilized in modifying the grasp or gripper type, if necessary. If the first suggestion of the grasp type does not produce effective results and the evaluation indicates a slightly different task requirement from initially accepted, the relative importance of the various task attributes for that particular job is varied. New choices of gripper type and grasp type are arrived at.

The evaluation of the grasp suggested by the planner is based on the heuristics triggered by the task attributes. A careful observation of the human way of accomplishing tasks brings forth numerous heuristics, some of which are described in [HLB89]. Four decision factors appear to be considered:

1. Distance of the grasping position from the axis of rotation.
2. Closeness of the grasping position to the center of gravity.
3. Preference for an approach direction parallel to the axis of assembly of the object which often coincides with the object axis.
4. Preference for a large contact area.

The task attributes associated with the given job determine how important each of these preferences are in choosing an apropriate grasp. An attribute may not be concerned with a criterion, in which case the factor is assigned a weightage of 0. For example, *torquability* does not recommend an approach direction parallel to the axis of the object. So this factor has zero weightage under the attribute of *torquability*. The importance of

Table 8.3 Weightage of factors influencing manipulability

Preference	Weightage
position close to c.g.	1
approach direction parallel to axis of assembly	2
large contact area	−2
large distance from the axis of the object	0

Table 8.4 Weightage of factors influencing torquability

Preference	Weightage
position close to c.g.	−2
approach direction parallel to axis of assembly	−2
large contact area	2
large distance from the axis of the object	0

each factor, if present, can be represented on a scale. An integer scale of −2 to 2 has been found to allow reasonable differentiation of the attributes. A negative weightage has been chosen to imply stress on the opposite preference. For example, a tag of −1 attached to the fourth factor in the list would indicate some preference for a small contact area. The following subsections discuss the heuristics triggered off by the task attributes and how the overall requirement of a criteria for the specified task is arrived at. These heuristics constitute the knowledge base concerning the mapping of task attributes onto the four factors. The relative importance of the decision factors is the basis of choosing grasp sites, approach directions, forces and torques involved in the grasp and in evaluating the choice.

8.5.1.1 Manipulability

A task requiring manipulability has been observed to be carried out with a grasp that shows the following:

- Preference for a grasping position close to the center of gravity.
- Preference for an approach direction parallel to the axis of assembly of the object which often coincides with the object axis.
- Tendency to have less contact area.

Table 8.3 shows how to quantify the above preferences in terms of the four factors mentioned earlier.

Equation (8.12) shows how the final weightage of a factor is found by considering all the task attributes:

$$w_i = \sum_{j=1}^{4} c_j f_{ij} \qquad (8.12)$$

where w_i is the final weightage for a factor i, c_j is the coefficient of weightage of the jth task attribute in the given job and f_{ij} is the weightage of the ith factor in the jth attribute. The linear model, though simple, represents the problem satisfactorily. The coefficients c_j are known deterministically. The proportionality requirement of a linear model is also satisfied by the weightage scheme. The task attributes are defined without any interaction effects with one another and therefore the additive property of the equation is also maintained. This total weightage for a factor is expressed as a fraction of the maximum weightage amongst all factors, as shown in equation (8.13).

$$w'_i = \frac{w_i}{\max_i w_i} \qquad (8.13)$$

Thus, the maximum value that can be attached to a criterion is 1.

Manipulation has a strong implication in the force domain and the force requirements can be modeled for some sample tasks like inserting a peg into a hole. Based on the features of the object and the task specification, an approximate set of wrenches (forces and torques) is taken on the basis of these sample tasks. While inserting a peg (see Figure 8.8), the forces in the X and Y directions should at least provide enough friction to counter the weight of the peg. In an ideal case with no interference during insertion, no other force would be required. But position and alignment errors necessitate some limits to forces and torques to enable manipulation. An estimate of the limits to torques about X and Y axes and forces along the three axes is made according to the fit, material hardness and the dimensional characteristics of the feature. The torque and force sensors can ensure that the actual values in each direction never exceed these limits. The contact configuration that allows the application of the set of wrenches is chosen.

A learning system can be designed with a provision for observing the change in performance with an alteration in the estimate of the wrenches required for the task. The resulting change can be fed back to the engine estimating the wrenches. The maximum positive change in performance due to change in each wrench can be stored

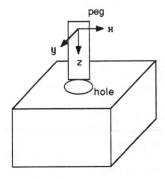

Fig. 8.8 A task requiring manipulability.

in the memory and various combinations of changes can be tried. This problem is beyond the scope of this thesis and has, therefore, not been addressed.

8.5.1.2 Torquability

The following heuristics are usually applicable to tasks requiring high torquability and have been expressed quantitatively in Table 8.4.

- Tendency to grasp near the end.
- Preference for large contact area.

A value of -2 assigned to the second factor implies that the approach direction of the grasp configuration should strive to be as perpendicular to the axis of assembly as is possible. In a case where there is no axis of assembly (for example, grasping a hammer), the axis of assembly is replaced by the long axis of the object. Torquability implies a torque about the supporting or contact point, which can be provided in the feature information. The long axis can also be specified in the feature information. Thus the torque direction can be calculated and the goal wrench specified on the basis of a suitable position chosen according to the heuristics above. A learning mechanism can be attached to this too. If the performance is not satisfactory, as inferred from the force, torque and proximity sensors which record any limit reached before the achievement of the goal state, a correction can be made in the wrench requirement in the proper direction.

8.5.1.3 Rotatability

The tendency to hold far from the axis of rotation is the most important heuristic observed while carrying out tasks involving rotatability. A big contact area is also preferable as indicated in Table 8.5.

The geometric specification about the axis of rotation can be obtained from the feature information. While estimating the necessary wrenches, it is important to remember that besides the torque in the direction of the required rotation, there has to be sufficient force along the axis of rotation to hold the object in contact with the fixed object without slippage. An example is shown in Figure 8.9.

Table 8.5 Weightage of factors influencing rotatability

Preference	Weightage
position close to c.g.	0
approach direction parallel to axis of assembly	1
large contact area	1
large distance from the axis of the object	2

Design and implementation

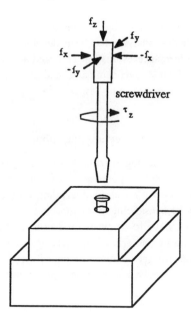

Fig. 8.9 Forces and torques in a task requiring rotatability.

8.5.1.4 Stability

The considerations usually made to have a stable grasp are given below. They can be initially quantified as shown in Table 8.6.

- Choosing a position close to the center of gravity.
- Having a large contact area.

8.5.1.5 Format for task description

The format fixed for the mid-level command is illustrated in the following example:
place_onto(*object type, movable object #, mating faces of movable object, fixed object #, mating faces of fixed object, location matrix of movable object*)

The object type indicates whether the object is a tool or a part to be assembled. The location matrix is a transformation matrix which provides the information about the

Table 8.6 Weightage of factors influencing stability

Preference	Weightage
position close to c.g.	2
approach direction parallel to axis of assembly	0
large contact area	2
large distance from the axis of the object	2

200   *Grasp planning*

current orientation and position of the object to be moved. The preliminary module has been implemented and a lexical analyzer has been developed in the process. The idea and guidelines for the performance evaluation and feedback have been developed in this research. Additions to the mid-level commands and modifications to the preliminary grasp type and gripper type choices can be easily made.

8.5.2 Feature reasoning

The primary idea of feature reasoning is to hasten the process of finding the optimal grasp. The reasoning is based on the purpose of a feature in the assembly. An index can be attached to each member of a feature, depending on the purpose of the feature, to indicate the probability of considering it for grasping. A set of heuristics, based on human response to features, have to be set up and probabilities assigned to the basic features. Initially all the grasp choices have an equal probability index, say 0.5. Every grasp choice passes through the set of rules that constitute the knowledge base of the feature reasoning engine and finally emerge with a probability index in the context of the current task.

The following is a list of heuristics which are provided in [ND] for parallel jaw grippers:

- The features *chamfer* and *fillet* are usually not grasped as their purposes are to ease the assembly of the part or to remove sharp corners. Hence a very low probability index is attached to the set of faces comprising the feature. The change may be brought about by multiplication with a number between 0 and 1.
- The feature *bevel* is also not usually the first choice while grasping an object. But if the beveled surfaces are large enough, they may be worth considering for grasping. Thus, an index whose value is proportional to the area, can be assigned to the *bevel*.
- A depression like the *slot* may provide a good grasping site. One of the sides of the slot, if not a mating surface, is often a member of a reasonable graspable pair as shown in Figure 8.10(a). The search for the candidate pairs can be started from here. If the resulting solution satisfies all the criteria, the geometric computation for all possible pairs can be avoided. Sometimes, the slot bottom has a very high surface finish and is not meant to be grasped, unless absolutely necessary. In such a case, the index associated with the pairs having this face as a member can be decreased by a suitable amount. If the slot is being used for the current assembly operation, the mating surface information will eliminate the entire feature.
- Protrusions are generally very suitable for grasping, unless they are directly involved in the mating operation. A *ridge* (see Figure 8.10(b)) or a *wedge* thus has a high probability index. If one of the consituent faces is a mating face, the index becomes very low as the feature cannot usually provide a good grasping possibility (see Figure 8.10(c)).
- A *cylindrical projection*, if not being used for the current mating operation, has a high index. The axis about which the projection has rotational symmetry can be extracted by the feature extractor. The grasp orientation and position can be directly derived from this information coupled with information about the projection geometry (diameter and length).
- The end feature information, for example, the *head* of a screw of the *handle* of a tool have a very high probability index because these features are usually designed for grasping and application of forces.

Design and implementation 201

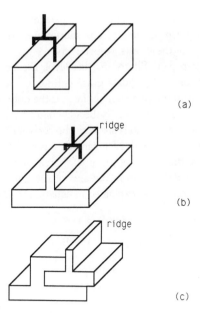

Fig. 8.10 Further examples in feature reasoning.

The index of each grasp candidate can be expressed in relation to the most favorable choice according to equation (8.14):

$$p'_i = \frac{p_i}{\max_i (p_i)} \qquad (8.14)$$

where p'_i is the relative probability index and p_i is the actual probability index of the ith candidate.

The symmetry information about the object to be grasped can be used very effectively. Rotational symmetry is defined as the angle, a rotation which results in the part being brought into coincidence with itself. The rotational symmetry axis of a tool or object defines the axis about which rotationability is measured and the axis, the turning of which measures the torquability. The insertion axis of the object can be derived from the rotational symmetry. This can provide a quick solution to the problem of choosing an approach direction. Tasks requiring manipulability usually prefer grasps parallel to the axis of insertion or movement of the object. This axis direction is given a high weightage during the computation of the optimal grasp, as described in section 8.5.1.1. Reflective symmetry, defined as a reflective transformation that brings a part into coincidence with itself, has not been found to be of great help in an assembly operation.

8.5.3 Geometrical constraints

The analysis of the object geometry for parallel jaw gripping has been implemented as described in the previous section. The procedure can be summarized in the following way:

1. The set of graspable entities are derived from the feature reasoning module. Mating faces are trimmed off from the set of graspable entities.

2. All the candidate grasp pairs are examined for parallelity, local accessibility, mutual visibility, and exterior grasp possibility. The subroutines *check_par*, *local_acc_of_face* and *check_dist* take care of these tests, respectively.
3. • For each of the grasp pairs filtered to this level, the general grasp planner will generate grasp sites on the candidate grasp entities. Currently, the centroid of the area of overlap between the two member faces is the only grasp site considered for each pair. More grasp sites can easily be generated by spatial resolution, i.e. by considering points at regular intervals from the centroid in chosen directions. The probability index associated with each candidate is modified according to the two factors that influence the choice of grasp site—preference for position close to the center of gravity and distance from the axis of rotation. The grasp sites are first sorted using the first criteria as explained in equation (8.15).

$$(d_g)_{ij} = \frac{dist(c.g, g_{ij})}{\max_i (dist(c.g, g_{ij}))}, \quad j = 1 \ldots n_i \qquad (8.15)$$

where $(d_g)_{ij}$ is the relative rating of a grasp site g_{ij} in the ith grasp candidate and n_i is the maximum number of grasp sites chosen for the ith candidate. The grasp sites are also evaluated using distance from the axis of rotation as the criterion and the corresponding relative rating $(r_g)_{ij}$ is determined.

Equation (8.16) shows how this modification is done

$$p_{ij} = p'_j [w_d(d_g)_{ij} + w_r(r_g)_{ij}] \qquad (8.16)$$

where p_{ij} is the new evaluation index associated with the jth grasp site in the ith grasp candidate and w_d and w_r are the weights associated with the two criteria used.
• The approach direction for every grasp site depends on two factors—contact area preference, preference for approach direction parallel to the axis of assembly. One has to ensure that there is no collision of the gripper with any part of the object to be picked up or the fixed object.

The subroutine **find_best_approach_for_point** considers collision of the jaws of the gripper with the object during its approach towards the grasp point as described in the previous chapter. The basic intersection algorithm being the same, this subroutine can easily include a test collision of the jaws with the fixed object provided the transformation matrix between the fixed and the movable object is known. At present, the only criterion considered for finding the optimum direction of approach is the contact area.

With two operating criteria, all the approach directions are rated relative to the best direction for each criterion. Equation (8.17) shows how the index a_{ijk} is calculated to rate the kth approach direction for the jth grasp site in the ith candidate according to the criteria of large contact area preference.

$$a_{ijk} = \frac{(a_r)_{ijk}}{\max_k (a_r)_{ijk}} \qquad (8.17)$$

where $(a_r)_{ijk}$ is the contact area between the jaw or finger and the object entity for the kth direction. Similarly, an index s_{ijk} for comparing the approach directions on the basis of parallelity with axis of assembly is calculated. The index evaluating the choices now takes the form shown in equation (8.18) which is modified from equation (8.16).

$$p_{ijk} = p_i[w_d(d_g)_{ij} + w_r(r_g)_{ij} + w_a a_{ijk} + w_s s_{ijk}] \tag{8.18}$$

4. The next stage in the procedure is the grasp force analysis and the grasp force required for translational stability (with no slippage) is calculated in the subroutine **find_grasp_force** as explained in section 4.4.1. The wrench estimate suggested by the task attributes can be calculated at this stage and a feedback of the performance can be used to modify the estimate if necessary, as described earlier in the chapter. The resulting set of grasps are then ordered according to the rotational stability in the subroutine **find_stability_parameter**. The expansion of the current program will include the calculation of overall rating t_{ijk} as expressed in equation (4.7) and the final rating f_{ijk} according to equation (8.19).

$$f_{ijk} = t_{ijk} p_{ijk} \tag{8.19}$$

The provision for soft contact analysis can also be included.

The current implementation considers only face-face-pairs. The mathematics required for the face-edge and face-vertex consideration have been developed in the previous chapter and can be easily coded into the existing program.

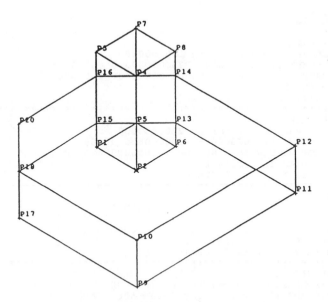

Fig. 8.11 Model of the die in the example.

```
the graspable pairs are 1, 2
the graspable pairs are 3, 4
the graspable pairs are 6, 7
the graspable pairs are 8, 9
-----------------------------------------------------------------
The GRASP PAIR is 1 2
The JAW SEPARATION (in mm) is 250.000000
THE BASE POINT IS 125.000000 375.000000 0.000000
THE APPROACH VECTOR IS -1.000000 0.000000 0.000000
THE GRASP FORCE IS 1.158223
The STABILITY INDEX is 0.521416
*****************************************************************
The GRASP PAIR is 3 4
The JAW SEPARATION (in mm) is 250.000000
THE BASE POINT IS 0.000000 625.000000 0.000000
THE APPROACH VECTOR IS 0.000000 -1.000000 0.000000
THE GRASP FORCE IS 1.211476
The STABILITY INDEX is 0.426742
*****************************************************************
```

(a)

```
the graspable pairs are 1, 2
the graspable pairs are 3, 4
the graspable pairs are 6, 7
the graspable pairs are 8, 9
-----------------------------------------------------------------
The GRASP PAIR is 1 2
The JAW SEPARATION (in mm) is 250.000000
THE BASE POINT IS 125.000000 375.000000 0.000000
THE APPROACH VECTOR IS -1.000000 0.000000 0.000000
THE GRASP FORCE IS 1.158223
The STABILITY INDEX is 0.521416
*****************************************************************
The GRASP PAIR is 3 4
The JAW SEPARATION (in mm) is 250.000000
THE BASE POINT IS 0.000000 625.000000 0.000000
THE APPROACH VECTOR IS 0.000000 -1.000000 0.000000
THE GRASP FORCE IS 1.211476
The STABILITY INDEX is 0.426742
*****************************************************************
The GRASP PAIR is 6 7
The JAW SEPARATION (in mm) is 1000.000000
THE BASE POINT IS 500.000000 0.000000 0.000000
THE APPROACH VECTOR IS -1.000000 0.000000 0.000000
THE GRASP FORCE IS 3.178213
The STABILITY INDEX is 0.109416
*****************************************************************
The GRASP PAIR is 8 9
The JAW SEPARATION (in mm) is 1000.000000
THE BASE POINT IS 0.000000 0.000000 -500.000000
THE APPROACH VECTOR IS 0.000000 0.000000 1.000000
THE GRASP FORCE IS 3.987761
The STABILITY INDEX is 0.082634
*****************************************************************
```

(b)

Fig. 8.12 Output file for the two cases in the example.

Summary

8.5.4 An example

Figure 8.11 shows a sample die with a protrusion as its prominent feature.

The object is created in IDEAS and an IGES translator converts it to an IGES file. The object description in terms of vertices, edges and faces is stored in a file. The file containing the object description is the input to the grasp planner. Another input is the file which contains feature information in terms of the faces which are highly probable candidates for grasping and the mating faces. Other inputs are the gripper size, maximum separation of the jaws and the frictional coefficient. The output file is shown in Figure 8.12(a) and the results are illustrated in Figure 8.13. The faces constituting the protrusion are very likely candidates for grasping and so the other faces are prevented from being considered by giving appropriate information in the file. If the feature information is not provided and all the faces are considered for grasping, the planner's suggestion is as shown in Figure 8.13(4). The actual output file is given in Figure 8.12(b).

8.6 SUMMARY

The focus of this chapter has been on the theoretical development of a grasp planner with a framework and a systematic methodology for a general purpose grasp planning that

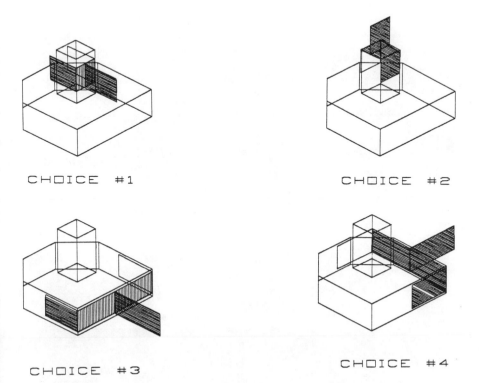

Fig. 8.13 Illustrations of the grasp planner's grasp choices.

uses the task description and feature information along with geometric reasoning to choose an optimum grasp and the consolidation of scattered works in grasping into an integrated, logical system.

It is evident that a thorough system with learning capability will be highly time consuming and will involve numerous trials and comparisons with human performance. Moreover there is no absolutely correct method of evaluating the performance as the various factors affecting it are not mutually independent. Only practical experiments and various case studies can provide insight into the empirical nature of performance evaluation. This evaluation is also necessary to resolve the cases when the grasp planner comes up with two suggestions of equal weightage. In case there is no difference in the performance or the cost of performance, either suggestion can be chosen.

A natural extension of this work is the inclusion of tasks that require dextrous hands. Some basic assembly operations like screwing and fitting keys are almost impossible to conduct with parallel jaw grippers. Thus the need for dextrous hands is now evident in the industrial scene. The development of an integrated sensor system for a suitable hand and a more detailed wrench estimator should be necessary.

CHAPTER 9

Trajectory planning and control

9.1 INTRODUCTION

Trajectory planning converts a description of the desired motion, such as *move* P_1, (Figure 9.1) to a trajectory defining the time sequence of intermediate configurations of the manipulator between the start point P_0 and the destination point P_1.

When the trajectory is executed, the reference point of the end-effector traces a curve in space and sometimes the end-effector changes its orientation. The space curve traced by the end-effector is the path of the trajectory. The result of the motion would be a new position and orientation – which we call location.

The trajectory planning system will produce a sequence of joint configurations for the manipulator $\{\theta_i\}$, $1 \leqslant i \leqslant I$. Also the first and second derivatives of the sequence $\{\theta_i\}$ will produce velocities and acceleration parameters necessary to control the servo mechanisms that perform the desired motion.

It has been seen that manipulator configuration can be expressed either in joint space or Cartesian space. It is easier to express the arm configuration in joint space to enable control and proper formulation of kinematics and dynamics. But when the arm is expressed in joint vectors, humans find great difficulty in relating them to locations in Cartesian space which are suitably described in orthogonal, cylindrical, or spherical coordinate frames [BHJ⁺83]. By using kinematics it is, however, relatively easy to compute the Cartesian space configuration from the joint vector. The inverse problem of relating the Cartesian location of the end-effector in space to joint configuration of the arm is called the inverse kinematic problem which is not only difficult but sometimes intractable. Sometimes, multiple solutions are obtained for the same Cartesian location. This can at times result in disastrous consequences for the task being performed. The careful design of the manipulator is one way of limiting some of these problems or being able to predict what the behavior of such a kinematic system will be.

Because of the problem of relating the Cartesian configuration to joint vectors, it is easier to perform trajectory calculations using joint space. Two popular methods for computing trajectories involve (a) using a suitably parametized class of functions, polynomials [BHJ⁺83], to provide the class of trajectories. The parameter values for each *move* are chosen to satisfy programmer defined constraints on position, velocity, and acceleration of the manipulator during the move process. (b) In the second method, a trajectory is obtained which constrains the manipulator end-effector to traverse a straight

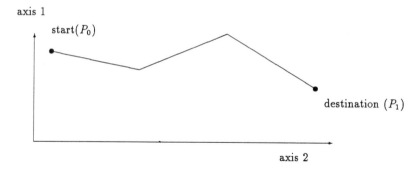

Fig. 9.1 A motion in the configuration space for a two-link manipulator.

line path in Cartesian space between the points P_0 and P_1. We now examine both the Cartesian and joint control approaches.

9.1.1 Cartesian space control

We have stated that it is more convenient to compute the manipulator trajectory in joint space. It is, however, useful to discuss the Cartesian motion, at least to expose the problems that can be inherent.

In Cartesian control, the configurations of the arm are specified in Cartesian coordinate frames at each point along the path. There are usually complex transformations necessary to achieve Cartesian control. It is possible to compute the path parameters off-line. This is accomplished by choosing the path in advance and at intervals of time, computing the Cartesian or joint configurations.

These computations can be in terms of hand location in space or the joint angles. The hand configuration can be $(x, y, z, \theta_x, \theta_y, \theta_z)$ where $\theta_x, \theta_y, \theta_z$ correspond to roll, pitch, yaw

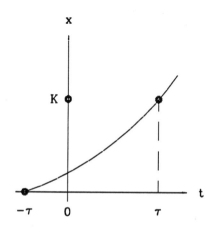

Fig. 9.2 Trajectory for $x(t)$ providing at $t = \tau$, $\dot{x}(\tau) = \dfrac{K}{T}$; and $\ddot{x}(T) = 0$.

Introduction

angles of the wrist. The joint angles are in terms of $(\theta_1, \theta_2, \ldots, \theta_n)$ each corresponding to the joint angle value in the computed configuration. But computations done this way invariably do not take accelerations into account and this can be disastrous especially when the system requires the robot to achieve accelerations beyond its limit.

Polynomial functions can be used to accomplish a smooth motion in Cartesian space continuous in position, velocity and acceleration. The path of each variable (x, y, z) can be defined to be a polynomial function of time with coefficients which we can find.

Imagine the simple case where the manipulator must move along the x-axis (Figure 9.2). Let $x(t)$ be the position of the hand in x-direction at time t. The arm can be accelerated from rest until time T when it uses the achieved velocity. At time τ, the end-effector will be at point $x = K$ and is moving at a constant velocity $\dot{x} = \frac{K}{T}$. For mathematical expediency, let us assume that the motion commences at $t = -\tau$. We desire to specify six parameters (position, velocity, and acceleration) at both ends of the motion in some n-order polynomial. Because of the symmetry of the problem, we choose a fourth order polynomial expression with unknown coefficients:

$$x(t) = a_4 t^4 + a_3 t^3 + a_2 t^2 + a_1 t + a_0 \tag{9.1}$$

We can differentiate equation (9.1) with respect to time to obtain $\dot{x}$ and $\ddot{x}$. We can specify boundary conditions as follows:

$$x(-\tau) = 0 \quad \dot{x}(-\tau) = 0 \quad \ddot{x}(-\tau) = 0 \tag{9.2}$$

$$x(\tau) = K \quad \dot{x}(\tau) = \frac{K}{T} \quad \ddot{x}(\tau) = 0 \tag{9.3}$$

By using the sets of equations obtained from x, $\dot{x}$ and $\ddot{x}$ with the boundary conditions, the coefficients of the polynomial can be obtained:

$$a_0 = \frac{3K}{16}; \quad a_1 = \frac{K}{2\tau}; \quad a_2 = \frac{3K}{8\tau^2} \tag{9.4}$$

$$a_3 = 0; \quad a_4 = \frac{-K}{16\tau^4} \tag{9.5}$$

In real time computation of the path, hand position relative to the robot base will need to be computed at each interval and the joint configurations obtained to compute the polynomial path. The chained expression which may relate the current location P_0 to the new location of P_1 will be found

$$R_{T_H} = R_{T_{P_0}} \cdot X_{1_{TP_0}} \cdots X_{n-1_{TX_n}} \cdots X_{n_{TP_1}} \tag{9.6}$$

where the point P_1 and the hand position H are the same.

9.1.1.1 Joint space control

Of course, the Cartesian space control approaches described above are not efficient and do not always guarantee the desired result. Computing the trajectory in joint space is therefore more convenient and accurate.

Consider the points P_0 and P_1 being the configuration of the two-link manipulator above. These points allow for positional constraints which can be satisfied by a trajectory of the nature:

$$\hat{\theta}(t) = g(t)\hat{\theta}_1 + (1 - g(t))\hat{\theta}_0 \tag{9.7}$$

where $g: [0, 1] \rightarrow [0, 1]$ is any continuous function satisfying $g(0) = 0$, $g(1) = 1$. It is clear that g converts the path traced by the hand into a trajectory. The simplest function which satisfies the above requirement is $g(t) = t$ which produces the linear combination:

$$\hat{\theta}(t) = t\hat{\theta}_1 + (1 - t)\hat{\theta}_0 \tag{9.8}$$

The problem with this function even with its advantage of simplicity is that joint space angular velocity is a constant: $\hat{\omega} = \hat{\theta}_1 - \hat{\theta}_0$. Therefore, the velocity cannot be specified independently at both ends of the trajectory and this requires "infinite" acceleration between movements. In addition to this, the joint solution $\hat{\theta}(t)$ cannot always be guaranteed to lie in the workspace.

It is however possible to constrain a trajectory by specifying its initial and final velocities such that $\dot{\hat{\theta}}(0) = \omega_0$ and $\dot{\hat{\theta}}(1) = \omega_1$. The four constraints will not be met by either a linear or a quadratic function of time, but by a cubic which has this symmetric form:

$$\hat{\theta}(t) = (1-t)^2\{\hat{\theta}_0 + (2\hat{\theta}_0 + \hat{\omega}_0)t\} + t^2\{\hat{\theta}_1 + (2\hat{\theta}_1 - \hat{\omega}_1)(1-t)\} \tag{9.9}$$

But the cubic is still an oversimplification when one considers that there is a maximum attainable velocity. In this approach, the acceleration, and therefore the torque cannot be independently specified at both ends of the trajectory since it grows linearly [BHJ$^+$83]. When the accelerations α_i are required to be specified at both ends of the trajectory along with the velocities and accelerations – a total of six constraints, then this fifth order polynomial suffices:

$$\hat{\theta}(t) = (1-t)^3\left\{\hat{\theta}_0 + (3\hat{\theta}_0 + \hat{\omega}_0)t + (\hat{\alpha}_0 + 6\omega_0 + 12\hat{\theta}_0)\frac{t}{2}\right\}$$
$$+ t^3\left\{\hat{\theta}_1 + (3\hat{\theta}_1 - \hat{\omega}_1)(1-t)\right. \tag{9.10}$$
$$\left. + (\hat{\alpha}_1 - 6\hat{\omega}_1 + 12\hat{\theta}_1)\frac{(1-t)^2}{2}\right\}$$

The problem with this quintic is that practical difficulties exist in general with trajectories planned by constraint satisfaction and with higher order polynomials in joint space. Even though the constraint satisfaction will work from simple descriptions of positions, velocities and accelerations, the main problem lies in the fact that it is weakly constrained. Specifically, the path and the rotation curve followed by the end-effector are not explicitly specified. Avoiding overshoot of the arm cannot be guaranteed because there is no constraint to prevent this.

The basic approach in trajectory planning is to treat the planner as a black box which accepts input of location, velocity, and acceleration expressed either in joint or Cartesian coordinates from starting location to destination location. Two approaches from planner perspective include, first, explicitly stating a set of constraints (e.g. continuity and smoothness) on position, velocity, and acceleration of the manipulator's generalized

Introduction

coordinates at selected locations — called *knot points* or *interpolation points* along the trajectory. The planner can then select a parametized trajectory from a class of functions (polynomial functions) within the interval $[0, \tau]$ that interpolates and satisfies the constraints at the interpolation and points. The second approach requires the user to explicitly specify the path which the end-effector must traverse by an analytical function and then the planner will determine a trajectory either in joint or Cartesian coordinates that approximates the desired path. As stated earlier, the first approach has problems because the end-effector is not constrained on the path, but in the second approach path constraints are specified in Cartesian coordinates, and the joint actuators are served in joint coordinates. This is why the problem is one of finding a parametized trajectory that satisfies the joint path constraints.

Many joint interpolation algorithms have been proposed to minimize the problems enunciated above. One such method is to use splines of low order polynomials (usually cubics) rather than higher order polynomials.

9.1.1.2 Joint interpolated control

The joint interpolation constitutes a compromise between point to point and continuous Cartesian path control. The goal is to provide a smooth and accurate path without incurring heavy computational burden. There are certain desirable characteristics in developing the strategy for path control:

1. Precise control of motion at or close to the workpiece.
2. Continuity of position, velocity, and acceleration.
3. Midrange motion which is predictable and only stops when necessary.

    (a) The first characteristic can only be achieved by using Cartesian control or joint control as discussed.
    (b) Characteristics 2 and 3 can be achieved by the following technique [Sny85]:
      i. Davide the path into a number of segments k.
      ii. For each segment, transform the Cartesian configuration at each end, and determine starting and ending angular positions for each joint.
      iii. Determine the time required to traverse each segment where ω_i is known (*a priori*) maximum velocity for joint i.

$$\tau_1 = \max \frac{\theta_{1i} - \theta_{2i}}{\omega_i^{max}} \qquad (9.11)$$

where

$$1 \leqslant i \leqslant n$$

   iv. Divide τ_1 into m equal intervals, $\Delta \tau$, where

$$m = \tau f$$

$$f = \text{sampling frequency} \qquad (9.12)$$

$$= 50 \text{ to } 60 \text{ Hz}$$

$$\Rightarrow \Delta \tau = \frac{1}{f}$$

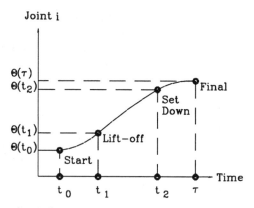

Fig. 9.3 Position conditions for a joint trajectory.

v. For each joint, determine the angular distance to be traveled during $\Delta \tau$

$$\Delta \theta_i = \frac{\theta_{1i} - \theta_{2i}}{m} \tag{9.13}$$

vi. When motion begins, at the nth sampling time, joint servo i receives set point $n\theta_i$.

- Using this algorithm, all joints begin and end their motion simultaneously.
- Path is smooth and predictable, though not necessarily along a straight line.

Another approach which is more commonly used is the one described by [Pau81] and which can also be found in [FGL87]. In this scheme, the path is divided into segments in which certain chosen polynomial functions are applied to satisfy the position, velocity and acceleration at the knot points of these segments. Using Figure 9.3, the knot points are chosen with the following considerations [Pau81], [FGL87]:

1. When picking up an object, the motion of the hand must be directed away from an object; otherwise, the hand may crash into the supporting surface of the object.
2. If we specify a departure position (lift-off point) along the normal vector to the surface out from the initial position and if we require the hand (i.e. the origin of the hand coordinate frame) to pass through this position, we then have an admissible departure motion. If we further specify the time required to reach this position, we could then control the speed at which the object is to be lifted.
3. The same set of lift-off requirements for the arm motion is also true for the set-down point of the final position motion (i.e. we must move to a normal point out from the surface and then slow down to the final position) so that the correct approach direction can be obtained and controlled.
4. From the above, we have four positions for each arm motion: initial, lift-off, set-down, and final.
5. Position constraints:

    (a) Initial position: velocity and acceleration are given (normally zero).
    (b) Lift-off position: continuous motion for intermediate points.

(c) Set-down position: same as lift-off position.
(d) Final position: velocity and acceleration are given (normally zero).

6. In addition to these constraints, the extrema of all the joint trajectories must be within the physical and geometric limits of each joint.
7. Time considerations:

   (a) Initial and final trajectory segments: time is based on the rate of approach of the hand to and from the surface and is some fixed constant based on the characteristics of the joint motors.
   (b) Intermediate points or mid-trajectory segment: time is based on maximum velocity and acceleration of the joints, and the maximum of these times is used (i.e. the maximum time of the slowest joint is used for normalization).

The following constraints apply to the planning of the joint interpolated trajectory [FGL87]:

Initial position:

1. Position (given).
2. Velocity (given, normally zero).
3. Acceleration (given, normally zero).

Intermediate positions:

1. Lift-off position (given).
2. Lift-off position (continuous with previous trajectory segment).
3. Velocity (continuous with previous trajectory segment).
4. Acceleration (continuous with previous trajectory segment).
5. Set-down position (given).
6. Set-down position (continuous with next trajectory segment).
7. Velocity (continuous with next trajectory segment).
8. Acceleration (continuous with next trajectory segment).

Final position:

1. Position (given).
2. Velocity (given, normally zero).
3. Acceleration (given, normally zero).

With the above constraints, some polynomial function of degree n or less can be chosen that satisfies the position, velocity and acceleration at the knot-joints (initial, lift-off, set-down, and final positions). An approach is to specify a seventh degree polynomial for each joint i:

$$q_i(t) = a_7 t^7 + a_6 t^6 + a_5 t^5 + a_4 t^4 + a_3 t^3 + a_2 t^2 + a_1 t + a_0 \qquad (9.14)$$

where the unknown coefficients a_j can be determined from the known positions and continuity conditions. But the use of such high degree polynomials to interpolate the given knot points can be problematic. It is difficult to find the extrema of this polynomial and it tends to have extraneous motion. Another approach is to split the entire joint trajectory into several trajectory segments so that different interpolating polynomials of a lower degree can be used to interpolate in each trajectory segment. There are many splitting approaches and each method has its attributes. The most common methods are the following [FGL87]:

4–3–4 Trajectory
Each joint has the following three trajectory segments: the first segment is a fourth-degree polynomial specifying the trajectory from the start position to the lift-off position. The second trajectory segment (or mid-trajectory segment) is a third-degree polynomial specifying the trajectory from the lift-off position to the set-down position. The last trajectory segment is a fourth-degree polynomial specifying the trajectory from the set-down position to the final position.

3–5–3 Trajectory
Same as the 4–3–4 trajectory, but uses polynomials of different degrees for each segment: a third-degree polynomial for the first segment, a fifth-degree polynomial for the second segment, and a third-degree polynomial for the last segment.

Cubic trajectory
Cubic spline functions of third-degree polynomials for five trajectory segments are employed here.

In summary, trajectory planning can be accomplished either in Cartesian space or in joint space. However, since manipulators are served in joint space and task specifications are accomplished in Cartesian space, the appropriate method to specify constraints is at some of the trajectory. These contraints relate to the position, velocity and acceleration. Using higher-order polynomials to solve the trajectory problem can be problematic due to the extraneous motion and finding the extrema. A more manageable approach is to split the trajectory into segments and apply lower degree polynomials.

9.2 EVALUATION OF TRAJECTORIES

Trajectory planning is an integral part of automatic robot planning or even robot design. As the number of different robots, and trajectories increase, some evaluation scheme to match a particular robot type with a particular trajectory becomes crucial. Having examined the trajectory planning problem in the previous section, we now examine how to evaluate these trajectories for different classes of robots.

An evaluation technique of trajectory paths based on joint torque, work, and power requirements is presented. Upon specification of a robot kinematic configuration and a set of trajectories, a systematic evaluation of each trajectory is performed to determine the optimal trajectory for the configuration. This allows for the determination of the optimal trajectory for each manipulator configuration.

The major goal in implementing automation is to improve or optimize some process. When robots are used in automation, they must somehow be evaluated if automation is to be used to its full potential. This evaulation must transcend the robot from design to implementation. In trajectory planning, the idea of optimization has been incorporated in design, but little attempt has been made at the implementation level.

Thus, research efforts in trajectory planning can be viewed as being single leveled. That is, most researchers have directed their efforts towards creating new trajectories. Here by necessity, different constraining criteria are incorporated. Thus, various parameters have been optimized in the trajectory generation process, and so, optimization is being achieved on the design level.

However, currently very little attention is directed towards the evaluation of the trajectories once they have been formulated. A few isolated attempts have been made to evaluate certain trajectories for specific robots, but no attempt has been made to derive a systematic evaluation process. It is evident that optimization at the implementation level has been neglected.

Clearly, one could use a number of criteria for evaluation. For instance Mujtaba [Muj77] has used such criteria as minimum time, minimum jerk (change in acceleration with respect to time), and time for computation, in the development and evaluation of various trajectories. One must ask if these criteria are valid.

A logical direction to turn to in order to answer this question is the biological case. Humans are thought to use a parabolic trajectory that is implemented to reduce joint torque, power, and work requirements. In the robotic analogy this seems reasonable. One does not want to load the manipulator with large clumsy motors needed to drive large torques.

Computationally, one can show that this is a reasonable choice as well. The tools needed to find these parameters lead to a systematic evaluation technique. Therefore, it may be beneficial to use other criteria than those used by Mutjaba. The work presented here makes use of this systematic process and incorporates three criteria for evaluation and derive largely from our early work [NA89]:

1. Total work done by the joints.
2. Individual peak joint torque.
3. Individual peak joint power requirements.

Certainly it is possible to consider an approach which takes into account the time to complete a trajectory path as well as the amount of jerk as important parameters. Yet that approach lends little insight to a systematic mapping process of trajectories to robots. It also only looks at the effects of the end-effector. Joint responses are not considered.

The work presented here, specifically considers the joint parameters. Therefore, information about both the feasibiltiy of a trajectory, and the resulting evaluation of that trajectory, is directly linked to a specific robot type. However, many of the results are general enough, so that much of the information drawn from one robot configuration, may be applied to other robot types. More importantly, the evaluation process is not robot or trajectory dependent.

9.3 OTHER TRAJECTORY EVALUATION APPROACHES

The task of trajectory planning has been a topic of concern in robotics for some time. Coiffet [Coi83], for example, introduces the underlying theory of trajectory planning for a general robot. He also presents some basic techniques currently being implemented for trajectory planning. Mujtaba [Muj77], in an in depth paper shows how five different trajectories can be derived for two different boundary conditions for the Stanford Blue Arm robot.

These authors, and others, focus on trajectory generation rather than on a method for evaluation. There have been a few isolated attempts towards evaluation. However, these consider specific trajectories for specific robot types. No attempt has been made to try to find a systematic evaluation process.

Mujtaba used time and jerk criteria to evaluate the trajectories presented in his work. He found for instance that the fastest path (Bang-Bang) leads to the greatest jerk. Other authors have touched on wandering (error) as a factor in evaluation of trajectories. Williams and Seireg, at the third symposium on Theory and Practice of Robots and Manipulators, presented a paper which used minimization of joint torques and forces to optimize the trajectory path. They use an example of a three jointed/three linked robot undergoing a spiral path.

Williams's and Seireg's research effort however, concentrates on optimization of actuation forces for a given trajectory path. In effect this is the antithesis of the evaluation process, which is presented in this chapter. Given a trajectory, Williams and Seireg are able to provide the optimal actuation forces (based on minimum power) to accomplish a given trajectory. They make no attempt to evaluate the trajectory for a given robot.

In total, there has been little done in the realm of evaluation of trajectories. Specifically, no one has tackled the fundamental problem presented here of trying to map an optimal trajectory, from a given set, to a given robot.

9.4 BACKGROUND MATERIAL

The goal in this presentation is to find an evaluation process that is not robot specific, but that will apply to all robot, and trajectory types. This implies the need for a method to classify robots. This system must be based on the kinematics of the robot, and not on its function. Nnaji [Nna86] developed a classification system that fits this specification, and which was used in this work. Other classification systems exist, however the majority of these are based on robot function, rather than kinematics.

For instance, Coiffet and Chirouze [CC82] list classification schemes that are used in three different industries; Japanese Industrial Robot Association (JIRA), Robotics Institute of America (RIA), and Association Francaise de Robotique Industrielle (AFRI). The classifications for JIRA are:

1. Manual handling device
2. Fixed sequence robot
3. Variable sequence robot
4. Playback robot
5. Numerical control robot
6. Intelligent robot

while RIA covers only 3, 4, 5, 6 of the JIRA classifications. The categories for AFRI are:

Type A: class 1 from JIRA
Type B: classes 2 and 3
Type C: classes 4 and 5
Type D: class 6

In contrast, Nnaji's system divides robots into five major classes based on the robot configuration. These are:

1. Cylindrical.
2. Jointed.
3. SCARA.
4. Rectilinear.
5. Spherical.

Four of the trajectories in this scheme are similar to Jensen's [Jen65] classification for cam design which resulted in the following classification system: "parabolic," "3-4-5 polynomial," "harmonic," and cycloidal cam curves respectively. If Nnaji's system is used, a few of the sub-classes can be ignored for the purposes of evaluation. This is solely due to the fact that consideration of certain robot types gives no new mappings of trajectories to robots. That is, the robot may be restricted to only a certain type of movement, or the robot sub-class may be a kinematic redundancy of another sub-class.

For instance, in all the classes, an extra DOF is added by placing the robot on a linear slide base. This does not change the basic kinematics or dynamics of the arm itself, but rather enlarges the work area for the robot.

Similarly, classifications that deal with double and triple end-effectors can be omitted from analysis. The addition of the "extra" end-effectors once again does not change any of the kinematics or dynamics that any one end-effector must satisfy. In effect all that has occurred is a superposition of two systems. Therefore, one can evaluate the system using the base components.

An additional set of classifications that is not considered for evaluation are rectilinear robots. The trajectories for these robot types are simple enough so that an evaluation is not necessary.

The last set of classifications that do not need to be evaluated are those that consider the addition of the end-effector DOF. Since this evaluation concerns gross movements; that is the actual trajectories of the manipulator, one is not concerned with orientation of the end-effector. Therefore the maximum number of DOF needed to be considered is three. There are a few exceptions to this whereby the robot is supplied with a redundant DOF. In all of these cases however links can be combined so that the three DOF torque equations can completely describe the system.

Once these unsuitable classes are eliminated, one is left with four classes which in total contain 40 different robot types.

Now that the robots can be classified, we will develop the tools needed for the evaluation process. In this work a Lagrangian formulation is used to determine the joint actuation forces required to undergo a given motion. This method shall not be explained in detail in this book; for further information the reader is referred to Paul [Pau81].

The basic concept of the Lagrangian formulation is the development of a single scalar

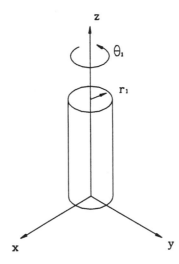

Fig. 9.4 Revolute joint without offset.

function, the Lagrangian L:

$$L = K - P$$

where L = Lagrangian, K = kinetic energy, P = potential energy.

The generalized forces can then be obtained by using the Lagrangian equation:

$$\frac{d}{dt}\left(\frac{\partial L}{\partial \dot{q}_n}\right) - \frac{\partial L}{\partial q_n} = Q_n$$

where Q = generalized force or torque, q is the generalized coordinate, and n = joint number. This gives n equations for an n DOF system.

The Lagrangian formulation is not the most computationally expediant technique available. However, it does present a systematic procedure which can be used to determine the joint torques and forces.

The Lagrangian has been computed for robots with up to 3 DOF.

A summary of this approach as detailed in [Asa87] and [NA89] is as follows.

Torque equations: List of variables.

T_{mn} torque at joint m due to link n
T^g_{nm} potential energy term of torque at joint n
K_e kinetic energy of link n
m_n mass of link n
r_n radius of link n
L_n length of link n
l_n radius of gyration of link n
q_n offset of link n
θ_n angle of rotation of link n
a_n acceleration of link n

v_n linear velocity of link n
$\dot{\theta}_n$ angular velocity of link n
$\ddot{\theta}_n$ angular acceleration of link n
C_n $\cos \theta_n$
S_n $\sin \theta_n$
C_{nm} $\cos(\theta_n + \theta_m)$
S_{nm} $\sin(\theta_n + \theta_m)$
g gravitational acceleration

The torque equations for both prismatic and revolute joints have been established for robots containing up to 3 DOF. Let joint 0 be revolute without offset. The torque required at this joint due to link 1 is computed in the following manner:

Clearly:

$$x = r\cos\theta_1 \qquad \dot{x} = \dot{\theta}_1 r \sin\theta_1$$
$$y = r\sin\theta_1 \qquad \dot{y} = \dot{\theta}_1 r \cos\theta_1$$

Therefore the kinetic energy of link 1 is:

$$K_e = \tfrac{1}{2} m v^2$$
$$= \tfrac{1}{2} m r^2 \dot{\theta}_1^2$$

Since the potential energy is not a function of θ, this will not contribute to the torque required at joint 0. Therefore the torque equation is:

$$T_{11} = \frac{d}{dt}\left(\frac{\partial K_e}{\partial \dot{\theta}_1}\right) - \frac{\partial K_e}{\partial \theta_1}$$
$$= \tfrac{1}{2} m r^2 \ddot{\theta}_1$$

The other torque equations are found in a similar fashion. The actual mathematics will not be carried out in full detail, but instead the final results are given. Note that the

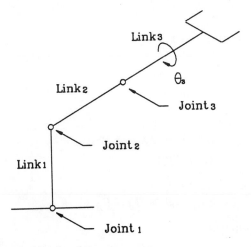

Fig. 9.5 Three degrees of freedom kinematic system.

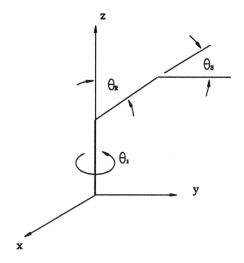

Fig. 9.6 Definitions of angular displacement.

kinetic energy and the potential energy terms of the torque equations are separable. Therefore, to avoid repetition the potential energy terms have been solved separately, and

$$T_{\text{total}} = T_{nm} - T_{nm}^g$$

It is important to know the definitions that were used to describe the angular displacements for each joint. Figure 9.6 contains a description of these definitions.

Revolute joints without offset:

$$T_{11} = \tfrac{1}{2}m_1 r_1^2 \ddot{\theta}_1$$

$$T_{12} = m_2 l_2^2 (2 S_2 C_2 \dot{\theta}_1 \dot{\theta}_2 + S_2^2 \ddot{\theta}_1)$$

$$T_{22} = m_2 l_2^2 (\ddot{\theta}_2 - S_2 C_2 \dot{\theta}_1^2)$$

$$T_{13} = \tfrac{1}{2}m_3 (2(L_2^2 S_2^2 + 2l_3 L_2 S_2 S_{23} + l_3^2 S_{23}^2)\ddot{\theta}_1$$
$$\quad + 2(2L_2^2 S_2 C_2 + 2l_3 L_2 (C_2 S_{23} + S_2 C_{23}) + 2l_3^2 S_{23} C_{23})\dot{\theta}_1 \dot{\theta}_2$$
$$\quad + 2(2l_3 L_2 S_2 C_{23} + 2l_3^2 S_{23} C_{23})\dot{\theta}_1 \dot{\theta}_3)$$

$$T_{23} = \tfrac{1}{3}m_3 (2(L_2^2 + 2l_3 L_2 C_3 + l_3^2)\ddot{\theta}_2$$
$$\quad + 2(l_3 L_2 C_3 + l_3^2)\ddot{\theta}_3$$
$$\quad - 2(2L_2^2 S_2 C_2 + l_3 L_2 (C_2 S_{23} + S_2 C_{23}) + l_3^2 S_{23} C_{23})\dot{\theta}_1^2$$
$$\quad - 2(l_3 L_2 S_3 \dot{\theta}_3^2) - 4l_3 L_2 S_3 \dot{\theta}_2 \dot{\theta}_3)$$

$$T_{33} = \tfrac{1}{3}m_3 (2(l_3 L_2 C_3 + l_3^2)\ddot{\theta}_2 + 2l_3^2 \ddot{\theta}_3$$
$$\quad - 2(l_3 L_2 S_2 C_{23} + l_3^2 S_{23} C_{23})\dot{\theta}_1^2 + 2l_3 L_2 S_3 \dot{\theta}_2^2)$$

Revolute joints with offset:

$$T_{11} = \tfrac{1}{2}m_1 r_1^2 \ddot{\theta}_1$$

$$T_{12} = \tfrac{1}{2}m_2(2(q_2^2 + l_2^2 S_2^2 - 4q_2 l_2 S_1 C_1 S_2)\ddot{\theta}_1$$
$$+ 2q_2 l_2(C_2 C_1^2 - S_1^2 C_2)\ddot{\theta}_2$$
$$+ 2(2l_2^2 S_2 C_2 - q_2 l_2 S_1 C_1 C_2)\dot{\theta}_1 \dot{\theta}_2$$
$$- 4q_2 l_2(s_1^2 S_2 - C_1^2 S_2)\dot{\theta}_1^2$$
$$+ 2q_2 l_2(s_1^2 S_2 - C_1^2 S_2)\dot{\theta}_2^2)$$

$$T_{22} = \tfrac{1}{2}m_2(2q_2 l_2(C_2 C_1^2 - S_2^2 C_2)\ddot{\theta}_1 + 2l_2^2 \ddot{\theta}_2 - 2l_2^2 C_2 S_2 \dot{\theta}_1^2)$$

$$T_{13} = \tfrac{1}{2}m_3(2(q_2^2 - 4q_2 L_2 S_1 C_1 S_2 - 4q_2 l_3 C_1 S_1 S_{23} + 2q_2 q_3$$
$$+ L_2^2 S_2^2 + 2L_2 l_3 S_2 S_{23} - 4q_3 L_2 S_1 C_1 S_2$$
$$+ l_3^2 S_{23}^2 - 4l_3 q_3 S_1 C_1 S_{23} + q_3^2)\ddot{\theta}_1$$
$$+ 2(2C_1^2 - 1)(q_2 L_2 C_2 + q_2 l_3 C_{23} + q_3 L_2 C_2 + q_3 l_3 C_{23})\ddot{\theta}_2$$
$$+ 2C_{23} l_3 (2C_1^2 - 1)(q_2 + q_3)\ddot{\theta}_3$$
$$+ 4(2S_1^2 - 1)(q_2 L_2 S_2 + q_2 l_3 S_{23} + q_3 L_2 S_2 + q_3 l_3 S_{23})\dot{\theta}_1^2$$
$$- 2(2C_1^2 - 1)(q_2 L_2 S_2 + q_2 l_3 S_{23} + q_3 L_2 S_2 + q_3 l_3 S_{23})\dot{\theta}_2^2$$
$$- 2l_3 S_{23}(2C_1^2 - 1)(q_2 + q_3)\dot{\theta}_3^2$$
$$+ 2(2L_2^2 S_2 C_2 + 2L_2 l_3 C_2 S_{23} + 2L_2 l_3 S_2 C_{23}$$
$$+ 2l_3^2 C_{23} S_{23} - 4q_2 L_2 C_2 C_1 S_1 - 4q_2 l_3 C_{23} S_1 C_1$$
$$- 4q_3 L_2 C_2 C_1 S_1 - 4q_3 l_3 C_{23} C_1 S_1)\dot{\theta}_1 \dot{\theta}_2$$
$$+ 2(2L_2 l_3 S_2 S_{23} + 2l_3^2 C_{23} S_{23} - 4q_2 l_3 S_1 C_1 C_{23}$$
$$- 4l_3 q_3 S_1 C_1 C_{23})\dot{\theta}_1 \dot{\theta}_3$$
$$- 4l_3 S_{23}(2C_1^2 - 1)(q_2 + q_3)\dot{\theta}_2 \dot{\theta}_3)$$

$$T_{23} = \tfrac{1}{2}m_3(2(2C_1^2 - 1)(q_2 L_2 C_2 + q_2 l_3 C_{23} + q_3 L_2 C_2 + q_3 l_3 C_{23})\ddot{\theta}_1$$
$$+ 2(L_2^2 + 2L_2 C_2 l_3 C_{23} + l_3^2 + 2L_2 S_2 l_3 S_{23})\ddot{\theta}_2$$
$$+ 2(L_2 C_2 l_3 C_{23} + l_3^2 + L_2 l_3 S_2 S_{23})\ddot{\theta}_3$$
$$+ 2(-2q_2 L_2 C_2 C_1 S_1 - 2q_2 l_3 C_{23} C_1 S_1 - 2q_3 L_2 C_2 C_1 S_1$$
$$+ 2q_3 l_3 C_{23} C_1 S_1 - L_2^2 C_2 S_2$$
$$- L_2 l_3 C_2 S_{23} - L_2 l_3 S_2 C_{23} - l_3^2 S_{23} C_{23})\dot{\theta}_1^2$$
$$+ 2(-L_2 C_2 l_3 S_{23} + L_2 l_3 S_2 C_{23})\dot{\theta}_3^2$$
$$+ 2(-2L_2 C_2 l_3 S_{23} + 2L_2 S_2 l_3 C_{23})\dot{\theta}_2 \dot{\theta}_3)$$

$$T_{33} = \tfrac{1}{2}m_3(2(q_2 l_3 C_{23}(2C_1^2 - 1) + q_3 l_3 C_{23}(2C_1^2 - 1))\ddot{\theta}_1$$
$$+ 2(L_2 C_2 l_3 C_{23} + l_3^2 + L_2 S_2 l_3 S_{23})\ddot{\theta}_2$$
$$+ 2l_3^2 \ddot{\theta}_3$$
$$+ 2(-2q_2 l_3 C_{23} S_1 C_1 - 2q_3 l_3 C_{23} C_1 S_1$$
$$- L_2 l_3 S_2 C_{23} - l_3^2 S_{23} C_{23})\dot{\theta}_1^2$$
$$+ 2(-L_2 l_3 S_2 C_{23} + L_2 l_3 C_2 S_{23})\dot{\theta}_2^2)$$

Prismatic joints with and without offset (note: a_n only denotes the component of the acceleration that the joint allows movement in. Therefore if an acceleration is perpendicular to the direction of motion, then $a_n = 0$):

$T_{11} = m_1 a_1$
$T_{12} = m_2(a_1 + a_2)$
$T_{22} = m_2(a_1 + a_2)$
$T_{13} = m_3(a_1 + a_2 + a_3)$
$T_{23} = m_3(a_1 + a_2 + a_3)$
$T_{33} = m_3(a_1 + a_2 + a_3)$

SCARA:

The SCARA robot requires a separate set of torque equations since joints 1 and 2 perform coplanar motion:

$T_{11} = m_1 l_1 \ddot{\theta}_1$

$T_{12} = \tfrac{1}{2} m_2 (2(L_1^2 + l_2^2 + 2l_2 L_1 C\theta_2) \ddot{\theta}_1$
$\quad\quad + 2(l_2^2 + L_1 l_2 C\theta_2) \ddot{\theta}_2$
$\quad\quad - 4 L_1 l_2 S\theta_2 \dot{\theta}_1^2 - 2 L_1 l_2 S\theta_2 \dot{\theta}_2^2)$

$T_{22} = m_2 (l_2^2 + L_1 l_2 C\theta_2) \ddot{\theta}_1 + l_2^2 \ddot{\theta}_2$
$\quad\quad - 2 L_1 l_2 S\theta_2 \dot{\theta}_1 \dot{\theta}_2$
$\quad\quad - L_1 l_2 S\theta_2 \dot{\theta}_1^2$

$T_{13} = \tfrac{1}{2} m_3 (2(L_1^2 + L_2^2 + 2 L_1 L_2 C\theta_2) \ddot{\theta}_1$
$\quad\quad + 2(L_2^2 + L_1 L_2 C\theta_2) \ddot{\theta}_2$
$\quad\quad - 4 L_1 L_2 S\theta_2 \dot{\theta}_1^2$
$\quad\quad - 2 L_1 L_2 S\theta_2 \dot{\theta}_2^2$

$T_{23} = m_3 (L_2^2 + L_1 L_2 C\theta_2) \ddot{\theta}_1 + L_2^2 \ddot{\theta}_2$
$\quad\quad - 2 L_2 L_1 S\theta_1 \dot{\theta} \dot{\theta} - L_1 L_2 S\theta_2 \dot{\theta}_1^2$

T_{33} = prismatic T_{33}

Potential energy terms for revolute joints:

$$T_{11}^g = 0$$
$$T_{12}^g = 0$$
$$T_{22}^g = -m_2 g l_2 S_2$$
$$T_{13}^g = 0$$
$$T_{23}^g = -m_3 g L_2 S_2 - m_3 g l_3 S_{23}$$
$$T_{33}^g = -m_3 g l_3 S_{23}$$

Potential energy terms for prismatic joints:

$$T_{11}^g = m_1 g$$
$$T_{12}^g = -m_2 g$$

$$T^g_{22} = -m_2 g C_2$$
$$T^g_{13} = -m_3 g$$
$$T^g_{23} = -m_3 g C_2$$
$$T^g_{33} = -m_3 g C_{23}$$

9.5 ROBOTS WITH MORE THAN 3 DEGREES OF FREEDOM

Most of the Nnaji's robot classifications have 3 or less DOF. However, in a few cases one or two redundant DOF are present. Therefore, either new torque equations must be derived to handle these extra joints, or the system must somehow be simplified so that the 3 DOF equations are sufficient to completely describe the system.

The second approach shall be used. Although this technique is intended for robots with more than 3 DOF, there are instances in 3 DOF systems where this method is applicable. The following is an example of such a case. This system has been chosen for its simplicity of mathematical manipulation.

Consider the following 3 DOF system.

T_{11}, T_{12}, T_{22} are found in the normal manner by using the appropriate torque equations. However, for link 3:

$$K = (K_1 + K_2) + K_3$$
$$= (ls_2^2 \dot{\theta}_1^2 + l^2 \dot{\theta}_3^2) + \tfrac{1}{4} m l_3^2 \dot{\theta}_3^2$$

where K = total kinetic energy, $K_1 K_2$ = kinetic energy of links 1, 2; l = length of link 2 + radius of gyration of link 3 and l_3 = radius of gyration of link 3.

The Lagrangian formulation yields:

$$T_{13} = \tfrac{1}{2} m L^2 (4 S_2 C_2 \dot{\theta}_1 \dot{\theta}_2 + 2 S_2^2 \ddot{\theta}_1)$$
$$T_{23} = \tfrac{1}{2} m L^2 (2 \ddot{\theta}_2 - 2 S_2 C_2 \dot{\theta}^2 - 1)$$
$$T_{33} = \tfrac{1}{2} m L^2 \ddot{\theta}_3$$

It is clear that this system can be considered to be a 2 DOF system with a 1 DOF system superimposed. However, one must remember that there must be a redefinition of the link length 1.

As stated above, this can be applied to systems with more than 3 DOF, thereby eliminating the need to develop new equations in many cases. In addition it is evident that robot configurations with redundant DOF do not need to be evaluated. These "parent" robot configurations can be broken down into simpler systems, "children", which are not as mathematically complex. Thus, evaluation can be carried out for the "children" rather than for the more complex "parent" systems.

9.6 EVALUATION AND ANALYSIS

The Lagrangian formulation also lends itself to the development of a work variable which is used in this analysis as an evaluation parameter. This parameter is found by taking the torque, or force at any segment in time and multiplying times the distance the joint moved in that time interval.

The absolute values of the work done in each segment are then added together to determine the work parameter. This is not the total work done by the system, in a classical physics sense. However, it presents an analogous work term, which is more suitable for evaluation purposes than a classical work term.

The work parameter has been developed because many of the torque and displacement profiles are symmetric about zero. Thus the net work done by a number of trajectories is zero. One would like to be able to differentiate between these trajectories, and so the classical work variable is not suitable.

First, we must decide which robot types are to be evaluated. In a practical application the decision is evident. However, the work described in this chapter is theoretical in nature. Therefore, a representative yet reasonable sized set of robots had to have been chosen.

We will illustrate this evaluation by using 8 robot types. The 8 robots were comprised of the entire cylindrical class, and one robot from each of the other classes. These robots are:

1. Cylindrical 1.1
2. Cylindrical 1.2
3. Cylindrical 2.1
4. Cylindrical 2.2
5. Cylindrical 5.1
6. Jointed 2.1
7. SCARA 2.1
8. Spherical 1.2

where the numbers (1.1, 1.2, etc.) refer to specific sub-classes.

This set of robots was chosen to allow comparison within a specific class, as well as comparison between classes. Thus both general and specific information about how the evaluation technique would behave for various robot classes was obtainable.

Next, decisions concerning trajectory types to be used must be made. For this analysis, 5 trajectory types were used. These trajectories, and the corresponding equations are as follows:

Bang–bang

$$\theta = 2t_n^2 \quad (0 \langle t_n \rangle 0.5)$$
$$\dot{\theta} = 4t_n$$
$$\ddot{\theta} = 4$$

$$\theta = (-1 + 4t_n - 2t_n^2) \quad (0.5 \langle t_n \rangle 1.0)$$
$$\dot{\theta} = (4 - 4t_n)$$
$$\ddot{\theta} = -4$$

Polynomial

$$\theta = 6t_n^5 - 15t_n^4 + 10t_n^3$$

$$\dot{\theta} = 30t_n^4 - 60t_n^3 + 30t_n^2$$
$$\ddot{\theta} = 120t_n^3 - 180t_n^2 + 60t_n$$

Exponential

$$\theta = (1.0 - (\exp(-\omega_d)))(1 + \omega_d t)$$
$$\dot{\theta} = \omega_d^2 t \exp(-\omega_d t)$$
$$\ddot{\theta} = \omega_d^2(\exp(-\omega_d t))(1 - \omega_d t)$$

where

$$\omega_d = \frac{9.0}{\text{(total time)}}$$

Cosine

$$\theta = 0.5(1 - \cos(\pi t_n))$$
$$\dot{\theta} = 0.5\pi \sin(\pi t_n)$$
$$\ddot{\theta} = 0.5\pi^2 \cos(\pi t_n)$$

Sine on ramp

$$\theta = t_n - \frac{(\sin(2\pi t_n))}{(2\pi)}$$
$$\dot{\theta} = 1 - \cos(2\pi t_n)$$
$$\ddot{\theta} = 2\pi \sin(2\pi t_n)$$

These equations are written for angular displacements. However, they also apply to the linear case. In addition, normalized time (t_n) and displacements are used (except in the exponential case where the time cannot be normalized). The exact form of each trajectory was developed by Mujtaba [Muj77], for the Stanford Blue Arm manipulator. These are the more common types of trajectories that one finds in the literature and in industry, and therefore are considered to be good representatives.

At this point, a few simplifying assumptions were needed to make the problem more tractable. These are:

1. Links are thin, straight rods:
   The thin rod assumption is incorporated into the torque equations and therefore is implicit in the evaluation. The basic assumption is that the mass of a link can be concentrated at the center of the mass, or at a point on the link which is located at a distance equal to the radius of gyration. Therefore a link length as well as an additional parameter is required for evaluation.
2. Each joint is restricted to be 1 DOF:
   This is to simplify the inverse kinematics. If a joint contains N DOF, this joint can be separated into N, 1 DOF joints, with zero link distance between them.
3. Orientation of the end-effector is not considered:
   The orientation of the end-effector is not considered to be important for this analysis. Only gross movement of the trajectory is considered. Therefore in general, wrist motions are neglected. One can postulate that only 3 DOF systems need to be evaluated, except in a few special cases when a redundant DOF would be present.

These assumptions do not adversely affect the evaluation process. Assumption 1 however does lead to an error in some of the torque equations. However, this will always lead to an overestimate, and is not trajectory dependent. Therefore, although the torque equation may not be exact, the evaluation is still valid (see Asano [Asa87]).

One may now develop the appropriate joint torque equations. The first step in doing this is to develop the transformation matrices which describe the joint coordinates relative to the base coordinate. These are then used to determine the inverse kinematics, and Jacobian which relate the joint displacements, velocities, and accelerations, to the hand displacements, velocities, and accelerations. The joint torques/forces are then found via the Lagrangian formulation as described in the previous section.

The work and power parameters are then formed. This requires no new information, but rather the application of the output from the inverse kinematics, Jacobian, and Lagrangian formulation.

The equations formed from the above work tend to be cumbersome. For this reason a computer program is needed to aid in the evaluation process.

The best trajectory can be determined by choosing the one that meets the torque and power constraints of the joint drivers, and has the lowest work parameter value.

9.6.1 Discussion

Power and torque were not used as evaluation criteria in these experiments. These parameters are only used as constraining criteria. If the maximum power, or torque required by a trajectory is greater than the maximum power, or torque available from the joint driver, then the trajectory is deemed inappropriate, and is immediately thrown out. Since this presentation is theoretical, there are no power or torque constraints imposed, and as such, these parameters are not used in evaluation.

The following notation shall be used for the rest of this section:

Robot 1 = cylindrical 1.1
Robot 2 = cylindrical 1.2
Robot 3 = cylindrical 2.1
Robot 4 = cylindrical 5.1
Robot 6 = jointed 2.1
Robot 7 = SCARA 2.1
Robot 8 = spherical 1.2

T-1 = bang–bang trajectory
T-2 = polynomial trajectory
T-3 = exponential trajectory
T-4 = cosine trajectory
T-5 = sine on ramp trajectory

In analyzing these robots with the trajectories shown some experiments can be conducted to determine the performance of each trajectory for each robot class [Asa87].

Consider the first experiment using a simple revolute planar motion of $45°$. The simplicity of the path enables a means of locating logical errors that may have been overlooked during program editing.

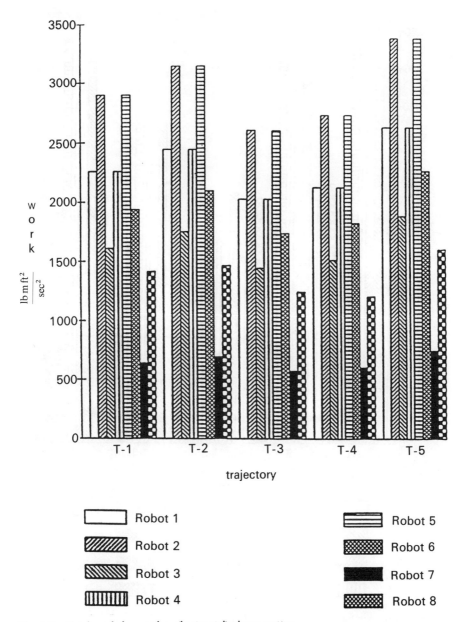

Fig. 9.7 Total work for revolute (horizontal) planar motion.

Trajectory planning and control

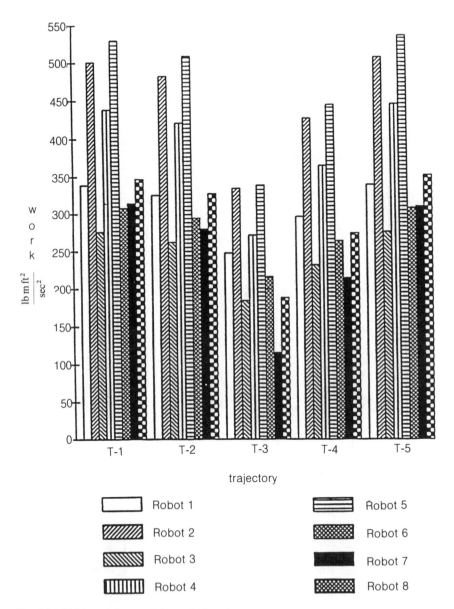

Fig. 9.8 Total work for revolute vertical motion.

The results of this trial can be plotted, and then analyzed. A characteristic shape of the graphs, for all 8 robots will show that for this simple planar motion the exponential trajectory is the best path out of the 5 considered, followed by cosine, polynomial, bang–bang and sine on ramp, respectively (Figure 9.7).

It can be found that the work parameter for this trial was controlled by the average acceleration. Thus, the bang–bang and sine on ramp trajectories which have the largest average accelerations, have the highest value for the work parameter. Similarly, exponential has the lowest average acceleration and thus has the smallest value for the work parameter.

Now consider a second trial involving an additional vertical motion which is perpendicular to the plane of motion used in trial 1. The general results will be the same. There is an overall increase in the work parameter, but the overall distribution remains the same (Figure 9.8).

To determine if these results are valid for any trajectory, or just the ones chosen, the problem can be reduced further. First a pure revolute path like the one in the first trial is considered for a wide range of angular displacements. Then, a purely prismatic trajectory is considered over a range of linear displacements.

In the first analysis, a surprising result will be discovered. For angles less than approximately $17°$, the results reveal that the expontial trajectory required the most work. The same type of results are obtained for all 8 robot types considered, which indicates that the response was not due to any particular kinematic set-up, but rather to the nature of the trajectories themselves.

In analyzing the acceleration equation for the exponential trajectory, it will be seen that the exponential trajectory is proportional to $(1/X)$, where X is the path length. If the acceleration is proportional to $(1/X)$ then so is the required torque. Thus, the torque actually decreases with an increase in displacement. This causes the work parameter for the exponential trajectory, in cases of small displacements, to become larger than one would expect. Next, the prismatic case is considered. If the explanations are correct, then one would expect to see similar results in the prismatic case. However, for the first prismatic trial, results different than those found for the revolute case can be seen.

In the initial prismatic trial, a motion of 10 vertical units can be prescribed. The general distribution will remain unchanged for a wide range of displacements. This indicates that it is not the robot configuration that is responsible for the variation in results, but rather differences in the trajectory equations. (Figure 9.9).

The vertical prismatic motion requires the effects of gravity to be considered. It was found that the gravity term affected the work parameter in different ways depending on the magnitude of the inertial term, and the trajectory involved.

For very short displacements, the inertial term contributes very little to the work parameter as compared to the gravitational term. Thus all of the trajectories required approximately the same amount of work.

For large displacements, the inertial term will become significant. As a result, the behavior will change. The addition of the gravitational term is equivalent to shifting the acceleration versus time graph upward, by a value equal to the gravitational constant. For the bang–bang trajectory (where the absolute value of the acceleration is equal to a constant), the work parameter becomes approximately equal to a constant (see Asano [Asa87]).

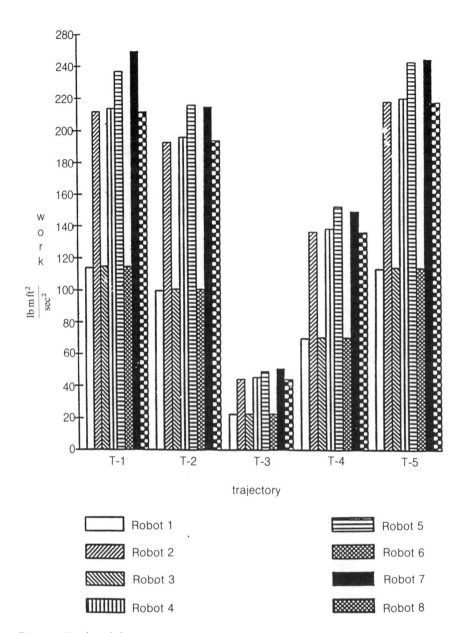

Fig. 9.9 Total work for prismatic motion.

The polynomial and sinusoidal trajectories use non-constant accelerations. For these trajectories an increase in the work parameter can be noted, which is controlled by the peak displacements.

The irregularity between the prismatic case and revolute case can thus be reasoned to be the result of the gravitational term. To check this, another trial can be initiated by using the same displacement as before, but by setting the gravitational constant to zero.

The resulting graph will take on the characteristic shape that is found for the revolute case. When the path is reduced, the exponential trajectory will experience an increase in the work parameter (relative to the other trajectories), and eventually surpass that of the other four trajectories.

Robot 8 will not be found to perform as expected in the prismatic experiments. This is due to the vastly different kinematic set-up of this robot relative to the other 7 types. Robot 8 has no vertical prismatic link. The robot reaches the vertical point which has been prescribed, through a rotation, and a translation which is in the plane of the rotational motion. Therefore, this is actually a combination of a prismatic and revolute motion.

An analysis of the trajectories shows that the linear displacement may affect the angular work done. This effect however, is minor, and does not substantially change the overall evaluation.

A final question that still remains unanswered is: What effect would changes in link masses, lengths, etc. have on the results? One would not expect changes in these parameters to change the overall distribution of the results, but rather scale the results up or down. This is because any changes in these parameters mathematically affect each trajectory in the same manner. The distribution was independent of changes in the robot geometry.

9.7 SUMMARY

An evaluation technique based on torque, work and power was shown to be feasible and efficient in determining the trajectory to use for different robot classes. As is evident from this presentation, it seems that most of the information about a trajectory can be obtained by using a basic set of test paths.

The exponential trajectory showed that a clear understanding of the robots working range may be needed. If small motions are used, then the exponential trajectory is not optimal, but if large paths are used, the exponential trajectory is clearly the best choice. Therefore, it may not be possible to find a trajectory that is optimal in all situations, but only for a certain working range of that robot.

One of the most promising results obtained from this evaluation approach is that the evaluation appears to be robot independent. Robots behave the same no matter what the configuration is, as long as the motion is relatively the same. Robot 8 did show an abnormality in evaluation for one test, but it must be remembered that this was due to the fact that the actual motion was vastly different for this particular case. In the same light, it was shown that changing robot parameters such as link masses, lengths, etc, does not affect the evaluation process.

It is clear that the exponential trajectory resulted in the lowest work parameter for large trajectories, and that the cosine trajectory resulted in the lowest for small trajectories.

Therefore, for all 8 robot configurations, the best trajectory, based on work requirements, is either exponential, or cosine.

This is a bit surprising since it is known that in the biological analogy, humans use a polynomial trajectory. The results however have been rationalized so the evaluation does not seem to be flawed. The reason for this result may lie in the fact that these trajectories were developed specifically for the Stanford Blue Arm robot. None of the robots evaluated here are very close kinematically to the Stanford Blue Arm.

Therefore, it is likely that the polynomial trajectory may be more robot specific than the others. In fact it is known that the exponential trajectory is really not robot specific at all since it is a general exponential form that was used. The coefficients were determined by an error criteria, not through robot characteristics.

In general, evaluation is situation specific to the extent that a table that ranks one trajectory as optimal in all cases is not feasible. Instead the analysis must be carried out for each differing case by use of a computer program.

CHAPTER 10

Considerations for generic kinematic structures

10.1 INTRODUCTION

It has been shown from earlier chapters that the goal in automatic robot programming is to generate a sequence of robot-level commands (MOVE, ALIGN, GRASP, etc) which are executable by all manipulators. But these commands cannot be executed directly by any manipulator. The ultimate objective is for the commands to be generated and transformed to joint motion by the computer. This chapter is intended to describe the principles necessary to transform these general robot-level commands into joint servo motion commands. The computer is used for computational as well as database management aspects of this transformation. In this chapter the generic robot kinematic structures (cylindrical, Cartesian, spherical, revolute, Scara) are discussed.

It may be worthwhile to describe how the present chapter fits into the overall strategy of RALPH.

The strategy of RALPH consists of [BNA88, Nna88, NC88, HLB89]:

1. Extracting and interpreting numerical CAD data for features through the standardized IGES interface format. This will enable a robot to have knowledge and understanding of its world. Alternatively, the features would have been implicitly specified using design with features.
2. Interpret and reason about the shapes and features in the robot's world to infer mating faces and grasping techniques to generate the desired robot independent plans.
3. The generated robot independent plans are passed to one of five kinematic structure interpreters (Cartesian, cylindrical, spherical, Scara or revolute). Each kinematic structure interpreter would then be able to translate the robot independent plans into its own executable plan.
4. For actual execution, these kinematically dependent plans would then be passed to a specific commercial robot which with the use of a language driver will implement the program.

Since parts 1 and 2 have already been described in previous chapters, we now concentrate on parts 3 and 4. The objective is to understand the kinematic behavior of manipulators in implementing a sequence of robot motions, which are general, robot independent and are not executable by a robot yet. A list of some of these commands is given below:

TRANS   Translate in x, y, z directions.

ALIGN   This requires re-orientation of the gripper so that the approach vector **a** is colinear with some entity's axis

GRASP   Close the gripper so that the distance between the jaws is the specified amount.

UNGRASP   Open gripper so that the distance between the jaws is the specified amount.

APPROACH   Move the gripper along the positive axis of the gripper frame.

DEPART   Move the gripper along one negative axis of the gripper frame.

LINEAR   Move gripper along an axis a specified amount.

ROT   Rotate about positive roll direction of the current gripper frame

These commands are generated by the computer after CAD data are extracted, and the objects and features interpreted and means of reaching, approaching and grasping an object have been decided upon.

10.2   KINEMATIC STRUCTURES

Colson and Perreira [J. C83] presented a survey of the kinematic arrangements used in industrial robots. They called the first three degrees of freedom of a manipulator (associated with the gross motion of the device), the body of the manipulator. Along with the robot bodies, they showed a number of wrists attached to the bodies. These bodies and wrists are shown in this section. There are six body types and seven wrist types that appear in industrial robots yielding 42 possible combinations. Here, all 42 kinematic structures are studied to assess their behavior in implementing general robot-level commands.

The following bodies and wrists shown here are taken directly from and with the permission of Colson and Perreira [J. C83]. The notation used for relating one joint to another is the same as Paul [Pau81]'s generalized **A** matrix. Links are related together in the following way:

$$\mathbf{A} = S_i \begin{pmatrix} d_i \\ \theta_i \\ a_{i,i+1} \\ \alpha_{i,i+1} \end{pmatrix}$$

where d_i, θ_i, α_i are the same as the generalized **A** matrix and $a_{i,i+1}$ is the same as a_n. With S_i, S can be either P or R depending on whether one has a revolute or prismatic joint. The subscript refers to the link number. An illustration of the bodies and wrists is provided at the end of this chapter. A list of link parameters for RPP (cylindrical) robots is included in section 10.5.

10.2.1   Inverse kinematic solution

Using the matrix transformation one can solve for the angles or displacements of each joint. For a 6 DOF system the following equation applies:

$$T_6 = A_1 A_2 A_3 A_4 A_5 A_6$$

One can obtain five more equations by premultiplying by A_n

$$^1T_6 = A_1^{-1}T_6 = A_2A_3A_4A_5A_6$$
$$^2T_6 = A_2^{-1}A_1^{-1}T_6 = A_3A_4A_5A_6$$
$$^3T_6 = A_3^{-1}A_2^{-1}A_1^{-1} = A_4A_5A_6$$
$$^4T_6 = A_4^{-1}A_3^{-1}A_2^{-1}A_1^{-1} = A_5A_6$$
$$^5T_6 = A_5^{-1}A_4^{-1}A_3^{-1}A_2^{-1}A_1^{-1} = A_6$$

T_6 is known since the position and orientation of the end-effector is known. Therefore since the matrix equality implies element by element equality, this gives 12 equations for each matrix equation. Through an inspection process one looks for entities that are only functions of one unknown variable. One then solves for this variable. This process is repeated until all variables are solved.

Note that in general unique solutions are not possible. One must look at geometric constraints which may restrict one solution. Also, one must remember to use Atan2 function to reduce ambiguity in angle rotations.

10.2.2 Jacobian

In cases where we want to transform differential changes in one coordinate frame into another, direct use of the Jacobian matrix is made. In robotics, one is interested in the differential changes in position and orientation of the coordinate frame attached to the end-effector T_6 as a function of the differential changes in a joint coordinate frame. Each column of the Jacobian consists of the differential translation and rotation vectors corresponding to differential changes of each of the joint coordinates.

A first step is to obtain expressions for differential rotations (differential translations are trivial to determine). By assuming small angle approximation ($\sin\theta = \theta$, $\cos\theta = 1$), one obtains:

$$\text{rot}(x, \delta_x) = \begin{pmatrix} 1 & 0 & 0 & 0 \\ 0 & 1 & -\delta_x & 0 \\ 0 & \delta_x & 0 & 0 \\ 0 & 0 & 0 & 1 \end{pmatrix}$$

$$\text{rot}(y, \delta_y) = \begin{pmatrix} 1 & 0 & \delta_y & 0 \\ 0 & 1 & 0 & 0 \\ \delta_y & 0 & 1 & 0 \\ 0 & 0 & 0 & 1 \end{pmatrix}$$

$$\text{rot}(z, \delta_z) = \begin{pmatrix} 1 & -\delta_z & 0 & 0 \\ \delta_z & 1 & 0 & 0 \\ 0 & 0 & 1 & 0 \\ 0 & 0 & 0 & 1 \end{pmatrix}$$

A general differential rotation matrix has become (neglecting higher order terms):

$$\text{rot}(x, \delta_x)\text{rot}(y, \delta_y)\text{rot}(z, \delta_z) = \begin{pmatrix} 1 & -\delta_z & \delta_y & 0 \\ \delta_z & 1 & -\delta_x & 0 \\ -\delta_y & \delta_x & 1 & 0 \\ 0 & 0 & 0 & 1 \end{pmatrix}.$$

As a result of the small angle approximation, and the fact that higher order terms were ignored, one can prove two important theorems:

Theorem 1:

A differential rotation w about an arbitrary vector $\mathbf{k} = (k_x, k_y, k_z)^T$ is equivalent to three differential rotations $\delta_x, \delta_y, \delta_z$ about the base axis x, y, z where:

$$k_x \delta_\theta = \delta_x$$
$$k_y \delta_\theta = \delta_y$$
$$k_z \delta_\theta = \delta_z$$

Theorem 2:

Differential rotations are independent of the order in which rotations are performed. If a frame undergoes a differential motion its next position is described by:

$$\mathbf{T} + \mathrm{d}T = [\mathrm{trans}(\delta_x, \delta_y, \delta_z)\mathrm{rot}(w, \mathrm{d}\theta)]T$$

where w = arbitrary axis of rotation and $\mathbf{T}$ = homogeneous transformation matrix. Solving for $\mathrm{d}T$, one obtains:

$$\mathrm{d}T = [\mathrm{trans}(\delta_x, \delta_y, \delta_z)\mathrm{rot}(w, \mathrm{d}\theta) - I]T$$

The differential operator, Δ, is defined as:

$$\Delta = [\mathrm{trans}(\delta_x, \delta_y, \delta_z)\mathrm{rot}(w, \mathrm{d}\theta) - I]$$

$$\begin{pmatrix} 0 & -\delta_z & -\delta_y & d_x \\ \delta_z & 0 & -\delta_x & d_y \\ \delta_y & \delta_x & 0 & d_z \\ 0 & 0 & 0 & 1 \end{pmatrix}$$

therefore:

$$\mathrm{d}T = \Delta T$$

The Δ operator represents differential translations and rotations. In other words, it is a differential translation and rotation transform, where the differential changes are with respect to the base frame.

One can solve for $\mathrm{d}T$ and in general one obtains the following equations:

$$^T\mathrm{d}_x = \mathbf{n} \cdot ((\delta \times \mathbf{p}) + \mathbf{d}) \tag{10.1}$$

$$^T\mathrm{d}_y = \mathbf{o} \cdot ((\delta \times \mathbf{p}) + \mathbf{d}) \tag{10.2}$$

$$^T\mathrm{d}_z = \mathbf{a} \cdot ((\delta \times \mathbf{p}) + \mathbf{d}) \tag{10.3}$$

$$^T\delta_x = \mathbf{n} \cdot \delta \tag{10.4}$$

$$^T\delta_y = \mathbf{o} \cdot \delta \tag{10.5}$$

$$^T\delta_z = \mathbf{a} \cdot \delta \tag{10.6}$$

where $\mathbf{n}, \mathbf{o}, \mathbf{a}$ = normal, orientation and approach vectors in frame T and $\mathbf{p}$ = translation vector in frame T.

One can similarly describe the translation and rotations of a frame via an operator that represents differential translations and rotations of that frame:

$$\Delta T = T^T \Delta$$

where $^T\Delta$ = operator with respect to displacement in frame T.

Solving for $^T\Delta$, one obtains:

$$^T\Delta = T^{-1}\Delta T$$

For a revolute joint, $d_i = 0$, and $\delta_i = 0i + 0j + 1k$. One can therefore show by applying equations (10.1)–(10.6) that:

$$^{T_6}d_i = (-n_x p_y + n_y p_x)i + (-o_x p_y + o_y p_x)j + (-a_x p_y + a_y p_x)k \quad (10.7)$$

$$^{T_6}\delta_i = n_z i + o_z j + a_z k \quad (10.8)$$

If the joint is prismatic $\delta = 0$, $d = 0i + 0j + 1k$, and the equation becomes:

$$^{T_6}d_i = n_z i + o_z j + a_z k \quad (10.9)$$

$$\delta_i = 0i + 0j + 0k \quad (10.10)$$

where vectors n, o and, a describe the end-effector orientation and vector d describes the end-effector position as described by the base coordinates.

Define an operator called the Jacobian that maps the differential changes in joint space to differential changes in Cartesian space:

$$D = J D_\theta$$

where J = Jacobian
D_θ = differential change in joint space
D = differential change in Cartesian space

10.2.3 Degeneracy

Whenever there is more than one set of joint variables that result in the same hand configuration, the arm is said to be degenerate. Several conditions may lead to degeneracy. For example, in using the Atan2 function two solutions may lie 180° apart. It must be emphasized that degeneracy is different from singularity, which is losing a degree of freedom in the Cartesian coordinate frame.

10.2.4 Singularity

When a Cartesian degree of freedom is lost, the manipulator is said to have reached a singular point. The directions in which motion is impossible are called *constraint directions*. This has special significance in implementing the APPROACH command. In some cases, it may be impossible for the robot to implement the APPROACH command because it has reached a *singular* point. Such a situation should be detected and the user should receive an appropriate message. If the Jacobian matrix is square, the robot is said to be *singular* where the determinant of the Jacobian matrix goes to zero. Under these circumstances, the Jacobian matrix is not invertible, and the constraint directions along which

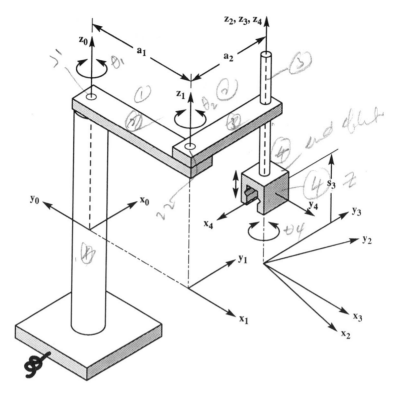

Fig. 10.1 Coordinate frames for the GE-A4 robot.

motion is impossible can be found. The ideas behind inverse kinematics analysis and finding the constraint directions are illustrated via an example described in the following section.

10.2.5 An example

One can derive the link parameters, relating one joint to another for the GE-A4 robot (Figure 10.1, Table 10.1).

For the GE-A4 robot:

From now on S_i will be substituted for $\sin(\theta_i)$, C_i for $\cos(\theta_i)$, S_{ij} for $\sin(\theta_i + \theta_j)$ and C_{ij} for $\cos(\theta_i + \theta_j)$.

Table 10.1 Link parameters for the GE-A4 robot

Link$_i$	Variables	α_i	a_i	d_i	$\cos(\alpha_i)$	$\sin(\alpha_i)$	θ_i
1	θ_1	0	a_1	0	1	0	θ_1
2	θ_2	0	a_2	0	1	0	θ_2
3	S_3	0	0	Z	1	0	0
4	θ_4	0	0	0	1	0	θ_4

If the values from Table 10.1 are inserted into the generalized **A** matrix, we get the following **A** matrices:

$$\mathbf{A}_1 = \begin{pmatrix} C_1 & -S_1 & 0 & a_1 C_1 \\ S_1 & C_1 & 0 & a_1 S_1 \\ 0 & 0 & 1 & 0 \\ 0 & 0 & 0 & 1 \end{pmatrix} \quad (10.11)$$

$$\mathbf{A}_2 = \begin{pmatrix} C_2 & -S_2 & 0 & a_2 C_2 \\ S_2 & C_2 & 0 & a_2 S_2 \\ 0 & 0 & 1 & 0 \\ 0 & 0 & 0 & 1 \end{pmatrix} \quad (10.12)$$

$$\mathbf{A}_3 = \begin{pmatrix} 1 & 0 & 0 & 0 \\ 0 & 1 & 0 & 0 \\ 0 & 0 & 1 & Z \\ 0 & 0 & 0 & 1 \end{pmatrix} \quad (10.13)$$

$$\mathbf{A}_4 = \begin{pmatrix} C_4 & -S_4 & 0 & 0 \\ S_4 & C_4 & 0 & 0 \\ 0 & 0 & 1 & 0 \\ 0 & 0 & 0 & 1 \end{pmatrix} \quad (10.14)$$

By multiplying the **A** matrices together, we get a relationship between the different frames. That is $^0T_4 = \mathbf{A}_1 \mathbf{A}_2 \mathbf{A}_3 \mathbf{A}_4$, is the relation between the base and the end-effector frame. The products of the **A** matrices for the GE-A4 robot starting at link four and working back, are:

the relationship between the second frame and the end-effector frame:

$$^2T_4 = \mathbf{A}_3 \mathbf{A}_4 = \begin{pmatrix} C_4 & -S_4 & 0 & 0 \\ S_4 & C_4 & 0 & 0 \\ 0 & 0 & 1 & Z \\ 0 & 0 & 0 & 1 \end{pmatrix} \quad (10.15)$$

the relationship between the first frame and the end-effector frame:

$$^1T_4 = \mathbf{A}_2 \mathbf{A}_3 \mathbf{A}_4 = \begin{pmatrix} C_{24} & -S_{24} & 0 & a_2 C_2 \\ S_{24} & C_{24} & 0 & a_2 S_2 \\ 0 & 0 & 1 & Z \\ 0 & 0 & 0 & 1 \end{pmatrix} \quad (10.16)$$

the relationship between the base frame and the end-effector frame:

$$^0T_4 = \mathbf{A}_1 \mathbf{A}_2 \mathbf{A}_3 \mathbf{A}_4 = \begin{pmatrix} C_{124} & -S_{124} & 0 & a_2 C_{12} + a_1 C_1 \\ S_{124} & C_{124} & 0 & a_2 S_{12} + a_1 S_1 \\ 0 & 0 & 1 & Z \\ 0 & 0 & 0 & 1 \end{pmatrix} \quad (10.17)$$

$^0T_4$, the relationship between the base frame and the end-effector frame is defined as:

$$^0T_4 = \begin{pmatrix} n_x & o_x & a_x & p_x \\ n_y & o_y & a_y & p_y \\ n_z & o_z & a_z & p_z \\ 0 & 0 & 0 & 1 \end{pmatrix} \quad (10.18)$$

Thus,

$$n_x = \cos(\theta_1 + \theta_2 + \theta_4) \quad (10.19)$$
$$n_y = \sin(\theta_1 + \theta_2 + \theta_4) \quad (10.20)$$
$$n_z = 0 \quad (10.21)$$
$$o_x = -\sin(\theta_1 + \theta_2 + \theta_4) \quad (10.22)$$
$$o_y = \cos(\theta_1 + \theta_2 + \theta_4) \quad (10.23)$$
$$o_z = 0 \quad (10.24)$$
$$a_x = 0 \quad (10.25)$$
$$a_y = 0 \quad (10.26)$$
$$a_z = 1 \quad (10.27)$$
$$p_x = a_2\cos(\theta_1 + \theta_2) + a_1\cos(\theta_1) \quad (10.28)$$
$$p_y = a_2\sin(\theta_1 + \theta_2) + a_1\sin(\theta_1) \quad (10.29)$$
$$p_z = Z \quad (10.30)$$

where n, o, a represent the orientation and p the translation.

Now one can proceed to solve for the joint angles in terms of the variables representing the orientation and position of the end-effector from equations (10.28) and (10.29):

$$p_x = a_2\cos(\theta_1 + \theta_2) + a_1\cos(\theta_1)$$
$$p_y = a_2\sin(\theta_1 + \theta_2) + a_1\sin(\theta_1)$$

Squaring both equations yields:

$$p_x^2 = a_2^2 C_{12}^2 + a_1^2 C_1^2 + 2a_1 a_2 C_1 C_{12}$$

and

$$p_y^2 = a_2^2 S_{12}^2 + a_1^2 S_1^2 + 2a_1 a_2 S_1 S_{12}$$

This yields:

$$C_2 = \frac{(p_x^2 + p_y^2 - a_1 - a_2)}{2a_1 a_2}$$

Now one can solve for S_2:

$$S_2 = \pm\sqrt{1 - C_2^2}$$

Obviously, we have a degeneracy here with the $\pm$ sign yielding two solutions. Using the atan2 function we get:

$$\theta_2 = \operatorname{atan}\left(\frac{S_2}{C_2}\right)$$

Solving the equations of p_x and p_y for C_1 and S_1 yields:

$$C_1 = \frac{(C_2 a_2 + a1)p_x + S_2 a_2 p_y}{(a_1^2 + a_2^2 + 2a_1 a_2 C_2)}$$

and:

$$S_1 = \frac{(C_2 a_2 + a1)p_y + S_2 a_2 p_x}{(a_1^2 + a_2^2 + 2a_1 a_2 C_2)}$$

Inserting in the atan2 function yields:

$$\theta_1 = \operatorname{atan2}(((C_2 a_2 + a_1)p_y + S_2 a_2 p_x), ((C_2 a_2 + a_1)p_x + S_2 a_2 p_y))$$

So far expressions for θ_1 and θ_2 have been gained. Equations (10.22) and (10.23) also give expressions for C_{124} and S_{124}:

$$\theta_{124} = \operatorname{atan}(-o_x, o_y)$$

Now the only remaining unknown is the variable dealing with the prismatic joint. This can be equated to equation (10.30).

For the GE-A4 robot which has four axes, the differential displacement of the end-effector as a vector can be defined as:

$$D = (\delta_x \delta_y \delta_z \Delta_z)^T$$

The differential displacement of the joint variables as a vector can be defined as:

$$D_{joint} = (\delta_{\theta_1}, \delta_{\theta_2}, \delta_{\theta_3}, \delta_{\theta_4})^T$$

The relationship between these two matrices is:

$$D = J D_{joint}$$

where J is the Jacobian matrix.

$$J = \begin{pmatrix} J_{11} & J_{12} & J_{13} & J_{14} \\ J_{21} & J_{22} & J_{23} & J_{24} \\ J_{31} & J_{32} & J_{33} & J_{34} \\ J_{41} & J_{42} & J_{43} & J_{44} \end{pmatrix}$$

For a revolute joint: $J_{1a} = -n_x p_y + n_y p_x$, $J_{2a} = -o_x p_y + o_y p_x$ and $J_{3a} = -a_x p_y + a_y p_x$, $J_{4a} = a_z$. For a prisimatic joint:

$$J_{1a} = n_z$$
$$J_{2a} = o_z$$
$$J_{3a} = a_z$$
$$J_{4a} = 0$$

where n, o, a and p come from: T_4^0 for the first joint, T_4^1 for the second, etc. For the first joint one gets:

$$= a_1 \sin(\theta_2 + \theta_4) + a_2 \sin\theta_4.$$

$$J_{11} = a_1 S_{24} + a_2 S_4$$
$$J_{21} = a_1 C_{24} + a_2 C_4 \quad a_1 \cos(\theta_2 + \theta_4) + a_2 \cos\theta_4$$
$$J_{31} = 0$$
$$J_{41} = 1$$

For the second joint one obtains:

$$J_{12} = a_2 S_4$$
$$J_{22} = a_2 C_4$$
$$J_{32} = 0$$
$$J_{42} = 1$$

For the third joint which is a prismatic, one obtains.

$$J_{13} = 0$$
$$J_{23} = 0$$
$$J_{33} = 1$$
$$J_{43} = 0$$

For the fourth joint one obtains:

$$J_{14} = 0$$
$$J_{24} = 0$$
$$J_{34} = 0$$
$$J_{44} = 1$$

Inserting in the Jacobian gives:

$$J = \begin{pmatrix} a_1 S_{24} + a_2 S_4 & a_2 S_4 & 0 & 0 \\ a_1 C_{24} + a_2 C_4 & a_2 C_4 & 0 & 0 \\ 0 & 0 & 1 & 0 \\ 1 & 1 & 0 & 1 \end{pmatrix}$$

In this case taking the inverse of the Jacobian leaves one with:

$$J^{-1} D = D_{\text{joint}}$$

which shows the differential displacement of the joint variables in terms of the differential displacement of the end-effector. Where J^{-1} does not exist, the J matrix is called "singular". It is known from matrix algebra that the J matrix is singular when $|J| = 0$ or the determinant of the Jacobian matrix is equal to zero. Physically this translates to losing a translational or rotational degree of freedom in the Cartesian coordinate frame of the end-effector. By definition, the inverse of a matrix is equal to the adjoint divided by the determinant of that matrix:

$$J^{-1} = \frac{J^+}{|J|}$$

Kinematic structures

where J^+ is the adjoint of matrix J. This immediately yields:

$$J^{-1}\left(\frac{J^+}{|J|}\right)D = D_{joint}$$

This directly implies that:

$$(J^+)D = |J|D_{joint}$$

Where $|J| = 0$ one obtains:

$$J_0^+ D = 0$$

where J_0^+ is the adjoint matrix evaluated at $|J| = 0$. For the GE-A4 robot the determinant of the Jacobian is given by:

$$|J| = a_1 a_2 S_2$$

From realizing that the determinant is equal to zero at $\theta_2 = 0$ one calculates the adjoint matrix as:

$$\text{adjoint}(J) = J^+ = \begin{pmatrix} a_2 C_4 & -a_2 S_4 & 0 & 0 \\ -a_1 C_4 - a_2 C_4 & a_1 S_4 + a_2 S_2 - 4 & 0 & 0 \\ 0 & 0 & a_1 a_2 S_2 & 0 \\ a_1 C_4 & -a_1 S_4 & 0 & a_1 a_2 S_2 \end{pmatrix}$$

Evaluating the adjoint matrix at $|J| = 0$ yields:

$$J_0^+ = \begin{pmatrix} a_2 C_4 & -a_2 S_4 & 0 & 0 \\ -a_1 C_4 - a_2 C_4 & a_1 S_4 + a_2 S_2 - 4 & 0 & 0 \\ 0 & 0 & 0 & 0 \\ a_1 C_4 & -a_1 S_4 & 0 & 0 \end{pmatrix}$$

Inserting expression for D one obtains:

$$\begin{pmatrix} a_2 C_4 & -a_2 S_4 & 0 & 0 \\ -a_1 C_4 - a_2 C_4 & a_1 S_4 + a_2 S_4 & 0 & 0 \\ 0 & 0 & 0 & 0 \\ a_1 C_4 & -a_1 S_4 & 0 & 0 \end{pmatrix} \begin{pmatrix} \delta_x \\ \delta_y \\ \delta_z \\ \Delta_z \end{pmatrix} = 0$$

From the above equality one can get four sets of equations. Row three consists of zeros and does not yield any information; rows 1, 2, 4 yield essentially the same information namely:

$$C_4 \delta_x - S_4 \delta_y = 0$$

From the above equation one can get the Cartesian constraint direction or the direction in which motion is impossible:

$$\text{at } \theta_4 = 0 \quad \delta_x = 0$$

or that translation in the x direction is impossible, which is the equivalent of losing a degree of freedom in the Cartesian coordinate frame. Also one can observe that:

$$\text{at } \theta_4 = 90 \quad \delta_y = 0$$

This information can be summarized as follows:
When

and

$\theta_2 = 0$ and $\theta_4 = 0$ then $\delta_x = 0$

$\theta_2 = 0$ and $\theta_4 = 90$ then $\delta_y = 0$

In some cases one may be interested in $\mathbf{J}_p$ or the positional Jacobian which deals only with differential translation ($\delta_x, \delta_y, \delta_z$). This can be obtained by eliminating the rows that have to do with differential rotation (in the case of GE-A4 this would be the fourth row). One also observes that in the case of GE-A4 the last joint does not impart any differential translational motion to the gripper (rows 1, 2, 3 of the last column of the Jacobian are equal to zero). This means that the last column can also be eliminated for obtaining $\mathbf{J}_p$ yielding a square matrix:

$$\mathbf{J}_p = \begin{pmatrix} a_1 S_{24} + a_2 S_4 & a_2 S_4 & 0 \\ a_1 C_{24} + A_2 C_4 & a_2 C_4 & 0 \\ 0 & 0 & 1 \end{pmatrix}$$

10.2.6 Quartenion representation of rotations

Even though in this presentation direct use is made of homogeneous matrix representations, it is important to mention a method of representing rotations that is less expensive for computation. This alternative method uses quartenions to represent rotations [YF86] [Ham69]. We briefly present the quartenion representation of rotations. We shall use it subsequently to implement linear motion.

In general, a quartenion Q consists of a scalar part s and a vector part $\mathbf{v}$ and here is written:

$$Q = [s + \mathbf{v}]$$

The multiplication rule is:

$$Q_1 \cdot Q_2 = [s_1 s_2 - \mathbf{v}_1 \cdot \mathbf{v}_2 + s_2 \mathbf{v}_1 + s_1 \mathbf{v}_2 + \mathbf{v1} \times \mathbf{v2}]$$

From this definition it follows that if:

$$s = \sin\frac{\theta}{2} \quad \text{and} \quad c = \cos\frac{\theta}{2}$$

then one can represent a rotation $\mathrm{rot}(\mathbf{n}, \theta)$ by a quartenion:

$$\left[\cos\frac{\theta}{2} + \sin\frac{\theta}{2} \cdot \mathbf{n}\right]$$

10.3 KINEMATIC IMPLEMENTATION

Realizing that general robot-level commands deal basically with moving the robot in a Cartesian coordinate frame one observes that the inverse kinematic solution for a robot provides a straightforward way of transforming Cartesian motion to joint motion.

Kinematic implementation

Realizing that the final position the robot has to move to is:

$$T = \begin{pmatrix} n_x & o_x & a_x & p_x \\ n_y & o_y & a_y & p_y \\ n_z & o_z & a_z & p_z \\ 0 & 0 & 0 & 1 \end{pmatrix}$$

One can solve for the joint variables.

Another important consideration in the kinematic implementation of the general robot-level commands is that of singularity. This occurs when one loses a Cartesian degree of freedom. The singular points lying within the working envelope of the robot can be found by determining where the determinant of the Jacobian matrix goes to zero. From that the constraint directions (the directions along which motion is impossible) can be determined. This is important in executing commands such as APPROACH and LINEAR. With these ideas in mind we consider all the above commands and show how they can be transformed to joint motion. In executing these commands one realizes that GRASP and UNGRASP are not kinematically significant as they represent only a closing of the gripper and do not depend on the kinematic arrangement used.

A list of general robot-level commands and their method of transformation to joint motion is given below:

TRANS: This represents a change in the last column of the transformation matrix relating the gripper to the base coordinate frame.

$$T = \begin{pmatrix} n_x & o_x & a_x & p_x \\ n_y & o_y & a_y & p_y \\ n_z & o_z & a_z & p_z \\ 0 & 0 & 0 & 1 \end{pmatrix}$$

Only p_x, p_y, p_z change in this case. One has to recalculate the inverse kinematic solutions with the new p_x, p_y, p_z to find the new values for the joint angles.

ALIGN: This shows a change in the first three columns of the transformation matrix in the T matrix shown. In some cases, it may be necessary to go through a translation to align the approach vector of the gripper with an object's axis. This means a change in all the columns of the transformation matrix (both the position and orientation of the gripper change). In either case, one has to recalculate the inverse kinematic solutions with the new values for variables in the transformation matrix to find the new values for the joint angles.

MOVE: This represents a change in all the variables of the transformation matrix since MOVE causes a change both in orientation and position.

APPROACH: This represents a change in the last column of the transformation matrix. The situation is depicted in Figure 10.2. One moves the robot from a position p_0 to a position p_1 along the positive axis of the gripper frame. We can solve for the new position vectorially by writing:

$$p_1 = p_0 + a$$

With the new position we can find the new joint angles necessary to attain this position by calculating the inverse kinematics solution with the new values for p_x, p_y, p_z. In this case the point from which the robot has to move may be a singular position, in which

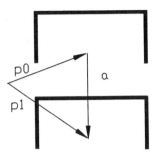

Fig. 10.2 The APPROACH command.

case it can be impossible for the robot to move along the *a* vector. Such a situation should be detected using the ideas introduced about singularities in section 5.2. If the distance to be covered is large, enough intermediate points should be interpolated so that the robot will not deviate from a linear path.

DEPART: The strategy used for implementing this command is similar to APPROACH except that one moves along the negative axis of the gripper frame.

LINEAR: This represents motion between two positions along a linear path. This motion from point 1 to point 2 was represented by Paul [Pau] as consisting of a translational motion and two rotational motions. The translational motion can be represented by a homogeneous transformation matrix $\mathbf{L}(t)$ and the motion is along a straight line:

$$\mathbf{L}(\Delta) = \begin{pmatrix} 1 & 0 & 0 & \Delta x \\ 0 & 1 & 0 & \Delta y \\ 0 & 0 & 0 & \Delta z \\ 0 & 0 & 0 & 1 \end{pmatrix}$$

The first rotational motion can be represented by a homogeneous transformation matrix and it serves to rotate the approach vector from position 1 to position 2. It is represented by:

$$R_A(\Delta) = \begin{pmatrix} S^2\psi V(\Delta\theta) + C(\Delta\theta) & -S\psi C\psi V(\Delta\theta) & C\psi S(\Delta\theta) & 0 \\ -S\psi C\psi V(\Delta\theta) & C^2\psi V(\Delta\theta) + C(\Delta\theta) & S\psi S(\Delta\theta) & 0 \\ C\psi S(\Delta\theta) & -S\psi S(\Delta\theta) & C(\Delta\theta) & 0 \\ 0 & 0 & 0 & 1 \end{pmatrix}$$

The second rotational motion serves to rotate the orientation vector from position 1 into the orientation vector at position 2. It can be represented by:

$$R_B(\Delta) = \begin{pmatrix} C(\Delta\phi) & -S(\Delta\phi) & 0 & 0 \\ S(\Delta\phi) & C(\Delta\phi) & 0 & 0 \\ 0 & 0 & 1 & 0 \\ 0 & 0 & 0 & 1 \end{pmatrix}$$

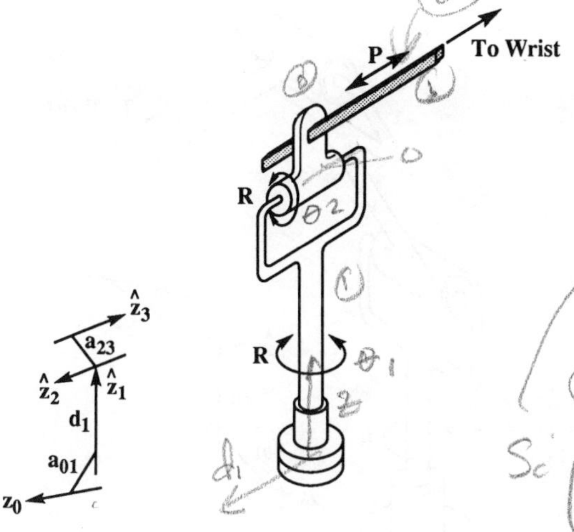

Kinematic implementation 247

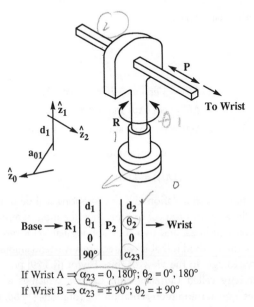

$$\text{Base} \rightarrow R_1 \begin{vmatrix} d_1 \\ \theta_1 \\ 0 \\ 90° \end{vmatrix} R_2 \begin{vmatrix} 0 \\ \theta_2 \\ a_{23} \\ 90° \end{vmatrix} P_3 \begin{vmatrix} d_3 \\ \theta_3 \\ a_{34} \\ \alpha_{34} \end{vmatrix} \rightarrow \text{Wrist}$$

If Wrist A $\Rightarrow \alpha_{34} = 0°, 180°; \theta_3 = 0°, 180°$
If Wrist B $\Rightarrow \alpha_{34} = \pm 90°; \theta_3 = 0°, 180°$

Fig. 10.3 RRP manipulator body.

$$\text{Base} \rightarrow R_1 \begin{vmatrix} d_1 \\ \theta_1 \\ 0 \\ 90° \end{vmatrix} P_2 \begin{vmatrix} d_2 \\ \theta_2 \\ 0 \\ \alpha_{23} \end{vmatrix} \rightarrow \text{Wrist}$$

If Wrist A $\Rightarrow \alpha_{23} = 0, 180°; \theta_2 = 0°, 180°$
If Wrist B $\Rightarrow \alpha_{23} = \pm 90°; \theta_2 = \pm 90°$

Fig. 10.4 RP manipulator body.

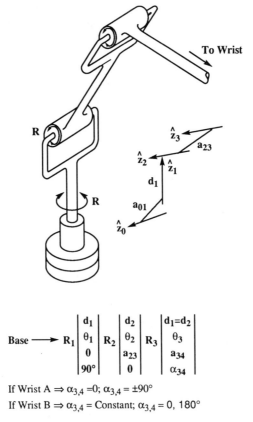

Base ⟶ $R_1 \begin{vmatrix} d_1 \\ \theta_1 \\ 0 \\ 90° \end{vmatrix} R_2 \begin{vmatrix} d_2 \\ \theta_2 \\ a_{23} \\ 0 \end{vmatrix} R_3 \begin{vmatrix} d_1=d_2 \\ \theta_3 \\ a_{34} \\ \alpha_{34} \end{vmatrix}$

If Wrist A ⟹ $\alpha_{3,4} = 0$; $\alpha_{3,4} = \pm 90°$
If Wrist B ⟹ $\alpha_{3,4}$ = Constant; $\alpha_{3,4} = 0, 180°$

Fig. 10.5 RRR manipulator body.

where

$$V(\Delta\theta) = Versine(\Delta\theta) = 1 - \cos(\Delta\theta)$$
$$C(\Delta\theta) = \cos(\Delta\theta) \text{ and } S(\Delta\theta) = \sin(\Delta\theta)$$
$$C(\Delta\phi) = \cos(\Delta\phi) \text{ and } S(\Delta\phi) = \sin(\Delta\phi)$$

Therefore the drive function for this linear motion consists of:

$$D(\Delta) = L(\Delta)R_A(\Delta)R_B(\Delta)$$

One realizes that both the translation and the rotations will be directly proportional to Δ and if Δ varies linearly with respect to time, then the motion represented by $D(\Delta)$ will correspond to a constant linear and two constant angular velocities.

The difficulty of this method is that it will require a wasteful amount of precomputation time and memory storage. In the view of this, Taylor [R.T79] proposed a joint variable space motion strategy called *Cartesian Path Control*, which is a refinement of Paul's technique but uses a quaternion representation for rotations [YF86, Ham69].

The motion along the linear path consists of translation of the tool frame's origin from p_0 to p_1 coupled with rotation of the tool frame from R_0 to R_1. Let $\Delta(t)$ be the remaining

Kinematic implementation

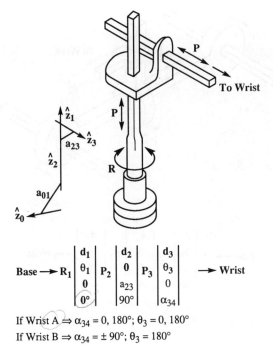

$$\text{Base} \rightarrow R_1 \begin{vmatrix} d_1 \\ \theta_1 \\ 0 \\ 0° \end{vmatrix} P_2 \begin{vmatrix} d_2 \\ 0 \\ a_{23} \\ 90° \end{vmatrix} P_3 \begin{vmatrix} d_3 \\ \theta_3 \\ 0 \\ \alpha_{34} \end{vmatrix} \rightarrow \text{Wrist}$$

If Wrist A $\Rightarrow \alpha_{34} = 0, 180°; \theta_3 = 0, 180°$
If Wrist B $\Rightarrow \alpha_{34} = \pm 90°; \theta_3 = 180°$

Fig. 10.6 RPP manipulator body.

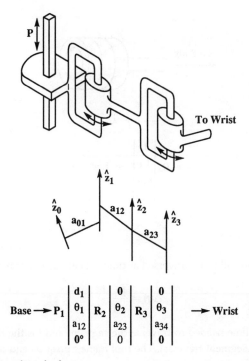

$$\text{Base} \rightarrow P_1 \begin{vmatrix} d_1 \\ \theta_1 \\ a_{12} \\ 0° \end{vmatrix} R_2 \begin{vmatrix} 0 \\ \theta_2 \\ a_{23} \\ 0 \end{vmatrix} R_3 \begin{vmatrix} 0 \\ \theta_3 \\ a_{34} \\ 0 \end{vmatrix} \rightarrow \text{Wrist}$$

Fig. 10.7 PRR manipulator body.

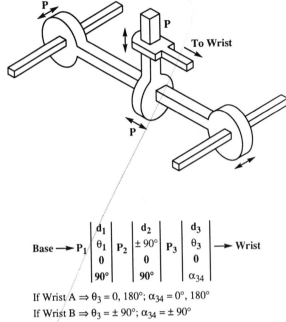

$$\text{Base} \rightarrow P_1 \begin{vmatrix} d_1 \\ \theta_1 \\ 0 \\ 90° \end{vmatrix} P_2 \begin{vmatrix} d_2 \\ \pm 90° \\ 0 \\ 90° \end{vmatrix} P_3 \begin{vmatrix} d_3 \\ \theta_3 \\ 0 \\ \alpha_{34} \end{vmatrix} \rightarrow \text{Wrist}$$

If Wrist A $\Rightarrow \theta_3 = 0, 180°; \alpha_{34} = 0°, 180°$
If Wrist B $\Rightarrow \theta_3 = \pm 90°; \alpha_{34} = \pm 90°$

Fig. 10.8 PPP manipulator body.

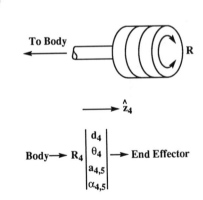

$$\text{Body} \rightarrow R_4 \begin{vmatrix} d_4 \\ \theta_4 \\ a_{4,5} \\ \alpha_{4,5} \end{vmatrix} \rightarrow \text{End Effector}$$

Fig. 10.9 Wrist 1A.

fraction of the motion still to be traversed at time t. Then for uniform motion we have

$$\Delta(t) = \frac{T-t}{T}$$

where T is the total time needed to traverse the segment and t is the time starting from the beginning of the segment traversal. The tool frame's position and orientation at time

Kinematic implementation   251

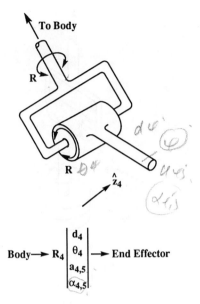

Body → $R_4 \begin{vmatrix} d_4 \\ \theta_4 \\ a_{4,5} \\ \alpha_{4,5} \end{vmatrix}$ → End Effector

Fig. 10.10   Wrist 1B.

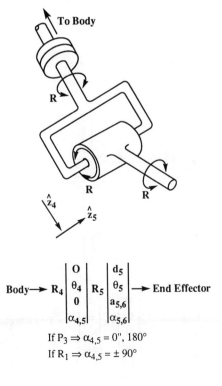

Body → $R_4 \begin{vmatrix} 0 \\ \theta_4 \\ 0 \\ \alpha_{4,5} \end{vmatrix} R_5 \begin{vmatrix} d_5 \\ \theta_5 \\ a_{5,6} \\ \alpha_{5,6} \end{vmatrix}$ → End Effector

If $P_3 \Rightarrow \alpha_{4,5} = 0"$, $180°$
If $R_1 \Rightarrow \alpha_{4,5} = \pm 90°$

Fig. 10.11   Wrist 2A.

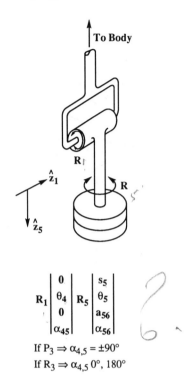

$$R_1 \begin{vmatrix} 0 \\ \theta_4 \\ 0 \\ \alpha_{45} \end{vmatrix} R_5 \begin{vmatrix} s_5 \\ \theta_5 \\ a_{56} \\ \alpha_{56} \end{vmatrix}$$

If $P_3 \Rightarrow \alpha_{4,5} = \pm 90°$
If $R_3 \Rightarrow \alpha_{4,5}$ 0°, 180°

Fig. 10.12 Wrist 2B.

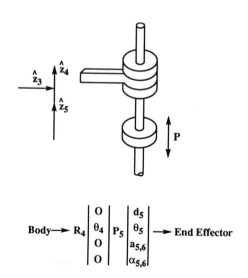

$$\text{Body} \rightarrow R_4 \begin{vmatrix} 0 \\ \theta_4 \\ 0 \\ 0 \end{vmatrix} P_5 \begin{vmatrix} d_5 \\ \theta_5 \\ a_{5,6} \\ \alpha_{5,6} \end{vmatrix} \rightarrow \text{End Effector}$$

Fig. 10.13 Wrist 2C.

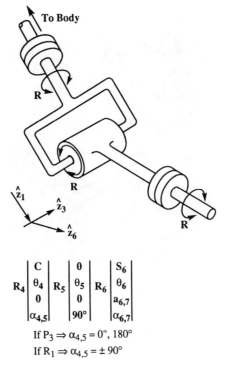

Fig. 10.14 Wrist 3A.

t are given, respectively, by:

$$P(t) = p_1 - \Delta(t)(p_1 - p_0)$$
$$R(t) = R_1 \text{rot}[n, -\theta\Delta(t)]$$

where rot(n, θ) is rotation by θ about an axis n to reorient R_0 into R_1:

$$\text{rot}(n, \theta) = R_0^{-1} R_1$$

where rot(n, θ) represents the resultant rotation of $R_0^{-1} R_1$ in a quaternion form. The approach proposed in this work is the same as Taylor's method of *Cartesian Path Control*.

ROT: This represents a rotation about the axis of the gripper frame. For the robots considered here, this is a roll motion for the manipulators having a roll motion as their last rotational joint. Referring to Figure OBs 10.3 to 10.15 at the end of this chapter showing the bodies and wrists for industrial robots, this means robots having a 1A, 2B, 2C, 3A, 3B wrist.

10.4 KINEMATIC ANALYSIS

As mentioned before, the goal here is to take a series of general, robot independent commands and study different kinematic structures to see how these commands can be

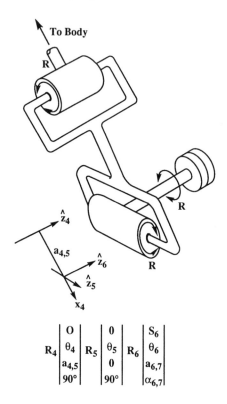

$$R_4 \begin{vmatrix} 0 \\ \theta_4 \\ a_{4,5} \\ 90° \end{vmatrix} R_5 \begin{vmatrix} 0 \\ \theta_5 \\ 0 \\ 90° \end{vmatrix} R_6 \begin{vmatrix} S_6 \\ \theta_6 \\ a_{6,7} \\ \alpha_{6,7} \end{vmatrix}$$

Fig. 10.15 Wrist 3B.

transformed to joint motion and also what sort of trends emerge in the kinematic behavior of robots implementing these commands. Methods of executing these commands in joint space using the inverse kinematic solution and the Jacobian matrix for the robot were described in the previous section of this chapter. It was decided that a total of 42 kinematic structures would be studied. Two basic assumptions were made regarding these robots.

The first one was that one would only consider solvable kinematic structures. The word "solvable" refers to structures whose joint solution can be found using inverse kinematic techniques and there is no need to resort to numerical solutions. We shall assume that all the robots with less than 6 DOF were solvable. For 6 DOF robots Yuchen and Wenhan

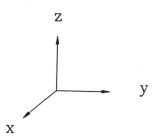

Fig. 10.16 Base coordinate frame for the robots.

Example

[Yuc86] listed the solvable structures of the 6 DOF manipulator. For the 6 DOF robots considered in this presentation, all the RRR structures turned out to be solvable regardless of the type of wrists attached to them. This matches the results obtained by Yuchen and Wenhan. For the other type of bodies, the primary requirement for solvability is for the last three joints to intersect at a point. For 3A wrists this is the case but for 3B wrists such a condition is lacking. To make the structure solvable for robots that do not have a RRR body but do have a 3B wrist, a_{45} is set equal to zero to make the kinematic structure solvable.

The second assumption deals with cases where the Cartesian degrees of freedom are not the same as the joint degrees of freedom yielding a Jacobian matrix that is not invertible (not a square matrix). In cases like this an attempt is made to isolate J_p or the positional Jacobian by eliminating the rows corresponding to rotational Cartesian degrees of freedom (refer to section 5.2). By using this technique one obtains J_p as a 3×3 matrix that has a determinant and is invertible. From this we can find the singularity for J_p. With these assumptions in mind we proceed to find the inverse kinematic solution for all the 42 kinematic structures. To solve this problem a software package that could multiply matrices symbolically may be used. Such a program could form the transformation matrices, multiply them together, simplify the results using algebraic and trigonometric rules, and form the Jacobian matrix and also find the determinant of the Jacobian matrix and its adjoint.

10.5 EXAMPLE

An example of the results that can be obtained is presented for the RPP (cylindrical) robots.

10.5.1 Link parameters for cylindrical robots

A list of link parameters for the RPP robots is given below. The coordinate frame for the base of the robot (world frame) is given in Figure 10.16. Using that and the link transformation matrix:

$$S_i \begin{pmatrix} d_i \\ \theta_i \\ a_{i,i+1} \\ \alpha_{i,i+1} \end{pmatrix}$$

one can obtain the coordinate frame for all the joints of a manipulator. The robots are grouped according to the body types used: Beginning with the base coordinate frame, the first two-link transformation matrices remain the same regardless of the wrist attached to the body. These are:

$$R_1 \begin{pmatrix} dd_1 \\ \theta_1 \\ 0 \\ 0 \end{pmatrix} \quad P_2 \begin{pmatrix} dd_2 \\ 0 \\ 0 \\ -90 \end{pmatrix}$$

The link parameters for the remaining joints are presented for every type of wrist:

RPP-1A:
$$P_3 \begin{pmatrix} dd_3 \\ 0 \\ 0 \\ 0 \end{pmatrix} R_4 \begin{pmatrix} 0 \\ \theta_4 \\ 0 \\ 0 \end{pmatrix}$$

RPP-1B:
$$P_3 \begin{pmatrix} dd_3 \\ 90 \\ 0 \\ 90 \end{pmatrix} R_4 \begin{pmatrix} 0 \\ \theta_4 \\ 0 \\ 90 \end{pmatrix}$$

RPP-2A:
$$P_3 \begin{pmatrix} dd_3 \\ 0 \\ 0 \\ 90 \end{pmatrix} R_4 \begin{pmatrix} 0 \\ \theta_4 \\ 0 \\ 90 \end{pmatrix} R_5 \begin{pmatrix} 0 \\ \theta_5 \\ 0 \\ 90 \end{pmatrix}$$

RPP-2B:
$$P_3 \begin{pmatrix} dd_3 \\ 90 \\ 0 \\ 0 \end{pmatrix} R_4 \begin{pmatrix} 0 \\ \theta_4 \\ 0 \\ 90 \end{pmatrix} R_5 \begin{pmatrix} 0 \\ \theta_5 \\ 0 \\ 0 \end{pmatrix}$$

RPP-2C:
$$P_3 \begin{pmatrix} dd_3 \\ 0 \\ 0 \\ 90 \end{pmatrix} R_4 \begin{pmatrix} 0 \\ \theta_4 \\ 0 \\ 0 \end{pmatrix} R_5 \begin{pmatrix} dd_3 \\ 0 \\ 0 \\ 0 \end{pmatrix}$$

RPP-3A:
$$P_3 \begin{pmatrix} dd_3 \\ 0 \\ 0 \\ -90 \end{pmatrix} R_4 \begin{pmatrix} 0 \\ \theta_4 \\ 0 \\ 90 \end{pmatrix} R_5 \begin{pmatrix} 0 \\ \theta_5 \\ 0 \\ 90 \end{pmatrix} R_6 \begin{pmatrix} 0 \\ \theta_6 \\ 0 \\ 0 \end{pmatrix}$$

RPP-3B:
$$P_3 \begin{pmatrix} dd_3 \\ 0 \\ 0 \\ 90 \end{pmatrix} R_4 \begin{pmatrix} 0 \\ \theta_4 \\ 0 \\ -90 \end{pmatrix} R_5 \begin{pmatrix} 0 \\ \theta_5 \\ 0 \\ -90 \end{pmatrix} R_6 \begin{pmatrix} 0 \\ \theta_6 \\ 0 \\ 0 \end{pmatrix}$$

10.5.2 Inverse kinematics results

The solution for the first three joints of the RPP robot is the same regardless of the wrist attached to the body:

$$\hat{\theta}_1 = \operatorname{atan}(-p_x/p_y) \qquad \theta_1 = \hat{\theta}_1 \text{ or } \theta_1 = \hat{\theta}_1 + 180$$
$$dd_2 = p_z - dd_1$$
$$dd_3 = p_y \cos(\theta_1) - p_x \sin(\theta_1)$$

The solution for the remaining joints is presented for every type of wrist:

RPP-1A:

$$\theta_4 = \operatorname{atan} \frac{o_y \sin(\theta_1) + o_x \cos(\theta_1)}{o_z}$$

RPP-1B:

$$\theta_4 = \operatorname{atan} \frac{a_z}{a_y \cos(\theta_1) - a_x \sin(\theta_1)}$$

RPP-2A:

$$\theta_4 = \operatorname{atan} \frac{o_y \sin(\theta_1) + o_x \cos(\theta_1)}{o_x \sin(\theta_1) - o_y \cos(\theta_1)}$$
$$\theta_5 = \operatorname{atan} \frac{(a_y \cos(\theta_1) - a_x \sin(\theta_1)) \sin(\theta_4) + (a_y \sin(\theta_1) + a_x \cos(\theta_1)) \cos(\theta_4)}{-a_z}$$

RPP-2B:

$$\theta_4 = \operatorname{atan} \frac{a_z}{a_y \cos(\theta_1) - a_x \sin(\theta_1)}$$
$$\theta_5 = \operatorname{atan} \frac{(o_y \cos(\theta_1) - o_x \sin(\theta_1)) \sin(\theta_4) - o_z \cos(\theta_4)}{o_y \sin(\theta_1) + o_x \cos(\theta_1)}$$

RPP-2C:

$$\theta_4 = \operatorname{atan} \frac{-o_y \sin(\theta_1) - o_x \cos(\theta_1)}{o_y \cos(\theta_1) - o_y \sin(\theta_1)}$$
$$dd_5 = p_z - dd_1 - dd_2$$

RPP-3A:

$$\theta_4 = \operatorname{atan} \frac{a_y \cos(\theta_1) - a_x \sin(\theta_1)}{a_y \sin(\theta_1) + a_x \cos(\theta_1)}$$
$$\theta_5 = \operatorname{atan} \frac{(a_y \cos(\theta_1) - a_x \sin(\theta_1)) \sin(\theta_4) + (a_y \sin(\theta_1) + a_x \cos(\theta_1)) \cos(\theta_4)}{-a_z}$$

$$A = -o_z \sin(\theta_5) - (o_y \cos(\theta_1) - o_x \sin(\theta_1)) \sin(\theta_4)$$
$$+ (o_y \sin(\theta_1) + o_x \cos(\theta_1)) \cos(\theta_4) \cos(\theta_5)$$
$$B = (o_y \sin(\theta_1) + o_x \cos(\theta_1)) \sin(\theta_4) - (o_y \cos(\theta_1) - o_x \sin(\theta_1)) \cos(\theta_4)$$

$$\theta_6 = \operatorname{atan} \frac{A}{B}$$

RPP-3B:

$$\theta_4 = \operatorname{atan} \frac{(a_x \sin(\theta_1) - a_y \cos(\theta_1))}{a_z}$$

$$\theta_5 = \operatorname{atan} \frac{(a_x \sin(\theta_1) - a_y \cos(\theta_1)) \sin(\theta_4) + a_z \cos(\theta_4)}{a_y \sin(\theta_1) + a_x \cos(\theta_1)}$$

$$A = (o_y \sin(\theta_1) + o_x \cos(\theta_1)) \sin(\theta_5)$$
$$+ ((o_y \cos(\theta_1) - o_x \sin(\theta_1)) \sin(\theta_4) - o_z \cos(\theta_4)) \cos(\theta_5)$$
$$B = O_z \sin(\theta_4) - (o_x \sin(\theta_1) - o_y \cos(\theta_1)) \cos(\theta_4)$$

$$\theta_6 = \operatorname{atan} \frac{A}{B}$$

10.5.3 Singular points for RPP (cylindrical) robots

It must be noted that in the case of cylindrical and other classes of robots, the singularities for the 2C wrist are ignored as this particular wrist is not a useful one. It must also be mentioned that in certain cases where the Jacobian matrix is large (6 DOF robot), the intermediate steps are not shown and only the final results are displayed for reference purposes. For this class of robots, the differential displacement of the end-effector as a vector can be defined as:

$$\boldsymbol{D} = (\delta_x, \delta_y, \delta_z, \Delta_x, \Delta_y, \Delta_z) \tag{10.31}$$

where $\delta_x, \delta_y, \delta_z$ represent differential translation and $\Delta_x, \Delta_y, \Delta_z$ represent differential rotation about the coordinate frame attached to the end-effector. This means that the Jacobian matrix has six rows. This implies that any cylindrical robot that has less than six degrees of freedom will yield a non-square and consequently non-invertible matrix. For cases where this is the case $\boldsymbol{J}_p$ or the positional Jacobian is isolated. The ideas behind finding the constraint directions are explained for the RPP-1A robot. The same techniques were applied for the other robots considered.

RPP-1A

For this robot, the differential displacement of the joint variables as a vector can be defined as:

$$\boldsymbol{D}_{\text{joint}} = (\delta\theta_1, \delta d_2, \delta d_3, \delta\theta_4) \tag{10.32}$$

The relationship between the differential displacement of the end-effector and $\boldsymbol{D}_{\text{joint}}$ is given by:

$$D = JD_{joint} \tag{10.33}$$

which shows the differential displacement of the joint variables in terms of the differential displacement of the end-effector. Where J^{-1} does not exist, the J matrix is called "singular". It is known from matrix algebra that the J matrix is singular when $|J| = 0$ or the determinant of the Jacobian matrix is equal to zero. Physically this translates to losing a translational or rotational degree of freedom in the Cartesian coordinate frame of the end-effector. By definition the inverse of a matrix is equal to the adjoint divided by the determinant of the matrix:

$$J^{-1} = \frac{J^+}{|J|} \tag{10.35}$$

where J^+ is the adjoint of matrix J. This immediately yields:

$$\frac{J^+}{|J|} D = D_{joint} \tag{10.36}$$

This directly implies that:

$$(J^+)D = |J| D_{joint} \tag{10.37}$$

Where $|J| = 0$ one obtains:

$$(J_0^+)D = 0 \tag{10.38}$$

where J_0^+ is the adjoint matrix evaluated at $|J| = 0$.

One observes that in the case of RPP-1A the last joint does not impart any differential translational motion to the gripper. This means that the least column can be eliminated for obtaining J_p, yielding a square matrix. The positional Jacobian is given by:

$$J_p = \begin{pmatrix} -dd_3 \cos(\theta_4) & -\sin(\theta_4) & 0 \\ dd_3 \sin(\theta_4) & -\cos(\theta_4) & 0 \\ a_{23} & 0 & 1 \end{pmatrix}$$

The determinant of J_p is given by:

$$|J_p| = dd_3$$

As described above, we evaluate the adjoint matrix at $dd_3 = 0$

$$\text{adjoint}(J_0) = \begin{pmatrix} -\cos(\theta_4) & \sin(\theta_4) & 0 \\ 0 & 0 & 0 \\ a_{23} \cos(\theta_4) & -a_{23} \sin(\theta_4) & 0 \end{pmatrix}$$

Utilizing equation (4.8), one obtains:

$$\begin{pmatrix} -\cos(\theta_4) & \sin(\theta_4) & 0 \\ 0 & 0 & 0 \\ a_{23} \cos(\theta_4) & -a_{23} \sin(\theta_4) & 0 \end{pmatrix} \begin{pmatrix} \delta_x \\ \delta_y \\ \delta_z \end{pmatrix} = 0$$

From this one obtains:

$$\cos(\theta_4)\delta_x - \sin(\theta_4)\delta_y = 0$$

which shows that:

$$\begin{aligned}\text{at} \quad dd_3 = 0 \text{ and } \theta_4 = 0 &\quad \delta_x = 0\\ \text{at} \quad dd_3 = 0 \text{ and } \theta_4 = 90 &\quad \delta_y = 0\end{aligned}$$

RPP-1B:

The positional Jacobian is given by:

$$J_p = \begin{pmatrix} -a_{23}\sin(\theta_4) & \cos(\theta_4) & \sin(\theta_4) \\ dd_3 & 0 & 0 \\ a_{23}\cos(\theta_4) & \sin(\theta_4) & -\cos(\theta_4) \end{pmatrix}$$

The determinant of J_p is given by:

$$|J_p| = dd_3$$

As described above, we evaluate the adjoint matrix at $dd_3 = 0$

$$\text{adjoint}(J_0) = \begin{pmatrix} 0 & 1 & 0 \\ 0 & 0 & 0 \\ 0 & a_{23} & 0 \end{pmatrix}$$

Utilizing equation (4.8) one obtains:

$$\begin{pmatrix} 0 & 1 & 0 \\ 0 & 0 & 0 \\ 0 & a_{23} & 0 \end{pmatrix} \cdot \begin{pmatrix} \delta_x \\ \delta_y \\ \delta_z \end{pmatrix} = 0$$

From this one obtains:

$$a_{23}\delta_y = 0$$

which shows that:

$$\text{at} \quad dd_3 = 0 \quad \delta_y = 0$$

RPP-2A:

The positional Jacobian is given by:

$$J_p = \begin{pmatrix} (a_{23}\sin(\theta_4) - dd_3\cos(\theta_4))\cos(\theta_5) & \sin(\theta_5) & \sin(\theta_4)\cos(\theta_5) \\ -(dd_3\sin(\theta_4) + a_{23}\cos(\theta_4)) & 0 & -\cos(\theta_4) \\ (a_{23}\sin(\theta_4) - dd_3\cos(\theta_4))\sin(\theta_5) & -\cos(\theta_5) & \sin(\theta_4)\sin(\theta_5) \end{pmatrix}$$

The determinant of J_p is given by:

$$|J_p| = dd_3$$

Example

So, as described above, we evaluate the adjoint matrix at $dd_3 = 0$

$$\text{adjoint}(J_0) = \begin{pmatrix} -\cos(\theta_4)\cos(\theta_5) & -\sin(\theta_4) & -\cos(\theta_4)\sin(\theta_5) \\ 0 & 0 & 0 \\ a_{23}\cos(\theta_4)\cos(\theta_5) & -a_{23}\sin(\theta_4) & a_{23}\cos(\theta_4)\sin(\theta_5) \end{pmatrix}$$

Utilizing equation (4.8), one obtains:

$$\begin{pmatrix} -\cos(\theta_4)\cos(\theta_5) & -\sin(\theta_4) & -\cos(\theta_4)\sin(\theta_5) \\ 0 & 0 & 0 \\ a_{23}\cos(\theta_4)\cos(\theta_5) & -a_{23}\sin(\theta_4) & a_{23}\cos(\theta_4)\sin(\theta_5) \end{pmatrix} \cdot \begin{pmatrix} \delta_x \\ \delta_y \\ \delta_z \end{pmatrix} = 0$$

From this one obtains:

$$-\cos(\theta_4)\cos(\theta_5)\delta_x - \sin(\theta_4)\delta_y - \cos(\theta_4)\sin(\theta_5)\delta_z = 0$$

which shows that:

at $dd_3 = 0$ and $\theta_4 = 0$ and $\theta_5 = 0$ $\quad \delta_x = 0$
at $dd_3 = 0$ and $\theta_4 = 0$ and $\theta_5 = 90$ $\quad \delta_z = 0$
at $dd_3 = 0$ and $\theta_4 = 90$ and $\theta_5 = 90$ $\quad \delta_y = 0$
at $dd_3 = 0$ and $\theta_4 = 90$ and $\theta_5 = 0$ $\quad \delta_y = 0$

RPP-2B:

The positional Jacobian is given by:

$$J_p = \begin{pmatrix} dd_3\sin(\theta_5) - a_{23}\sin(\theta_4)\cos(\theta_5) & \cos(\theta_4)\cos(\theta_5) & \sin(\theta_4)\cos(\theta_5) \\ dd_3\cos(\theta_5) + a_{23}\sin(\theta_4)\sin(\theta_5) & \cos(\theta_4)\sin(\theta_5) & -\sin(\theta_4)\sin(\theta_5) \\ a_{23}\cos(\theta_4) & \sin(\theta_4) & -\cos(\theta_4) \end{pmatrix}$$

The determinant of J_p is given by:

$$|J_p| = dd_3$$

We evaluate the adjoint matrix at $dd_3 = 0$

$$\text{adjoint}(J_0) = \begin{pmatrix} \sin(\theta_5) & \cos(\theta_5) & 0 \\ 0 & 0 & 0 \\ a_{23}\sin(\theta_5) & a_{23}\cos(\theta_5) & 0 \end{pmatrix}$$

Utilizing equation (4.8), one obtains:

$$\begin{pmatrix} \sin(\theta_5) & \cos(\theta_5) & 0 \\ 0 & 0 & 0 \\ a_{23}\sin(\theta_5) & a_{23}\cos(\theta_5) & 0 \end{pmatrix} \cdot \begin{pmatrix} \delta_x \\ \delta_y \\ \delta_z \end{pmatrix} = 0$$

From this one obtains:

$$\sin(\theta_5)\delta_x + \cos(\theta_5)\delta_y = 0$$

which shows that:

at $dd_3 = 0$ and $\theta_5 = 90$ $\quad \delta_x = 0$
at $dd_3 = 0$ and $\theta_5 = 0$ $\quad \delta_y = 0$

RPP-3A:

Unlike the robots considered so far in this class (cylindrical), this robot has a square Jacobian matrix, the determinant of which is:

$$|J| = dd_3 \sin(\theta_5)$$

From the adjoint matrix evaluated at $dd_3 = 0$, we obtain the following equation:

$$(\sin(\theta_4)\sin(\theta_6) + \cos(\theta_4)\cos(\theta_5)\cos(\theta_6))\delta_x + (\cos(\theta_4)\cos(\theta_5)\sin(\theta_6)$$
$$- \sin(\theta_4)\cos(\theta_6))\delta_y + \sin(\theta_5)\cos(\theta_4)\delta_z = 0$$

From which we get:

at $dd_3 = 0$ and $\theta_4 = 90$ and $\theta_5 = 0$ and $\theta_6 = 90$ $\delta_x = 0$
at $dd_3 = 0$ and $\theta_4 = 0$ and $\theta_5 = 90$ and $\theta_6 = 0$ $\delta_z = 0$
at $dd_3 = 0$ and $\theta_4 = 90$ and $\theta_5 = 90$ and $\theta_6 = 0$ $\delta_y = 0$
at $dd_3 = 0$ and $\theta_4 = 90$ and $\theta_5 = 90$ and $\theta_6 = 90$ $\delta_x = 0$
at $dd_3 = 0$ and $\theta_4 = 0$ and $\theta_5 = 0$ and $\theta_6 = 90$ $\delta_y = 0$
at $dd_3 = 0$ and $\theta_4 = 90$ and $\theta_5 = 0$ and $\theta_6 = 0$ $\delta_y = 0$

Now we evaluate the adjoint matrix at $\theta_5 = 0$ and get with $dd_3 \neq 0$:

$$\cos(\theta_6)\Delta_x - \sin(\theta_6)\Delta_y = 0$$

where Δ_x and Δ_y represent differential rotation about x, y of the end-effector frame. We finally obtain:

$$dd_3 \neq 0 \quad \text{and} \quad \theta_6 = 0 \quad \Delta_x = 0$$
$$dd_3 \neq 0 \quad \text{and} \quad \theta_6 = 90 \quad \Delta_y = 0$$

RPP-3B:

As may be expected, the results are quite similar to RPP-3A. The determinant of the Jacobian matrix is:

$$|J| = dd_3 \sin(\theta_5)$$

At $\theta_5 = 0$, from the adjoint matrix we get:

$$dd_3 \cos(\theta_6)\Delta_x - dd_3 \sin(\theta_6)\Delta_y = 0$$

with $dd_3 \neq 0$ and $\theta_6 = 0$ $\Delta_x = 0$
with $dd_3 \neq 0$ and $\theta_6 = 90$ $\Delta_y = 0$

with $dd_3 = 0$, one obtains:

$$-\sin(\theta_5)\cos(\theta_6)\delta_x - \sin(\theta_5)\sin(\theta_6)\delta_y - \cos(\theta_5)\delta_z$$

at $\theta_5 = 0$ $\delta_z = 0$
at $\theta_5 = 90$ and $\theta_6 = 0$ $\delta_x = 0$
at $\theta_5 = 90$ and $\theta_6 = 90$ $\delta_y = 0$

10.6 PATTERN OF KINEMATIC BEHAVIOR

Once all the solutions for the inverse kinematics of the 42 structures are determined and the singularities found, one can proceed to determine the pattern that emerges in these solutions. The primary aim of such effort is to:

1. Find out which joints are involved in executing a general robot-level command.
2. Take steps towards developing a system that can find all the joint solutions when a specific kinematic structure is given. For example once a body type is specified (for instance RRR) such a system would know what the solution for joints one to three are. This would result ultimately in a program in Prolog [CM81] [MM87] that contains all the information regarding industrial robots.
3. Realize what the kinematic constraints (singularities) for each class of robots are. (This classification is done according to the body types used; for example, RRR is articulate, PPP is Cartesian, RPP is cylindrical and PRR is spherical.)

To understand which joints are involved in executing a command the following guidelines are used:

One begins by looking at the general level robot commands:

TRANS   Joints that contain p_x, p_y, p_z in their solution play a role in implementing this command.

ALIGN   Any joint for which n_x, n_y, n_z, o_x, o_y, o_z, a_x, a_y, a_z are non-zero in its solution has a role in executing the ALIGN command. For instance, in the case of a RRR, 3B manipulator, every joint except the first joint has a role in this. In cases where ALIGN includes translational motion, every joint of the robot may have a role in executing the ALIGN command.

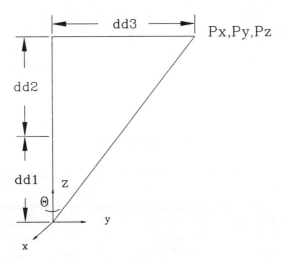

Fig. 10.17   Geometric solution for RPP robot.

APPROACH The same principles that were used for TRANS can be applied. Every joint in the robot will have a role in this as a MOVE consists of a change both in position and orientation.

In developing the generic level program, one makes the basic assumption that all the wrist joints are located on the same point. That way, location and orientation are separated. In particular rather than solving for the tool origin position as a function of the link displacements, we can instead solve for the wrist origin position as a function of only the first three link values. The joints that come after this (wrist joints) change only the orientation of the gripper. Such an approach is reasonable because the only class of robots for which the last three joints do not have to be at the same point for the robot to be solvable is the articulate (RRR). To verify the results, we will solve for the first three joints geometrically. The situation for the cylindrical (RPP) is shown in Figure 10.17.

Obviously, the sum of dd_1 and dd_2 gives us the coordinates of p_z or:

$$dd_1 + dd_2 = p_z$$

Since the variable θ_1 allows the robot to rotate in the x, y envelope, we obtain:

$$\theta_1 = \operatorname{atan} \frac{-p_x}{p_y}$$

Solving for dd_3, we obtain:

$$dd_3 = p_y \cos(\theta_1) - p_x \sin(\theta_1)$$

The only case where orientation and position are not separate (a robot with RRR, 3B structure) is handled individually.

In order to find the singularities for different classes of robots, we take the approach of realizing whether robots with the same body but different wrists can suffer from the same constraints (singularities). An example of such a generalization is described here for illustrative purposes. For the robots having a RPP body (cylindrical), it can be seen that the determinant of the positional Jacobian J_p is given by dd_3. This is the variable dealing with the third joint of the robot (a prismatic joint). This means that when $dd_3 = 0$, the robot loses a translational degree of freedom. If one is working with a cylindrical robot and a position is reached where $dd_3 = 0$, the user can be informed that a translational degree of freedom has been lost and the APPROACH command in that direction will not work.

10.7 SUMMARY

In this chapter we have defined a systematic way of transforming general robot-level commands (MOVE, ALIGN, etc.) to joint motion for different classes of robots. A total of 42 robots have been considered and all generic robot kinematic structures (cylindrical, Cartesian, spherical, jointed, Scara) were described. The steps taken to make the transformation of these commands to joint motion possible included: (1) Development of forward and inverse kinematics solutions for all the manipulators considered. (2) Realizing where the singular points for a robot lie using the Jacobian transformation. (3) Observing the patterns that emerge in the location of singular points for different classes of robots. (4) Writing statements that make the information gained in the above steps part of the database of RALPH.

One of the benefits of the issues addressed in this chapter is the fact that RALPH can now take a series of general, robot independent commands (MOVE, ALIGN, etc.) and transform them directly to joint motion using the ideas presented here.

In determining where the singular points for a robot lie and from that finding the constraint directions (the directions along which motion is impossible), the basic objective should be to realize the constraints a kinematic structure suffers from, depending on the kind of body and wrist used.

A high level of automation has been achieved as a result of this research. This conclusion is substantiated by the fact that once the body type and wrist type are known, one can immediately transform general robot-level commands to joint motion.

Primary attention has been focused in this chapter on industrial robots. The methods described in this chapter can also be applied to other kinematic structures. As far as the first three joints (body) of a robot are concerned, the configurations not found in industrial robots include RPR, PRP, PPR. Although, these bodies are not particularly useful ones, the techniques described here can be used to analyze them.

An important point to consider in using the ideas presented in this chapter is that of workspace singularities. In general, singularities can be classified into two broad categories [Cra86, Wol87]:

1. *Workspace boundary singularities:* are those which occur when the manipulator is fully stretched out or folded back on itself such that the end-effector is near or at the boundary of the workspace.
2. *Workspace interior singularities:* are those which occur away from the workspace boundary and generally are caused by losing a degree of freedom in Cartesian coordinate space.

Workspace boundary singularities however do not always make the determinant of the Jacobian matrix go to zero. This makes the problem of kinematic control difficult. The safest way is to operate within the working envelope and never be on it. Although the subject of workspace design is outside the objectives of this chapter, a scheme by Sugimoto, Duffy and Thomas [K.S81b, K.S81a] is suggested for finding the extreme reaches of a robot hand. It is reasonable to assume that as long as the manipulator is not on the boundary of the workspace (it has not reached the extreme distance along either x, y, z-axis), that any singularities can be detected by realizing whether the determinant of the Jacobian matrix has gone to zero as explained in this chapter.

It should be noted that all the manipulator configurations associated with the working envelope present singular configurations. However, such (workspace boundary) singularities will not always correspond to Jacobian singularities.

It may be useful to assume that a portion of the robot workspace (close to the boundary of robot working envelope) should be ignored in implementing a task. That makes the problem of kinematic control more manageable.

In terms of the computational requirements, issues of importance include computational speed and also pattern matching capabilities that allow one to have easy access to the information regarding a specific type of manipulator.

CHAPTER 11

Program synthesis and other planners

11.1 INTRODUCTION

In the previous chapters we have developed the principles necessary to generate a robot program automatically. We can now model the world and adequately specify tasks to the robot. We have also developed some crucial planning strategies such as those for obstacle avoidance, trajectory planning, grasp planning, etc. In this chapter, we will discuss some other planning issues that are important for automatic robot planning. These include the precedence generator and the approach directions generator. We will then discuss fine motion planning and finally program synthesis.

11.2 SPANNING VECTOR FOR ASSEMBLY DIRECTIONS AND OTHER APPLICATIONS

One of the major obstacles in process planning for automatic robot assembly has been providing the robot with sufficient information needed for task execution. In general, assembly task consists of automating part acquisition, part handling, part positioning, and part mating. In this section we focus our attention on the part mating process. Even for a simple assembly, such as a peg-and-hole insertion, the assembly or part mating task can be complicated. It consists of a combination of several complex operations. Some of the major operations in an automatic mating process can include determining the mating faces of the parts, identifying the faces and the area of the faces which can be gripped by the end-effector, finding the center of gravity of the parts, determining the axis of rotation, and finding the directions along which the parts can be put together. Each of these complex issues has to be carefully addressed, if one tries to automate the mating task in an assembly process.

We will specifically address the problem of finding the directions along which parts can be put together without interference by other portions of the mating parts. We develop an algorithm to determine the set of assembly directions and describe a representation for this set of directions. We also present a method to obtain this set of vectors directly from the geometric data produced by the CAD system.

To describe assembly directions, consider the two objects which are to be assembled in Figure 11.1. The directions along which part A can be put together with part B, and C

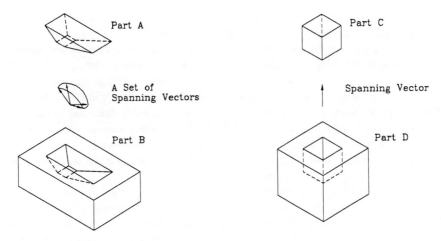

Fig. 11.1 Spanning vector for assembly directions.

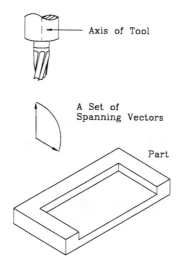

Fig. 11.2 Spanning vector in machining.

with D respectively without interference are the assembly directions, and can be represented by a set of spanning vectors as shown.

Thus, a spanning vector is a vector which can be projected from an internal feature of a part into outside space without interference with any portion of the part. For some parts, such as a blind square hole, there can only be one such vector. For others, there can be an infinite number of vectors. The angle created by any two spanning vectors, which do not lie on the same surface, can be viewed as an *angle of span*.

As in the case of generating a process plan for automatic robot assembly, generating a process plan for automatic machining requires a set of directions from which a part can be approached by the machine tool. This set of directions is also known as tool-approach directions. In this process, the part is machined with the axis of the machine tool (the z-axis) inside the angle of span of the spanning vectors of the feature to be machined. This way, the part can be machined in such a way that no interference occurs between the machine tool and the surfaces of the part that are not currently being machined. In other words, the set of tool-approach directions can help determine the position and orientation of the part, so that no interference will occur during machining [TJ91]. Since the concept of tool-approach directions is the same as assembly directions, in the sense that we can think of the machine tool as the other half of a two-item assembly, the set of tool-approach directions can be found by using the same algorithm we use for determining assembly directions. Figure 11.2 shows a part to be machined, and the set of spanning vectors representing the tool-approach directions.

In injection molding or die casting process, the design of molds and dies directly affects productivity, quality of product, and cost of manufacturing. The most significant aspect of the design is the choice of the *parting surface* location, that is the location of the surface separating the two halves of the mold or die. A combination of several mechanical, metallurgical and process parameters influences the choices of the parting surface [RS90]. The algorithm presented in this section can be used to assess the feasibility of parting surface locations of a mold or die, by looking at the directions of mold closure. This enables design engineers to have choices of parting surface location alternatives. In doing this, we consider both halves of the die or mold as a final assembly of two parts, and the contact area of both halves as the parting surface. If a disassembly direction exists for the halves, it also means that the location of the parting surface is valid, and the disassembly direction becomes the direction of mold closure. Figure 11.3 illustrates both halves of a die with a possible location of parting surface. Note that in Figure 11.3 the die can only be opened along one direction. If we think of the process as an assembly process, the direction along which the dies can be separated is the assembly direction.

Many researchers have contributed to a better understanding of concepts related to assembly directions and tool-approach directions. For example, T.C. Woo (1987) developed a procedure to generate a sequence of motions for removing parts from a final assembly. In that work, Woo uses the notion of *monotonicity* for generating disassembly direction

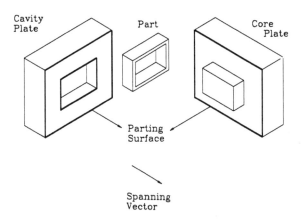

Fig. 11.3 Spanning vector in injection molding and die casting.

[Woo87]. In general, disassembly directions are the same as assembly directions since the problem of finding how to assemble a given set of parts can be converted into an equivalent problem of finding how the same parts can be disassembled.

Sedas and Talukdar (1987) also developed an algorithm for a disassembly planner. One of the steps in their algorithm is *Picking a Promising Direction* for disassembly using a trial and error approach [ST87]. The concept they use for picking a promising direction for disassembly can also be used for picking a promising direction for assembly.

Hoermann and Rembold (1988) described a hierarchical robot action planning system for automatic assembly. In their work, they introduce translation motions in assembly, which are computed from the type of assembly contact using some heuristics [HR88].

For automatic process planning for machining, Tseng and Joshi (1991) developed two algorithms to determine feasible tool-approach directions for machining Bezier curves and surfaces. In their algorithm, the surfaces are approximated by many small faces, which are treated as planes, and map onto a 3D sphere. They use this sphere to represent the set of tool-approach directions [TJ91].

Unfortunately, except for very few cases such as the above scheme, many of the algorithms cannot be applied in three-dimensional space. In addition, many algorithms use a trial and error approach. This approach is not only inefficient, but also not feasible if seen from a computational viewpoint. Another deficiency of some of the existing approaches is that given a direction, the algorithm only determines if the direction is a valid direction. To obtain an *optimal direction* from the set of feasible directions, the algorithm should not only test the feasibility of a direction, but it must provide all of the feasible directions. In this chapter we present an algorithm for determining the maximum possible range of assembly directions and tool-approach directions, which can be applied in 3D space, and are also feasible to compute. The algorithm [BNYew] we are presenting here requires that only polyhedral parts are considered. However, this is not undue restriction, since parts with curvilinear surfaces can be approximated.

11.2.1 Determining control faces

Having represented the parts using boundary representation, the next step is to determine which faces of the part will be involved in the reasoning process. These faces are called control faces. Control faces are important in the algorithm, since these faces determine the set of directions we are interested in. In machining, the control faces are those which need to be machined in one machining set up. In injection molding and die casting, the faces of the cavity determine the directions of the mold closure. In assembly process, the mating faces are the control faces. Mating faces are the area where the boundaries of the parts overlap after the assembly takes place. In the following subsections, we will describe how we determine the mating faces of a two-item assembly using our spatial relationship engine.

11.2.2 Finding mating faces

In order to automatically find the mating faces in an assembly, the relative position and orientation of the parts in the final assembly have to be known, since mating faces are determined by the contact area of the parts after the assembly process takes place. Spatial relationship about faces of the mating parts can be used to determine the relative position and orientation of the mating parts. By specifying the relationship among the faces, such as *parallel, against, parax*, etc., the degree of freedom of the parts is reduced. To reach a fixed position in an assembly, the number of degrees of freedom left on the parts have to be zero. When it has not reached zero, more spatial relationships are needed. In other words, spatial relationships of the parts can be interpreted as a constraint imposed on the degrees of freedom between relative mating or interacting faces of the parts [LN91]. In Figure 11.4, the task is to assemble part A and part B. If an *against* relationship is given between face 1 of part A and face 4 of part B, we impose a constraint on the degree of freedom between the two faces. If the same type of relationship is given between face 2 of part A and face 6 of part B, and between face 3 of part A and face 5 of part B, then the degrees of freedom left on the parts is zero, and hence, the parts are in a fixed position.

Based on these spatial relationships, a coordinate frame can then be attached to each part. These coordinate frames are attached such that when the parts are put together according to the spatial relationships given to the parts, these frames would coincide with each other. These frames are called mating frames. Assembly between two parts can then be expressed by these mating frames. A mathematical model is applied to determine a collection of spatial relationships to configure the final and fixed location of the parts after the assembly process takes place [NL90b]. As an example, Figure 11.5 shows the assembly relations between $Part_a$ and $Part_b$. The origin coordinate frame of $Part_a$ is O_a and the mating frame of $Part_a$ is M_a. O_b is the origin coordinate frame of $Part_b$ and M_b is the mating frame of $Part_b$. The origin coordinate frame is automatically created by the CAD system which may lie inside or outside the body. The mating frames were created by the spatial relationships of the faces of the parts. In order to assemble $Part_a$ with $Part_b$, one has to match the position and orientation of both mating frames M_a and M_b. If the transformation matrix for $Part_b$ with respect to $Part_a$ is $^{O_a}T_{O_b}$, then

$$^{O_a}T_{O_b} = {}^{O_a}T_{M_a} \cdot {}^{O_b}T_{M_b}^{-1}.$$

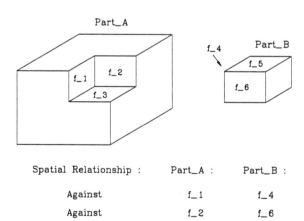

Fig. 11.4 Spatial relationships of an assembly.

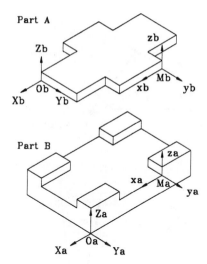

Fig. 11.5 Original frames and mating frames.

This equation gives us the relative position and orientation of the parts in the final assembly. A spatial relationship engine has been developed by other researchers in our laboratory to perform this type of operation.

If we know the fixed position and orientation of the parts after the assembly takes place, we can determine the area where the boundaries of the parts overlap, and we know the mating faces of the parts. The unit normal vectors of these mating faces are the basis of our algorithm to determine the *spanning vectors* of the parts.

11.2.3 Spanning vector

11.2.3.1 Mathematical background

A set X in E^n is called a convex set if given any two points x_1 and x_2 in set X, then $rx_1 + (1 - r)x_2 \in X$ for each $r \in [0, 1]$.

$rx_1 + (1 - r)x_2$ for r in the interval $[0, 1]$ represents a point on the line segment joining x_1 and x_2.

The boundary of a convex set can be classified as extreme points or regular points. A point x in a convex set X is called an extreme point of set X, if x cannot be represented as a linear combination of two distinct points in set X. An example of extreme and non-extreme points can be seen in Figure 11.6.

Given a convex set X, a non-zero vector $\boldsymbol{d}$, which is called the *direction* of the convex set, is inside the set X, if for each x in the set X, the ray $\{x + r\boldsymbol{d} : r \geq 0\}$ also belongs to the set X. Hence, starting at any point x in the convex set X, one can recede along $\boldsymbol{d}$ for any step length $r \geq 0$, and remain within the convex set X. Note, if the convex set is bounded, then no directions exist for the convex set.

The notion of extreme directions is similar to the notion of extreme points. An extreme direction of a convex set is the direction of the set that cannot be represented as a linear

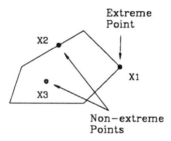

Fig. 11.6 Extreme points.

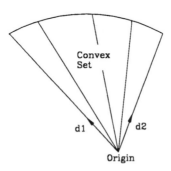

Fig. 11.7 Extreme directions.

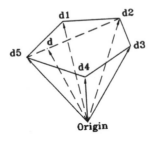

Fig. 11.8 Polyhedral convex cone.

combination of two distinct directions of the convex set. Directions d_1 and d_2 are called distinct directions if d_1 cannot be represented as a positive multiple of d_2 [BJS90]. Figure 11.7 shows an example of extreme directions of a convex set.

The concept of convex sets in 2D can be extended to convex sets in 3D. A special class of convex sets in 3D is convex cones. A convex cone is a convex set that consists entirely of rays emanating from the origin. Since as convex sets in 2D, a non-extreme direction can be represented as a linear combination of extreme directions, a convex cone can be represented by its extreme directions. In this presentation, we are interested in a special case of convex cones called polyhedral convex cones. Figure 11.8 shows a polyhedral convex cone with its extreme directions. In Figure 11.8 we have five extreme

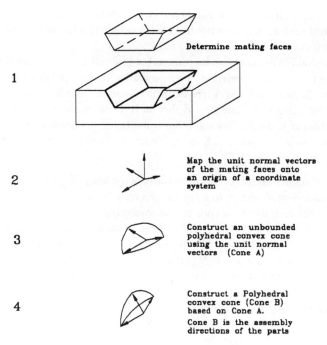

Fig. 11.9 Spanning vector algorithm.

directions, namely d_1, d_2, d_3, d_4, d_5. Note that any direction emanating from the origin of the polyhedral convex cone can be represented as a linear combination of these five extreme directions. To illustrate, choose a direction d shown in the figure. It can be seen that d can be represented as a linear combination of d_2 and d_5. That is $\boldsymbol{d} = rd_2 + (1-r)d_5$ $r \in [0, 1]$.

The concept of convex set and convex cone is essential in the methodology for finding spanning vectors of a part, since we use polyhedral convex cone to represent the spanning vectors of a part.

11.2.3.2 Algorithm for finding spanning vector

Orientation of a face can be determined by the unit normal vector of that face. We have discussed how we can get such information from the CAD system. The first step in the algorithm is to map all the unit normal vectors of the control faces of the feature to an origin of a coordinate system. Then, we check if an unbounded polyhedral convex cone can be constructed using the unit normal vectors of the control faces. We do this by letting the origin of the coordinate system be the intersection of a set of planes $\boldsymbol{P}$. If there is at least one plane p_i in the set of planes $\boldsymbol{P}$ for which all of the unit normal vectors lie above or below plane p_i, then an unbounded polyhedral convex cone can be constructed using the unit normal vectors of the control faces. In other words, if the unit normal vectors of the feature lie in one halfspace, then a polyhedral convex cone exists for the feature. This step is needed, since an unbounded polyhedral convex cone should exist for a feature to have a direction. If an unbounded polyhedral convex cone cannot be constructed using the unit normal vectors of the control faces, we conclude that the feature does not have spanning vectors.

If a polyhedral convex cone can be constructed (call this Cone A), then the spanning vector of the feature can be determined by generating another polyhedral convex cone (call this Cone B), based on Cone A. The extreme vectors of Cone B are the cross products of the extreme vectors of Cone A, and they have to lie in the same halfspace as the extreme vectors of Cone A. This algorithm is illustrated in Figure 11.9.

From a computational viewpoint, the procedure of finding spanning vectors can be done in the following way:

1. Get the unit normal vectors of the control faces of the feature from the CAD system.
2. Map the unit normal vectors to an origin of a coordinate system.
3. Check if a polyhedral convex cone can be constructed using the unit normal vectors of the feature. Let $P = \{f_1, f_2, f_3, \ldots, f_i\}$, where f_i is face i of part P, and N_i is the normal vector of face i. Then let $V = N_j \times N_k$ for $j, k = 1, 2, 3, \ldots, i$ and $j \neq k$. If there is at least one V which follows: $V \cdot N_j \geq 0$ for $j = 1, 2, 3, \ldots, i$ $\forall N_j$ or $V \cdot N_j \leq 0$ for $j = 1, 2, 3, \ldots, i$ $\forall N_j$, then a polyhedral convex cone can be constructed using the unit normal vectors of the feature.
4. Determine the extreme vectors of the polyhedral convex cone. Let $V = N_l \times N_m$ for $l, m = 1, 2, 3, \ldots, i$ and $l \neq m$. If V follows $V \cdot N_j \geq 0$ for $j = 1, 2, 3, \ldots, i$ $\forall N_j$ or $V \cdot N_j \leq 0$ for $j = 1, 2, 3, \ldots, i$ $\forall N_j$. Then N_l and N_m are the extreme vectors of the polyhedral convex cone.
5. If a polyhedral convex cone can be constructed using the unit normal vectors of the control faces (Cone A), then create another polyhedral convex cone (Cone B) based on Cone A. Let C_i be the extreme vectors of (Cone A), and let $V = C_k \times C_l$ for $k, l = 1, 2, \ldots, i$ and $k \neq l$. V is an extreme direction of Cone B if: $V \cdot N_j \geq 0$ for $j = 1, 2, 3, \ldots, i$ $\forall N_j$ or $V \cdot N_j \leq 0$ for $j = 1, 2, 3, \ldots, i$ $\forall N_j$. Then, Cone B becomes the set of spanning vectors of the feature.

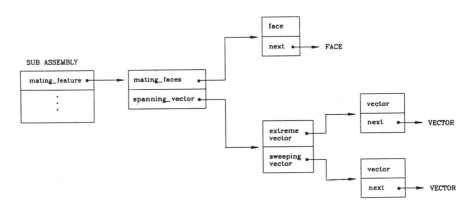

Fig. 11.10 Spanning vector data structure.

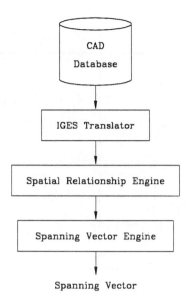

Fig. 11.11 The procedure for obtaining spanning vectors.

11.2.4 Representation of spanning vector

Since the spanning vector of a part can be described by a polyhedral convex cone, we use polyhedral convex cone representation to represent a set of spanning vectors.

If a polyhedral convex cone is not empty, this cone has a finite number of extreme vectors, say $d_1, d_2, d_3, \ldots, d_k$. If a direction d is in the set $\mathbf{X}$, where $\mathbf{X}$ is a polyhedral convex cone, d can be represented as a linear combination of the extreme vectors: $d = \Sigma_{j=1}^{k} r_j d_j$, where $r_j \geq 0$. Therefore, using the extreme vectors of the polyhedral convex cone, we can represent every vector in the cone.

In Figure 11.10, we show the data structure that we use in our spanning vector engine.

11.2.5 Some examples of spanning vector

The final output of the resulting spanning vector engine is the set of spanning vectors of the part. Figure 11.11 shows the procedure for obtaining the spanning vector, and Figure 11.12 shows some of the parts which have been processed in the spanning vector engine.

Using spanning vector algorithm, we can generate a set of assembly directions, tool approach directions, or check the validity of parting surface location. However, in some cases, it is necessary to choose a direction among the directions in the feasible set. Each application has its own criteria to determine the *optimal direction*. Machine constraint, collision avoidance, tolerance, and travel time of the robot arm are some of the things that need to be considered when picking a direction from the feasible set. With this set of spanning vectors, an algorithm can now be employed to pick a direction which is optimal to a particular environment, based on various criteria.

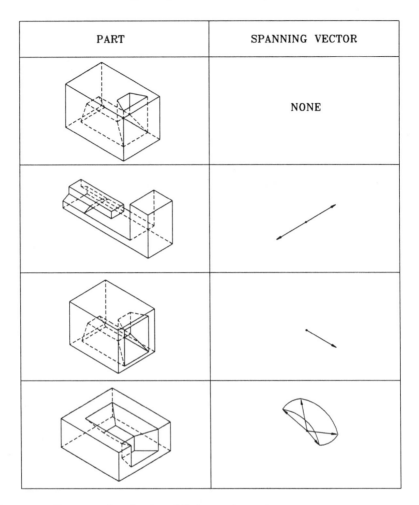

Fig. 11.12 Four examples of parts and their spanning vectors.

11.3 PRECEDENCE GENERATION

Two important aspects of the research in assembly planning are the representation of final or goal assemblies and the representation of the precedence constraints. They are important because the representation scheme determines the ease with which precedence knowledge can be automatically acquired and then efficiently manipulated. We briefly enumerate the different approaches taken by researchers in tackling the problem of precedence generation in assembly planning.

Bourjault [Bou84a] defined a *liaison model* or a graph of connections as a diagram representing all possible topological relations and connections between all the parts in an assembly. He then proposed a method in which all the precedence knowledge about an assembly is obtained by answering a set of structured questions based on this *liaison model* of the assembly. These questions were typically YES–NO questions which determined the ability or inability of assembling a part to another part or subassembly. The main drawback of this approach was that the number of questions increased exponentially as the number of parts. Thus the user was faced with a large number of questions to answer. De Fazio and Whitney [FD87] followed on the path of Bourjault and refined the method. They reduced the number of questions from being exponentially explosive to $2l$ where l represented the number of liaisons or connections in the assembly. This was achieved by reasoning about previously answered questions to deduce answers to other questions. The number of liaisons is usually greater than the number of parts in the assembly, so it was found that it is still difficult for a human to answer all these questions correctly and completely.

Homem de Mello and Sanderson [dMS89] have used the method of relational models for representing assemblies. They represented all the assembly plans explicitly using an AND/OR graph structure. Their model includes the types of entities, their contacts and attachments and a set of relationships between entities. From this relational model a graph of connections is established. Using the cut-set method, the assembly is decomposed into subassemblies. Knowing the fact that the geometry of the parts and their relative positions are the same in the whole assembly and in the subassembly, the relational model of the subassemblies is deduced by knowing the subassembly's set of parts. The geometric feasibility of the decomposition is checked and AND/OR graph representations of all feasible assembly sequences is generated.

Huang and Lee [HL90] have arrived at a symbolic representation of precedence knowledge. They describe assemblies in the form of a feature mating operations graph (FMOG). In this FMOG graph there are two subsets of vertices one subset represents the components and the other subset represents the mating operations. The state of all the feature mating operations whether it is done or undone describes the state of the assembly. The precedence constraint knowledge for each mating operation is represented by them in the form of a *pre_condition* and a *post_condition* for that operation. This *pre_condition* of an operation tells you which operations must be completed before performing this operation. Similarly the *post_condition* of an operation is defined as the set of all operations which have to appear after completing this operation. They then develop a geometric mating graph to include all the geometric and topological information. They also use the method of cutsets to simulate the disassembly and arrive at the precedence constraints.

In almost all the existing work it is not clear how the liaison or connectivity relations are automatically generated. It is also not an easy task for the user to answer all the questions pertaining to geometric interference. In recent work of Japtap and Nnaji [JN91] assemblies are defined as a set P containing parts which are bound by spatial relationships. In this approach, it is not necessary to specify all the mating spatial relationships in order to define the part in its goal position. We have the user define only enough spatial relationships to constrain the part uniquely in its assembly position. We then automatically determine all the contact mating face equations and thereby construct the component connectivity graph. Subsequently we have chosen to only use geometric reasoning to determine all the feasibly assembly plans. We present this method in detail in the next subsections.

11.3.1 Concept of assembly precedence

Assembly is an *inherently integrative* process [NW89] fundamental in the manufacture of many products. It has been defined as an act of putting together individual parts and subassemblies so as to achieve the design specified final configuration of a given product. Traditionally assembly operations were segregated into two broad classes. (1) Those that required highly skilled workers (fitters) to perform complex assembly operations on mating parts with varying nominal dimensions. (2) Those that required simple pick and place or insert type assembly operations. Now with the advent of the concept of *design for assembly* (DFA), *unidirectional assemblies* and *concurrent engineering* the skill requirement for assemblies is fast diminishing. In the case of high volume and stable market products, dedicated assembly line machines have been used successfully to maintain consistent quality and high production rates. In today's competitive market, product life cycles have become shorter and shorter thus requiring fast, accurate assemblies and the ability to assemble with the same setup or minimum setup times, different kinds of assemblies.

To use the product model for generating assembly plans, it is essential to study all possible alternative assembly sequences. Different assembly sequences mean different assembly operations, different setups, different assembly times and result in different assembly costs [LC90]. In this chapter any set of parts that are joined to form a stable unit is called an assembly. Subassembly refers to an assembly that is part of another assembly.

In this research the specific task addressed is that of determining the feasible assembly seuqences based on geometric precedence constraints.

The user constructs the final assembly by defining *spatial relationships* between various parts of the assembly. These relationships are of the type *against, aligned, parallel-offset*, etc. and they are just enough to uniquely constrain the parts in their goal positions in the assembly. The Product Assembly Modeler (PAM) provides the interface to the user to interactively define the above relationships. PAM then outputs the final assembly representation of all the parts. Techniques in *graph theory* are then used to manipulate the final assembly structure to deduce the precedence constraints automatically. First an undirected graph is used to represent the part connectivity relationships in the product. Spatial relationships input in PAM are used in determining the mating frames of the various component parts of the product. The disassembly directions are determined from the spanning vector (see section 11.3) for each component part. These disassemblibility directions are then tested to determine whether there is collision with any other component in the assembly.

The input to the precedence algorithm is primarily from PAM. This input is in the form of an assembly structure or model of the final assembly. The objective of the precedence algorithm is:

- Find all the direct mating faces for each part in the assembly with every other part in the assembly. Store this information in the form of a component connectivity graph.
- Find the disassembly directions for every part using the spanning vector.
- Use heuristic procedures to reduce the set of disassembly directions to within manageable limits to make the algorithm computationally efficient.
- Use a disassembling algorithm to check for parts which can be removed without collision. Based on this procedure form an AND/OR graph structure which details all the possible feasible sequences for that assembly.

A precedence generator interacts with other planners which have the responsibility for such tasks as grasp planning, gross motion planning and fine motion planning. A simulation of generated assembly sequence can be carried out on any desired robot. This checks if the motion path generated is collision free. The simulation of the various feasible sequences can provide the user valuable information about requirement of jigs and fixtures to hold intermediate subassemblies and the possibility of requiring different tools and robot configurations to perform the same assembly. The choice of the assembly sequence also determines the layout of factory cell and determines the time and costs of the assembly operation. This information can be used in an iterative design cycle to optimize the design of products. The user now chooses the final and actual sequence of assembly and the robot on which to perform that assembly once the feasible set of sequences have been generated.

11.3.2 Assumptions for precedence

From a given set of parts P there can be many different assembly configurations which can be formed. Thus if one is using a forward search algorithm to find feasible subassemblies in the process of generating a final known assembly, the branching factor becomes high enough to make the algorithm inefficient. The problem of finding how to assemble a given product can thus be converted to an equivalent problem of finding how the same product can be disassembled. It is not always that assembly operations are reversible but we shall restrict ourselves presently to the reversible ones. In this presentation, we assume that all the parts are rigid and there is no internal energy stored when the assembly is done. An example of such a part having nonrigid or variable geometry is a *spring*. Subassemblies are also assumed to be inherently stable for the present work. We concentrate primarily on geometric reasoning of the final assembly model to extract the precedence constraints. We initially constrain ourselves to polyhedral approximation of objects. This is not a major constraint considering the fact that we are only interested in identifying contact surfaces and not the nature of the surfaces. The system automatically finds disassembly directions in the cases when the disassembly direction is linear. It is also possible to consider cases of complex disassembly motions as are involved in disassembling parts mating by snap fit, screws, rotate and lock mechanisms. This is possible because in PAM when the spatial relationships are being assigned the designer can input process information associated with each mating entity. Since screws and rotate and lock mechanisms involve rotational motions these can be specified explicitly in the form of number of complete revolutions and the direction of motion as either clockwise or anti-clockwise.

11.3.3 Spatial relationships and precedence

As mentioned in Chapter 3, PAM is a subset of a larger system Promod [LN91] primarily intended to facilitate the user to represent general product specifications, tolerance information, process information and the designers intention in general. This system encompasses all manufacturing processes like machining, sheet metal fabrication, assembly etc. PAM provides a graphics interface to interactively assign assembly mating relations to the parts. These spatial relationships describe the final desired relationship among component parts of an assembly. Design with spatial relationships is used not only for

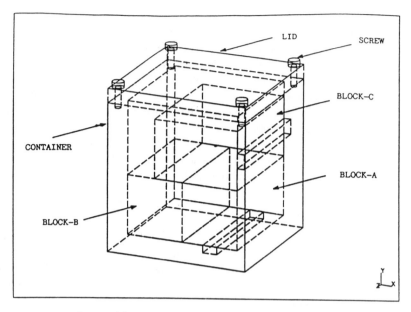

Fig. 11.13 Example assembly.

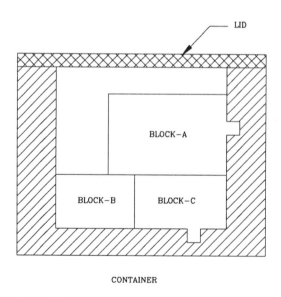

Fig. 11.14 Example assembly C/S view.

inferring the assembly positions of parts but also as a model to capture the designer's intent.

The relationships can be of the type *against, parallel-offset, parax-offset, aligned, inclined-offset* and *include-angle* [LN91]. Attributes of mating relationships can also be defined between various mating faces, like

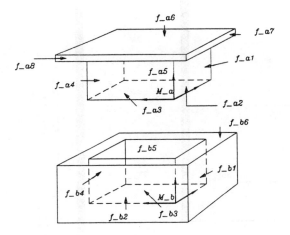

Fig. 11.15 Mating frames.

- mating process, whether welding, glueing, riveting, pressure fit etc.,
- contact forces, torques,
- relative motion parameters between two mating parts, rotational motion for screws and rotate and lock mechanisms, complex linear motions required for snap fits, etc.

Consider the five item example assembly shown in Figure 11.13. The assembly operations involved in assembling this product are typical of those encountered in any robotic assembly operation. A cross-section of this assembly is shown in Figure 11.14 to show the mating surfaces and types clearly. For creating an assembly model of this assembly, the user may assign on an average three mating relationships for each of the parts to constrain it uniquely in its assembly position. The degree of freedom of the part (DOF) is used to determine the uniqueness of the assembly position and to determine if the part is constrained completely. Essentially each spatial relationship is interpreted as a constraint imposed on the DOF between the relative mating features. The concept of mating frames is used for this purpose. In Figure 11.15 a very simple assembly of two parts Part a and Part b, is used to illustrate this concept more lucidly. Let $F_a\{f_{a1}, f_{a2}, \ldots, f_{an}\}$ and $F_b\{f_{b1}, f_{b2}, \ldots, f_{bm}\}$ be the face set for these two parts.

These two parts can be constrained in their assembly positions by defining relationships like:

- three *against* relationships between faces $f_{a3} :: f_{b3}, f_{a3} :: f_{b2}$ and $f_{a1} :: f_{b1}$ or
- two *against* and one *parallel-offset* relationship between faces $f_{a4} :: f_{b4}, f_{a5} :: f_{b5}$ and $f_{a6} :: f_{b6}$ respectively.

There could also be other combinations of the different mating primitives that could be used to achieve the same thing. The users need use only the ones which are easy for them to assign naturally and easily. These mating relationships determine the mating frames as shown in Figure 11.15.

Let the origin coordinate frame of Part a be O_a and the mating frame be M_a and similarly for Part b let it be, O_b and M_b. The origin coordinate frames are created by

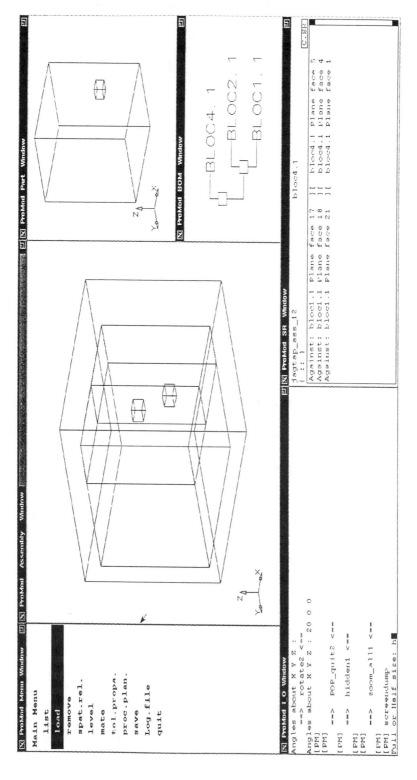

Fig. 11.16 Spatial relationships.

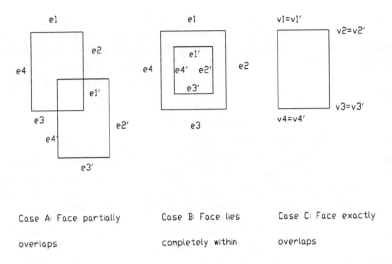

Fig. 11.17 Planar mating types.

the CAD modeler and all the vertices of the parts are located relative to these frames. In order to assemble these two parts the mating frames of the two parts should exactly match. Now if the transformation matrix for Part b with respect to Part a is $^{O_a}T_{O_b}$, then:

$$^{O_a}T_{O_b} = {}^{O_a}T_{M_a} \cdot {}^{O_b}T_{M_b}^{-1}$$

PAM thus generates these mating frames for all the parts in the assembly and transforms them into their assembly position. The output of PAM is a final assembly tree structure consisting of parts, their spatial relations, the transformation matrices for each part, etc. This representation provides all useful assembly information and also stores the designers intention. The specific assembly contacts, conditions of fastening or joining information can also be stored in this representation. This representation of the final assembly is then used by the precedence algorithm to extract the precedence relationships in assembly. Figure 11.16 shows the five-item example assembly being assembled in PAM. As can be seen in the figure the screen interface is so designed that it allows the user to see the completed subassembly or assembly in one window and the next part to be assembled in the parts window. The status of the degree of freedom of the object as it is being restricted by the assignment of spatial relationships is also shown in a window and a BOM assembly tree shows the various parts which are already present in the assembly.

11.3.4 Find all mating faces

In PAM the user assigns only the necessary mating relationships to constrain a part uniquely in its assembly position. As a result of this PAM outputs the transformation matrices for each part which enable it to achieve this assembly position. Using the above transformation matrices and the face equations which are already provided by the IGES translator from the raw IGES data we find out all the contact mating faces for each part with every other part in the assembly.

Geometrically speaking we have to first find all faces of a part which lie in the same infinite plane as that of a face of another part and their normals should be in the opposite directions. The normalized face equations are of the form $F(x, y, z) = 0$. If two faces, face A and face B of parts A and part B have l_i and m_i as the normalized coefficients then they are coplanar and have normals in opposite directions if the following condition is satisfied.

$$l_i + m_i = 0; \text{ for all } i$$

To find whether face A and face B actually intersect each other, the edges and the boundary loops of the faces are used. The boundary loops [Yeh89] in conjunction with the plane equations represent the actual faces of the parts. They are defined by the edges and the vertices of the parts. The check for intersection is done in three stages because as shown in Figure 11.17 there are three cases of planar mating.

- Case A: In this first case where there is partial overlap of two faces, the edge intersection is checked i.e. whether the edge of one part intersects with the edge of the other part.
- Case B: Here one face completely lies within the other face, thus it is checked whether all the vertices of face A lie within the loop of face B or vice-versa. This is done by the ray casting method.
- Case C: In this case of exact overlap of two faces, it is checked if all the vertices of one face are equal to all the corresponding vertices of the other face.

This method guarantees to find all possible mating faces for the parts. In the implementation for making the algorithm computationally efficient Case C was checked first, then every face which did not satisfy Case C was checked for Case A and then all the remaining faces were checked for Case B. This algorithm, Find_Mating_face, is shown below:

```
LET P = {P₁, P₂, P₃,..., Pₙ};
FOR:Pᵢ, i = 1...n;
    F = {F₁, F₂, F₃,..., Fₖ};
    FOR:Fⱼ, j = 1...k;
        CALL:Function(check_vertices(Fⱼ));
        IF:return = NIL;
            CALL:Function(check_edge_intersection(Fⱼ));
            IF:return = NIL;
                CALL:Function(ray_casting(Fⱼ));
return P = {P₁, P₂, P₃,..., Pₙ};
```

Proceeding in this way we get a linked list of all the mating faces of a part.

11.3.5 Determining the disassembly direction

After all the mating face equations for all the component parts with every other part are generated, the feasible disassembly directions for each part in the assembly can be determined. The mating face equations help in determining the *spanning vector*, **r**. This spanning vector shows the direction or set of directions in which the part can be assembled.

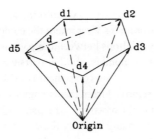

Fig. 11.18 Polyhedron convex cone.

The determination of the spanning vector is done as follows. Whenever there is a planar contact between two faces of two parts, the parts can be removed in any direction in the halfspace created by the mating face. Thus if $F = f_1, f_2, \ldots, f_j$ represents a face set of all planar mating faces for a part then the set of removable directions due to the ith mating face is given by:

$$R_i = \{r | r \cdot n_i \geq 0; \quad i = 0, 1, 2 \ldots j\}$$

Here n_i is the unit normal vector to the ith mating face. The actual or set of resultant removable directions due to all the j mating faces is:

$$R = \bigcap_{i=1}^{j} R_i$$

The set R can be seen as a polyhedral convex cone [BNYew]. In Figure 11.18 we can see a polyhedron convex cone, with its extreme directions, namely d_1, d_2, d_3, d_4, d_5. Here the extreme directions of a convex polyhedron are defined as the direction of the set that can be represented as a linear combination of two distinct directions of the convex set. Note that any direction emanating from the origin of this polyhedral cone can be represented as a linear combination of these five extreme directions. For example, choose a direction d shown on the figure. It can be seen that d can be represented as a linear combination of d_2 and d_5. That is $d = rd_2 + (1 - r)d_5$ $r \in [0, 1]$. Thus r can be chosen as a vector sum of R or can be set to the vertical direction or to the x, y, and z directions.

11.3.6 Precedence algorithm

We shall now look closely at the disassembly algorithm which generates the feasible assembly sequences. We now already have the assembly representation of the product from the output of PAM. The mating face information of all the parts is now generated as well as the disassembly directions.

1. Deduction of precedence constraints.
   The first thing to be done here is to analyze the mating face information. This is done by constructing the component connectivity graph (Figure 11.19). This undirected graph is constructed to give the user an idea of the connectivity relations of the various parts in the assembly. It also serves as an efficient representation of the mating information which can be easily manipulated by the dissassembly algorithm. Let

$P\{P_1, P_2, P_3, \ldots P_n\}$ be a set of parts which go into the assembly. An undirected graph $G(V, E)$ in which the vertices or nodes V represent the component parts P and the edges E represent the connectivity relations between the parts is constructed. The degree of each vertex $d(v_i)$ determines the number of other parts with which each part has direct mating contacts. The usefulness of the component connectivity graph is in helping to find out which parts can be removed from the assembly without interfering with other parts in their goal positions in the assembly. The part with minimum degree of vertex is generally the first choice but many times by the virtue of its position in the assembly it might not be possible to remove it first without either making the remaining subassembly unstable or causing geometric interferences. For the example assembly the component connectivity graph is shown in Figure 11.19.

2. Assembly decomposition algorithm.

This algorithm derives the precedence constraints based on geometric interferences from the component connectivity graph. The method of cutsets adapted in part from [dMS89] is used to decompose the assembly structure. From the component connectivity graph we get $P = V$, the set of parts which are present in the final goal assembly. Let $d(v_i)$ be the degree of the vertex(part) v_i. The following steps enumerate the disassembly algorithm:

> Choose part v with minimum $d(v_i)$
> Let $V_1 = v$, $V_2 = \overline{V}_1$ and $V_1, V_2 \subset V$.
> Call Function(Disassembly_direction(v));
> return(R),
> choose r_1, $r_1 \subset R$;
> Call Function(check_intersection(v, r_1));
> IF return = true;
> > then choose another r_2, $r_2 \subset V$;
> > Call Function (distance(v, face of v_2));
> > return = d;
> > Call Function(translate(v, d));
> > Call Function (check_intersection (v, r_2);
> > > IF return = NIL;
> > > delete v from $G(V, E)$;
> choose another v with minimum $d(v_i)$;

Utilizing this method it is now possible to simulate the disassembly of the given product assembly. If it is possible to remove a part in the disassembly direction r, then it has no predecessors in the disassembly procedure. This simply means that no other part has to be removed before. If on the other hand the part is not removable in any disassembly direction then the set of predecessors contains all the parts which are an obstacle for that part when moving in the disassembly direction r.

Once the precedence algorithm is executed an AND/OR graph representation of all the possible and feasible sequences is constructed. The AND/OR graph for the example assembly is shown in Figure 11.20. Associated with each node in the graph is a subassembly that might be reached in the process of building the final product. An AND-arc represents the operation of putting two child subassemblies together to make the parent, while OR-arcs give different ways of creating the same parent subassembly. Thus each AND

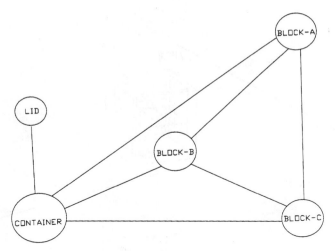

Fig. 11.19 Component connectivity graph.

subtree of a full AND/OR graph represents one order of assembly for the root of the tree with the geometric relations in performing the disassembly. As can be seen from the AND/OR graph there are two feasible sequences for the example assembly.

Since the Block-A (Figure 11.20) has a protrusion feature which mates with the hole feature in the container, it might be argued that Block-A should go before Block-B so that it has more maneuverability. These kinds of common assembly heuristics can be used to decide upon an optimal and preferable sequence from the feasible sequences generated by the disassembly algorithm. The above example also illustrates how complex, albeit linear disassembly motions can be deduced only from the geometry of the assembly.

We have talked so far about testing for object interference in the direction of the disassembly, in the following description we shall see an algorithm for doing it.

3. Testing the disassembly direction.

A geometric precedence constraint between two parts or subassemblies can result if an object in its goal position becomes an obstacle for the assembly of another object. A search algorithm takes each part and checks whether it can be removed without interference. If the part can be removed then it is removed from the component connectivity graph and the algorithm is applied again. As the parts are removed from this component connectivity graph another directed precedence graph is constructed which shows the sequence in which the parts are to be assembled. As explained in step 2, given a disassembly direction r the algorithm first finds a convex polygon perpendicular to this direction which encloses the part completely. Rays are then shot [PS85b] from points on this polygon and their intersection with other parts is detected. This is done as follows.

- Find faces of the parts in V_2 whose surface normals are not orthogonal with the disassembly direction i.e.

$$n_j^k \cdot r \neq 0$$

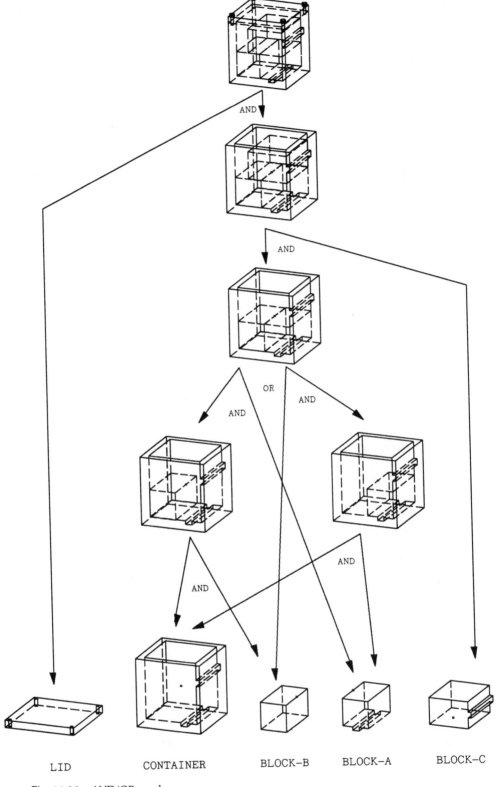

Fig. 11.20 AND/OR graph.

Here n_j^k represents the surface normal for the jth face and the kth part in the set V_2 formed by the cutset procedure.
- The intersection of an infinite line in the disassembly direction and an infinite plane containing the face as determined above, is tested.
- Check whether this point of intersection lies within the face of that part as bounded by the loop.

If there is intersection then it is first checked whether the part being disassembled has a disassembly direction in a direction parallell to this intersection face. If there is then this new direction is checked again for collision. If the part being disassembled does not have a disassembly direction parallel to this intersection face then the distance d between the intersecting faces of these two parts is calculated and the part is linearly translated in the disassembly direction by d.

The new mating faces of this part are re-evaluated and the disassembly direction recomputed. The part is now checked again for collision in the new disassembly direction by the same method. If there is still collision then the part with which the collision occurs is moved from set V_1 to set V_2; and the whole algorithm begins again with this new cutset.

11.3.7 Interpretation of results

The output of the precedence algorithm is the AND/OR graph which represents all the feasible assembly sequences for the given assembly. Each sequence may result in different assembly times, different subassemblies hence different fixtures, different layout of the assembly line etc. The net effect is that different sequences result in different costs. It is possible to determine an optimum sequence from the set of given feasible sequences. This has to be done at a later stage in automatic planning when more information is available about:

- robot type,
- gripper type for the robot,
- the robot environment or robot world,
- type and cost of fixtures required for various subassemblies,
- assembly line layout,
- decision criteria, cost of assembly or time of assembly.

11.4 FINE MOTION PLANNING

Fine motion planning is usually performed in the vicinity of the task which a manipulator is expected to perform. For example, if a manipulator picks up a bolt to screw onto a nut, then fine motion must be executed near the point of pick up of the bolt and when the principal axis of the bolt is to be aligned to the axis of the hole. These fine motions are typically planned for approach and departure to the tasks.

While gross motion planning ensures a feasible path from an initial configuration to near final configuration, the *grasp planner* ensures feasible grasp. There is no assurance that the final motions will actually accomplish the task.

At regions near the initial and final robot configurations, the manipulator is usually constrained by external positional constraints which will require motion control different from that the of the gross motion. There are basically two types of motions involved in external position constraints: *guarded motion*, when the manipulator is about to contact a surface; and *compliant motion*, used when the robot is in continuing contact with a surface [BHJ⁺ 83].

The typical motion required for an assembly operation has four parts: a guarded departure from current configuration, a free motion towards the destination configuration of the task step, a guarded approach to contact at the destination, and a compliant motion to achieve the task configuration [BHJ⁺ 83]. The problem of free motion planning has been addressed in Chapter 7.

Guarded and compliant motions are both sensor-based motion strategies used to achieve or maintain relative configurations among objects even in the presence of uncertainty in the actual configurations. In compliant task, the legal configurations of the robot are constrained by the kinematics of both the task and the robot. The legal robot configurations are constrained by the task kinematics to lie on a surface within the robot's configuration space. This type of surface is called *C-surface* [BHJ⁺ 83].

The goal in guarded motion is to achieve a manipulator configuration on an actual C-surface while avoiding excessive forces. It is necessary to avoid the range of forces which can damage the robot or the task objects. Since the exact configuration of the objects and the robot are usually not known, there is a need to rely on sensors (force, touch or combined) to guarantee that contact is achieved without excessive forces. Also, the speed of the arm must be relatively low.

A viable procedure is for the planner to specify the approach path to the C-surface including the maximal forces to be exerted for guarded motions. One way to compute the approach path is by finding a path which avoids collisions with nearby objects and at the same time guaranteeing collision with the desired C-surface. This can be seen to be a special case of the collision avoidance problem discussed in Chapter 7. With the new methods of designing products for manufacturability, many of the problems inherent in assembly can be minimized. Problems such as jamming due to lack of chamfer can easily be avoided. Sensor-tuned planners also help in providing robust guarded and compliant plans.

11.5 PROGRAM SYNTHESIS

Program synthesis encompasses the tasks of grasp planning, motion planning, and error detection. The resulting output of this program is usually several grasp commands, motion specifications and error tests. The program is also robot-level commands which can be executed by any robot. We have discussed how to transform the general robot-level commands into generic robot commands for the five general classes of robot (Cartesian, cylindrical, etc.) in Chapter 10.

The grasp planner has been discussed in detail in Chapter 9. In general, the grasp planner chooses grasp configurations and ensures that the grasp can withstand the forces generated during motion and contact with other objects. It is important that the grasping does not increase any certainty in the configuration of the object being grasped.

Program synthesis

After grasping the object, then the planner must synthesize reliable motions to achieve the desired goal of the task. At this stage, all the motion plans which we have discussed come into play: gross motion, collision avoidance, fine motion (guarded and compliant motions) plannings.

Guarded and compliant motions planning ensures that the actual state of the world will differ from the world model but only within known bounds. But this may not be the case. The actual state may be well outside the known boundaries. For example, objects can be the wrong size or they can be absent. In such cases, the program should be able to detect the failure and correct it or stop the motion. Because of this, a very important part of a program will be sensory tests (discussed in Chapter 5) for error detection and correction.

References

[ACP83] A.P. Ambler, D.F. Corner, and R.J. Popplestone. Reasoning about the spatial relationship derived from a rapt program for describing assembly by robot. *IJCAI*, pp. 824–844, 1983.

[AH] P. Adolphs and G. Höhn. Transformation of obstacles into configuration space by using a special look-up-table for collision free on-line path planning.

[All88] Peter K. Allen. Integrating vision and touch for object recognition tasks. *The International Journal of Robotics Research*, 7(6):15–33, December 1988.

[AO87] H. Asada and K. Ogawa. Manipulator dynamics analysis using the generalized centroid and virtual mass for task planning and end-effector design. In *The Winter Annual Meeting of ASME*, pp. 101–110, 1987.

[AP75] A.P. Ambler and R.J. Popplestone. Inferring the positions of bodies from specified spatial relationships. *Artificial Intelligence*, 6(2):157–174, 1975.

[Asa87] D. Asano. Evaluation of trajectories for different classes of robots. Master's thesis, University of Massachusetts Mechanical Engineering Program, Amherst, MA, May 1987.

[BBM82] Thomas Beckwith, N. Lewis Buck, and Roy Marangoni. *Mechanical Measurements*. Addison-Wesley, 1982.

[BC82] Robert C. Bolles and Ronald A. Cain. Recognizing and locating partially visible objects: The local-feature-focus method. *The International Journal of Robotics Research*, 1(3):57–82, Fall 1982.

[BF82] A. Barr and E.A. Feigenbaum. *The Handbook Of Artificial-Intelligence*, volume 1-3. William Kaufmann Inc., 1981–82.

[BHJ$^+$83] M.J. Brady, J.M. Hollerbach, T.L. Johnson, T. Lozano-Perez, and M.T. Mason. *Robot Motion – Planning and Control*. MIT Press, 1983.

[BJS90] M.S. Bazaraa, J.J. Jarvis, and H.D. Sherali. *Linear Programming and Network Flows*. John Wiley & Sons, Inc., 1990.

[BLL89] Jérôme Barraquand, Bruno Langlois, and Jean-Claude Latombe. Numerical potential field techniques for robot path planning. Technical Report STAN-CS-89-1285, Department of Computer Science, Stanford University, October 1989.

[BLP83] R.A. Brooks and T. Lozano-Pérez. A subdivision algorithm in configuration space for find-path with rotation. In *Proceedings of the Eighth International Conference on Artificial Intelligence*, Karlsruhe, Germany, 8–12 August 1983. Also in: *IEEE Transactions on Systems, Man, and Cybernetics*. SMC-15(2) March/April 1985.

[BNA88] J.Y. Chu, B.O. Nnaji and M. Akrep. A schema for cad-based robot assembly task planning for csg-modeled objects. *Journal of Manufacturing Systems*, 7(2), 1988.

[BNRew] H.C. Liu, B.O. Nnaji and U. Rembold. A product modeler for discrete components. *IJPR* (in press).

[BNYew] J.B. Sadrach, B.O. Nnaji and H. Yeh. Spanning vector for assembly directions and other applications. *Computer-aided Design* (accepted).

[Boi82] J. Boissonnat. Stable matching between a hand structure and an object silhouette. *IEEE Transactions on Pattern Analysis and Machine Intelligence*, pp. 603–612, November 1982.

[Bou84a] A. Bourjault. Contribution a une approache methodologique de l'assemblage automatise: Elaboration automatique des sequences operatoire. PhD thesis, Universite de Franche-Comte, Besançon November 1984.

[BPM82] G. Boothroyd, C. Poly, and L.E. Murch. *Automatic Assembly*. Marcel Dekker, Inc., 1982.
[Bro83a] R.A. Brooks. Planning collision-free motions for pick-and-place operations. *International Journal of Robotics Research*, 1(4):19–44, Winter 1983.
[Bro83b] R.A. Brooks. Solving the find-path problem by good representation of free space. *IEEE Transactions on Systems, Man, and Cybernetics*, SMC-13(3):190–197, March/April 1983.
[Buc87] C.E. Buckley. A foundation for the "flexible-trajectory" approach to numeric path planning. In Ulrich Rembold and Klaus Hörmann, (eds), *Languages for Sensor-Based Control in Robotics*, pp. 389–424. Springer-Verlag, 1987.
[BX87] A. Barraco and D.M. Xing. Determination of trajectories with obstacle avoidance. In Ulrich Rembold and Klaus Hörmann (eds), *Languages for Sensor-Based Control in Robotics*, pp. 361–388. Springer-Verlag, 1987.
[Cam89] Stephen Cameron. Efficient intersection tests for objects defined constructively. *The International Journal of Robotics Research*, 8(1):3–25, 1989.
[Can87] John Francis Canny. *The Complexity of Robot Motion Planning*. PhD thesis, Massachusetts Institute of Technology, 1987.
[CC82] P. Coiffet and M. Chirouze. *An Introduction to Robot Technology*. Mc-Graw Hill, New York, 1982.
[CH88] M.R. Cutkosky and R.D. Howe. Human grasp choice and robotic grasp analysis. In S.T. Venkatraman and T. Iberall (eds), *Dextrous Robot Hands*. Springer-Verlag, 1988.
[Clas] CAM-I. Cam-i's illustrated glossary of workpiece form features. *CAM-I Inc.*, R-80-PPP-02.1, Arlington, TeX.
[CLT90] A. Ijel C. Laugier and J. Troccaz. Combining vision-based information and partial geometric models in automatic grasping. In *IEEE International Conference on Robotics and Automation*, pp. 676–682, 1990.
[CM81] W.F. Clocksin and C.S. Mellish. *Programming in Prolog*. Springer-Verlag, Berlin, Heidelberg, 1981.
[Coi83] P. Coiffet. *Robot Technology Vol 1 – Modeling & Control*. Prentice Hall, Englewood Cliffs, NJ, 1983.
[CP86] M.R. Cutkosky and P.K. Wright P.K. Friction, stability and the design of robotic fingers. *The International Journal of Robotic Research*, pp. 20–37, 1986.
[Cra86] J.J. Craig. *Introduction to Robotics: Mechanics and Control*. Addison-Wesley, Reading, MA, 1986.
[Cut85] M.R. Cutkosky. *Robotic Grasping and Fine Manipulation*. Kluwer Academic, 1985.
[DH55] J. Denavit and R.S. Hartenberg. A kinematic notation for lower-pair mechanisms based on matrices. *Journal of Applied Mechanics*, 22(2), June 1955.
[Dix88] J.R. Dixon. Designing with features: Building manufacturing knowledge into more intelligent cad systems. In *Proceedings of ASME Manufacturing International*, Atlanta, GA, April 1988. ASME.
[DL82] B. Dufay and C. Laugier. Geometrical reasoning in automatic grasping and contact analysis. In T.M.R. Ellis and O.J. Semenkov (eds), *Advances in CAD/CAM*. 1982.
[DLL$^+$87] J.R. Dixon, E.C. Libardi, S.C. Luby, M. Valghul, and M.K. Simmons. Examples of symbolic representations of design geometric engineering with computers. *Expert Systems for Mechanical Design*, 2:1–10, 1987.
[dM89] L.S. Homem de Mello. *Task Sequence Planning for Robotic Assembly*. PhD thesis, Carnegie Mellon University, 1989.
[dMS89] L.S. Homem de Mello and A.C. Sanderson. A correct and complete algorithm for the generation of mechanical assembly sequences. *International Conference On Robotics and Automation*, 1, 1989.
[DW88] Hugh F. Durrant-Whyte. *Integration, Coordination and Control of Multi-Sensor Robot Systems*. Kluwer, 1988.
[EK88] Peter A. Eggleston and Charles A. Kohl. Symbolic fusion of mmw and ir imagery. In *Sensor Fusion: Spatial Reasoning and Scene Interpretation*, pp. 20–27. SPIE, November 1988.
[Fav84] Bernhard Faverjon. Obstacle avoidance using an octree in the configuration space of a manipulator. In *Proceedings of the 1984 IEEE International Conference on Robotics*, pp. 504–512, 1984.

[FD87] T. De Fazio and Whitney D.E. Simplified generation of all mechanical assembly sequences. *EEE Journal of Robotics and Automation*, **RA-3**(6), 1987.

[Fea86] R.S. Fearing. Simplified grasping and manipulation with dextrous robot hands. *IEEE Journal of Robotics and Automation*, pp. 188–195, 1986.

[FGL87] K.S. Fu, R.C. Gonzales, and C.S.G. Lee. *Robotics: Control, Sensing, Vision, and Intelligence.* McGraw-Hill, New York, 1987.

[Fly88] A.M. Flynn. Combining sonar and infrared sensors for mobile robot navigation. *The International Journal of Robotics Research*, 7(6):5–14, December 1988.

[Gar87] Thomas D. Garvey. A survey of ai approaches to the integration of information. *SPIE Infrared Sensors and Sensor Fusion*, **782**:68–82, Fall 1987.

[Gat85] L.B. Gatrell. Cad-based grasp synthesis. Master's thesis, University of Utah, 1985. Dept. of Computer Science.

[GL88] Robert R. Goldberg and David G. Lowe. Hessian methods for verification of 3d model parameters from 2d image data. In *Sensor Fusion: Spatial Reasoning and Scene Interpretation*, pp. 63–67. SPIE, November 1988.

[Gla90] Bernhard Glavina. Solving findpath by combination of goal directed and randomized search. In *Proceedings of the 1990 IEEE International Conference on Robotics and Automation*, 1990.

[GT78] D.D. Grossman and R. H. Taylor. Interactive generation of object models with a manipulator. *IEEE Trans. Systems, Man, Cybernetics SMC*, **8**(9):667–679, 1978.

[GW90] R.A Grupen and R.S. Weiss. Sensor-based planning for grasping and manipulation with multifingered robot hands. Technical Report 90-58, COINS, University of Massachusetts at Amherst, Laboratory for Perceptual Robotics, July 1990.

[GWNO86] M.P. Grover, M. Weiss, R.N. Nagel, and N. Odrey. *Industrial Robotics:Technology, Programming, and Applications.* McGraw-Hill, New York, 1986.

[HA77] H. Hanafusa and H. Asada. Stable prehension by a robot hand. In *Proceedings of the 7th International Symposium on Industrial Robots*, pp. 323–335, 1977.

[Ham69] W.R. Hamilton. *Elements of Quaternions.* Chelsa Publishing, 1969.

[HdMS86] L.S. Homen de Mello and A.C. Sanderson. AND/OR Graph Representation of Assembly Plans. In *Proceedings of the Fifth National Conference on Artificial Intellgence*, pp. 1113–1119, 1986.

[Her78] J.M. Hervé. Analysis structurelle des mécanismes par groupe des déplacements. *Mechanism and Machine Theory*, **13**(4), 1978.

[HFH84] Thomas C. Henderson, Wu So Fai, and Charles Hansen. Mks: A multisensor kernel system. *IEEE Transactions on Systems, Man, and Cybernetics*, SMC-14, September/October 1984.

[HHB85] Thomas Henderson, Charles Hansen, and Bir Bhanu. The specification of distributed sensing and control. *Journal of Robotic Systems*, **2**(4):387–396, Winter 1985.

[HL90] Y.F. Huang and C.S.G Lee. An automatic assembly planning system. *IEEE International Conference on Robotics and Automation*, 1990.

[HLB89] T. Iberall H. Liu and G.A. Bekey. The multidimensional quality of task requirements for dextrous robot hand control. *IEEE Journal of Robotics and Automation*, **1**:452–457, 1989.

[Hoe87] Klaus Hoermann. A Cartesian approach to findpath for industrial robots. In Ulrich Rembold and Klaus Hömann (eds), *Languages for Sensor-Based Control in Robotics*, pp. 425–450. Springer-Verlag, 1987.

[HR88] K. Hoermann and U. Rembold. Robot action planner for automatic parts assembly. *Proceedings: IEEE International Workshop on Intelligent Robots and Systems*, Japan, October/November 1988.

[HS84] Thomas Henderson and Esther Shilcrat. Logical sensor systems. *Journal of Robotic Systems*, **1**(2):169–193, Summer 1984.

[Hun78] K.H. Hunt. *Kinematic Geometry of Mechanisms.* Oxford University Press, Oxford, 1978.

[HWMH88] T. Henderson, E. Weitz, A. Mitiche, and C. Hansen. Multisensor knowledge systems: Interpreting 3rd structure. *The International Journal of Robotics Research*, **7**(6):114–137, December 1988.

[Inc84] Unimation Inc. Programming manual user's guide to valii, document 398t1. Technical report, Westinghouse Co., Danbury, CT, 1984.

References

[IRF89] M. Sudhaker I.N. Rao and M.A. Faruqi. Octree modelling for cad/cam interface and robotic path generation. In *Proceedings of CAD, CAM, Robotics and Factories of the Future Conference*, Delhi, Dec. 19–22 1989. IIT.

[Jam85] J.W. Jameson. Analytic techniques for Automated Grasps. PhD thesis, Stanford University, 1985.

[JBP85] J.P. Hermann, J. Bach, S. Herard and P. Pardo. Bin-picking using a 3-d sensor and a special gripper. In *Proceedings of the 5th International Conference on Robot Vision and Sensory Controls*, pp. 529–540, 1985.

[J.C83] D. Perreira and J. Colson. Kinematic arrangements used in industrial robots. In *Proceedings of the 13th International Symposium on Industrial Robots*, 1983.

[Jen65] P.W. Jensen. *Cam Design and Manufacture*. Industrial Press, New York, 1965.

[JLP90] J.L. Jones and T. Lozano-Perez. Planning two fingered grasps for pick and place operations on polyhedra. In *IEEE International Conference on Robotics and Automation*, pp. 683–688, 1990.

[JN91] P. Jagtap and B.O. Nanji. Automatic precedence plan generation with spatial relationships in assembly. In *SME Technical Paper MS91-387*, 1991.

[JWW85] R.A. Volz, J.D. Wolter and A.C. Woo. Automatic generation of gripping positions. *IEEE Transaction on Systems, Man and Cybernetics*, pp. 204–213, 1985.

[Kan91] Tzong-Shyan Kang. Interpretation of CAD Models for an automatic Machine programming planner. PhD thesis, University of Massachusetts At Amherst, 1991.

[Kha86] O. Khatib, Real-Time obstacle Avoidance for Manipulators and Mobile Robots. *International Journal of Robotics Research*, **5**(1):90–98, Spring 1986.

[KL85] T.R. Kane and D.A. Levinson. *Dynamics*. McGraw-Hill, 1985.

[K.S81a] J. Duffy and K. Sugimoto. Determination of extreme distances of a robot hand. 2. *Journal of Mechanical Design*, pp. 776–783, 1981.

[K.S81b] J. Duffy K. Sugimoto. Determination of extreme distances of a robot hand. 1. *Journal of Mechanical Design*, pp. 631–636, July 1981.

[KVC$^+$] A.C. Kak, A.J. Vayda, R.L. Cromwell, W.Y. Kim, and C.H. Chen. *Knowledge-based robotics*. Robot Vision Lab, Purdue University.

[LA86] Y. Liu and M.A. Arbib. A robot planner in the assembly domain. Technical Report 86-36, Dept. COINS, University of Massachusetts at Amherst, Amherst, MA 01003, July 1986. COINS Technical Report.

[Lau86] C. Laugier. A program for automatic grasping of objects with a robot arm. In *Robot Grippers*. 1986.

[Lau88] Christian Laugier. Planning robot motions in the SHARP system. In B. Ravani (ed), *CAD Based Programming for Sensory Robots*. Springer-Verlag, 1988.

[LC90] Alan C. Lin and T.C. Chang. Automated process planning for 3-dimensional mechnaical assemblies. *Proceedings of NSF Design and Manufacturing systems Conference*, pp. 633–640, 1990.

[LD86] E.C. Libardi and J.R. Dixon. Designing with features: Design and analysis of extrusions as an example. In *Spring National Design Engineering Conference and Show*, pp. 24–27, Chicago, March 1986.

[Lin91] E. Lin. Sensor representation for automatic robotic assembly. Master's thesis, University of Massachusetts At Amherst, 1991.

[Liu90] Y. Liu. Symmetry Groups in Robotic Assembly Planning. PhD thesis, University of Massachusetts at Amherst, 1990.

[LN91] H. Liu and B.O. Nnaji. Design with Spatial Relationships. *Journal of Manufacturing Systems*, **10**(6), 1991.

[LP80] T. Lozano-Perez. Automatic planning of manipulator transfer movements. *IEEE Trans. Sys., Man., Cybern,.* SMC-**11**(10):681–689, 1980. also in MIT AI Memo 606 and 'Robot Motion'. Brady *et al* (eds). Series:MIT Press.

[LP83a] T. Lozano-Perez. Robot programming. *Proceedings of the IEEE*, **71**:821–841, 1983.

[LP83b] T. Lozano-Pérez. Spatial Planning: A Configuration Space Approach. *IEEE Transactions on Computers*, **C32**(2), 1983.

[LP83c] T. Lozano-Perez. *Task Planning in Robot Motion. Planning and Control*. MIT Press, Cambridge, MA, 1983.

References

[LP87] T. Lozano-Pérez. A Simple Motion-Planning Algorithm for General Robot Manipulators. *IEEE Journal of Robotics and Automation*, **RA-3**(3):224–238, June 1987.

[LPB85] T. Lozano-Pérez and R.A. Brooks. An Approach to Automatic Robot Programming. Technical Report A.I. Memo No. 842, Massachusetts Institute of Technology, Artificial Intelligence Laboratory, MIT, April 1985.

[LPW77] T. Lozano-Perez and P.H. Winston. Lama: A language for automatic mechanical assembly. In *Proceedings of the Fifth International Joint Conference on Artificial Intelligence*, Cambridge, MA, August 1977. MIT.

[LPW79] T. Lozano-Perez and M.A. Wesley. An algorithm for planning collision free paths among polyhedral obstacles. *Comm. ACM*, **22**(10):560–570, October 1979.

[LS88] Z. Li and S.S. Sastry. Task oriented optimal grasping by multifingered robot hands. *IEEE Journal of Robotics and Automation*, pp. 32–44, 1988.

[LW77] L. Lieberman and M. Wesley. AUTOPASS: An automatic programming system for computer controlled mechanical assembly. *IBM Journal of Research and Development*, 1977.

[Lyo85] D. Lyons. A simple set of grasps for a dextrous hand. Technical Report 85–37, University of Massachusetts at Amherst, Department of Computer and Information Science, 1985.

[Mas81] M.T. Mason. Compliance and force control for computer controlled manipulators. *IEEE Transactions on Systems, Man, and Cybernetics*, SMC-1(6), June 1981. Also in *Robot Motion* (ed. Brady et al.) Series: MIT Press.

[MG91] W. Meier and J. Graf. A two-arm robot system based on trajectory optimization and hybrid control including experimental evaluation. In *Proceedings of the 1991 IEEE International Conference on Robotics and Automation*, 1991.

[MHL88] Amar Mitiche, T.C. Henderson, and R. Laganière. Decision networks for multisensor integration in computer vision. In *Sensor Fusion: Spatial Reasoning and Scene Interpretation*, pp. 291–299. SPIE, November 1988.

[MM87] J.F. Wilson, M. McCord, M. Sowa. *Knowledge Systems and Prolog*. Addison-Wesley, 1987.

[Muj77] M. Mujtaba. Exploratory study of computer integrated assembly systems. Technical report, Discussion of Trajectory Calculation Methods, Stanford Artifical Intelligence Laboratory Memo AIM-285, Stanford Computer Science Report Stan-6S-76-568, March 1977.

[NA88] B.O. Nnaji and M. Akrep. The relationship between form and function in robotic manipulation and assembly. Technical Report 8, Automation and Robotics Laboratory, University of Massachusetts at Amherst, Amherest, MA 01003, 1988. (Memo.)

[NA89] B.O. Nnaji and D.K. Asano. Evaluation of trajectories for different classes of robots. *Journal of Robotics and Computer Integrated Manufactuctuing*, **6**(1):25–35, 1989.

[NC88] B.O. Nnaji and J.P. Chen. Feature and shape classification and coding of csg-based models for robot assembly language. Technical Report 11, Automation and Robotics Laboratory, University of Massachusetts at Amherst, Amherst, MA 01003, 1988. (Memo.)

[NC90] B.O. Nnaji and J. Chu. Ralph static planner: Cad based robotic assembly task planning for csg-based objects. *Internatiol Journal of Intelligent Systems*, **5**(2):153–181, 1990.

[ND92] B.O. Nnaji and A. Dubey. Task oriented feature-based grasp planning. *Journal of Robotics and Computer-Integrated Manufacturing*, **9**(6):471–491.

[Nil80] N.J. Nilsson. *Principles of Artificial Intelligence*. Palo Alto, Tioga, California, 1980.

[NK88] B.O. Nnaji and T.S. Kang. A solid model interpretation and unification of csg-based models using principal axes. Technical Report 10, Automation and Robotics Laboratory, University of Massachusetts at Amherst, Amherst, MA 01003, 1988. (Memo.)

[NK90] B.O. Nnaji and T. Kang. Interpretation of cad models through iges interface. *Artifical Intelligence for Design, Analysis and Manufacturing*, **4**(1):15–45, 1990.

[NL90a] B.O. Nnaji and H. Liu. Feature reasoning for automatic robotic assembly. *International Journal of Production Research*, **28**(3):517–540, 1990.

[NL90b] B.O. Nnaji and H. Liu. A product assembly modeler. In *1st International Conference on Automation Technology*, Taipei, Taiwan, R.O.C., July 1990.

[Nna86] B.O. Nnaji. *Computer-Aided Design, Selection and Evaluation of Robots*. Elsevier Science, New York, 1986.

[Nna88] B.O. Nnaji. A framework for cad-based geometric reasoning for robot assembly planning. *The International Journal of Production Research, Special Issue*, **26**(5):735–764, 1988.

[NNA90] Hiroshi Noborio, Tomohide Naniwa, and Suguru Arimoto. A Quadtree based Path-Planning Algorithm for a Mobile Robot. *Journal of Robotics Systems*, **7**(4):555–574, 1990.

[NP90] B.O. Nnaji and R.J. Popplestone. A generalized shape and feature descriptor from models on a cad/cam system for automatic assembly: An approach to: Form, function, design & manufacture. In *NSF Engineering Conference*, Arizona, 1990.

[NW89] James L. Nevins and Daniel Whitney, *Concurrent Design of Product and Processes—A strategy for the next generation in manufacturing*. McGraw-Hill, 1989.

[Nyu88] Van-Duc Nyugen. Constructing force closure grasps. *The International Journal of Robotics Research*, **7**(3), June 1988.

[OLP89] Patrick A. O'Donnell and Tomás Lozano-Pérez. Deadlock free and collision free coordination of two robot manipulators. In *IEEE Transactions on Robotics and Automation*, 1989.

[OSY87] C. Ó'Dúnlaing, M. Sharir, and C.K. Yap. Retraction: a new approach to motion planning. In J.T. Schwartz, M. Sharir, M. Sharir, and J. Hopcroft (eds), *Planning, Geometry and Complexity of Robot Motion*. Ablex, 1987.

[Owe85] Tony Owens. *Assembly with Robots*. Prentice-Hall, 1985.

[PAB78] R.J. Popplestone, A.P. Ambler, and I.M. Bellos. Rapt: A language for describing assemblies. *The Industrial Robot*, 1978.

[PAB80a] R.J. Popplestone, A.P. Ambler, and I.M. Bellos. An efficient and portable implementation of rapt. In *Proceedings of the First ICAA*, Bedford, UK, March 1980. ICAA.

[PAB80b] R.J. Popplestone, A.P. Ambler, and I.M. Bellos. An interpreter for a language for describing assemblies. *Artificial Intelligence*, **14**(1):79–107, 1980.

[PAD85] Production Automation Project, University of Rochester. *PADL-2 User Manual*, December, 1985.

[Pau] R.P. Paul. Manipulator Cartesian path control. *IEEE Trans. System, Man, Cybern.*, **SMC-11**(6):449–455.

[Pau81] R.P. Paul. *Robot Manipulators: Mathematics, Programming and Control*. MIT Press, Cambridge, MA, 1981.

[PDE88] National Institute of Standard and Technology. *Product Data Exchange Specification: The First Working Draft*, nistir 88-4004 edition, February, 1988.

[Pie68] D. Pieper. The kinematics of manipulators under computer control. PhD thesis, Stanford University, 1968.

[PJ87] F. Pfeiffer and R. Johanni. A concept for manipulator trajectory planning. *IEEE Journal of Robotics and Automation*, **RA-3**(2), April 1987.

[Pop84] R.J. Popplestone. Group theory and robotics. In *Robotics Research: The First International Symposium* (eds Brady, M. and Paul, R.), Cambridge, MA and London, 1984. MIT Press.

[Pop87] R.J. Popplestone. The edingburgh designer system as a framework for robotics. In *Proceedings of the IEEE Intertnational Conference on Robotics and Automation*, volume 3, 1987. IEEE.

[PS85a] F.P. Preparata and M.I. Shamos. *Computational Geometry: An Introduction*. Springer Verlag, 1985.

[PS85b] Franco Preparata and Michael Ian Shamos. *Computational Geometry: An Introduction*. 1985.

[PT88] J. Pertin-Troccaz. Geometric reasoning for grasping: A computational point of view. In Bahram Ravani (ed), *CAD-based Programming for Sensory Robots*, pp. 397–423. Springer-Verlag, 1988.

[PW85] M.J. Pratt and P.H. Wilson. Requirements for support of form features in a solid modelling system. Report R-85-ASPP-01, CAM-I, Arlington, TX, June 1985.

[Rei86] John H. Reif. Complexity of the Generalized Mover's Problem. In J.T. Schwartz, M. Sharir, and J. Hopcroft (eds), *Planning, Geometry and Complexity of Robot Motion*. Ablex Series in AI, 1986.

[Rem89] Ulrich Rembold. Autonome mobile Roboter (in German). *Robotersysteme*, **4**(1):17–26, 1989.

[Req83] A.A.G. Requicha. Toward a theory of geometric tolerancing. *The International Journal of Robotics Research*, **2**(4):45–60, Winter 1983.

[RK88] Elon Rimon and Daniel E. Koditschek. The construction of analytic diffeomorphismus for exact robot navigation on star worlds. Technical Report 8809, Department of Electrical Engineering, Yale University, August 1988.

[RS90] B. Ravi and M.N. Srinivasan. Decision criteria for computer-aided parting surface design. *CAD*, **22**(1):11–18, 1990.

[RSF89] S. Rao, M. Sudhakar, and M.A. Faruqi. Octree modelling for CAD/CAM interface and robotic path generation. In *CAM/CAM, Robotics and Factories of the future: Fourth International Conference*, pp. 125–137, 1989.

[R.T79] R. Taylor. Planning and execution of straight line manipulator trajectories. *IBM Journal of Research and Development*, **23**:424–436, 1979.

[SD89] Zvi Shiller and Steven Dubowsky. Robot path planning with obstacles, accurator, gripper, and payload constraints. *International Journal of Robotics Research*, **8**(6), December 1989.

[SDR86] User Mannual for GEOMOD. *SDRC I-IDEAS*, 1986.

[Sha88] J.J. Shah. Feature transformations between application-specific feature spaces. *Computer-Aided Engineering*, **5**(6):247–255, 1988.

[SJ86] S.C. Jacobsen, E.K. Iversen. et al. Design of the utah/mit dextrous hand. In *Proceedings of the IEEE Conference on Robotics and Automation*, April 1986.

[Sma88] James R. Smart. *Mordren Geometries*. Brooks/Cole, 1988.

[Sny85] Wesley E. Snyder. *Industrial Robots: Computer Interfacing and Control*. Prentice-Hall, 1985.

[SS83] Jacob T. Schwartz and Micha Sharir. On the piano movers' Problem II: general techniques for computing topological properties of real algebraic manifolds. *Advances in Applied Mathematics*, **4**:298–351, 1983.

[ST87] S.W. Sedas and S.N. Talukdar. A disassembly planner for redesign. *Proceedings: The Winter Annual Meeting of The American Society of Mechanical Engineers, Boston*, December 1987.

[Sta88] S.A. Stansfield. Integrating multiple views into a single representation of a range imaged object. In *Sensor Fusion: Spatial Reasoning and Scene Interpretation*, pp. 52–58. SPIE, November 1988.

[TJ91] Y.J. Tseng and S. Joshi. Determining feasible tool-approach directions for machining bezier curves and surface. *CAD*, **23**(5):367–379, 1991.

[TM49] S. Timeshenko and G.H. MacCullough. *Elements of Strength of Materials*. Van Nostrand, 1949.

[TS89] Osamu Takahashi and R.J. Schilling. Motion planning in a plane using generalized Voronoi diagrams. *IEEE Transations on Robotics and Automation*, **5**(2), April 1989.

[TT88a] F. Thomas and C. Torras. Group-theoretic approach to the computation of symbolic part relations. *IEEE Journal of Robotics and Automation*, **4**(6), 1988.

[TT88b] S.C.A. Thomopoulos and R.Y.J. Tam. A study of isotropic and anisotropic GIE robot manipulators under optimal path control. In *Proceedings of the U.S.A.–Japan Conference on Flexible Automation*, pp. 249–256, 1988.

[VA86] R. Vijaykumar and M.A. Arbib. A task-level robot planner for assembly operations. Technical Report 86-31, Dept. COINS, University of Massachusetts at Amherst, Amherst, MA 01003, July 1986. COINS Technical Report.

[VNss] A. Vishnu and B.O. Nnaji. Analysis of sub-assembly stability for automated assembly. *International Journal of Robotics and CIM* (in press).

[VRH$^+$78] H.B. Voelker, A. Requicha, E. Hartquist, W. Fisher, R. Metzger, R. Tilove, N. Birrel, W. Hung, G. Armstrong, T. Check, R. Moote, and J. McSweeney. The padl-1.0/2 system for defining and displaying solid objects. *ACM Computer Graphics*, August 1978.

[WA88] Y.F. Wang and J.K. Aggarwal. Geometric modeling using both active and passive sensing. In *Sensor Fusion: Spatial Reasoning and Scene Interpretation*, pp. 12–19. SPIE, November 1988.

[War89] C.W. Warren. Global path planning using artifical potential fields. In *Proceedings of the 1989 IEEE International Conference on Robotics and Automation*, pp. 316–321, 1989.

[WCBW86] Jr. William C. Burns and Janet Evans Worthington. *Practical Robotics: Systems, Interfacing, and Application*. Prentice-Hall, 1986.

[WDM89] C.W. Warren, J.C. Danos, and B.W. Mooring. An approach to manipulator path planning. *International Journal of Robotics Research*, **8**(5):87–95, October 1989.

[Wol87] W.A. Wolovich. *Robotics: Basic Analysis and Control*. Hort, Rinehart and Winston, New York, 1987.

[Wol88] J.D. Wolter. On the automatic generation of plans for mechanical assembly. PhD thesis, Computer, Information and Control Enginnering in the University of Michigan, 1988.

[Woo87] T.C. Woo. Automatic disassembly and total ordering in three dimensions. In *Proceedings: The Winter Annual Meeting of The American Society of Mechanical Engineers*, December 13–18, Boston, MA, pp. 291–303, 1987.

[Yap87] Chee-Keng Yap. Algorithmic motion planning. In J.T. Schwartz and C.K. Yap (eds), Algorthmic and Geometric Aspects of Robotics, volume 1 of *Advances in Robotics*. Lawrence Erlbaum Associates, 1987.

[Yeh89] S. Yeh. Translation of iges data from cad into a structure for object reasoning. Master's thesis, University of Massachusetts at Amherst, 1989.

[YF86] A.T. Yang and F. Freudenstein. Application of dual-number quaternion algebra to the analysis of spatial mechanisms. *J. Applied Mechanics*, pp. 300–308, 1986.

[YTA87] K. Youcef-Toumi and H. Asada. The design of arm linkages with decoupled and configuration-invariant inertia tensors, *Journal of Dynamic Systems, Measurement, and Control*, **109**:268–275, September 1987.

[Yuc86] Wenhan Yuchen. Solvability of practical 6r manipulator. *Recent Trends in Robotics*. Elsevier Science, 1986.

[ZL89] David Zhu and Jean-Claude Latombe. New heuristic algorithms for efficient hierarchical path planning. Technical Report STAN-CS-89-1279, Department of Computer Science, Stanford University, August 1989.

Index

Abel 180
Adjacency graph 159
A* algorithm 159
ALIGN 233–4, 245, 263–4
Allen 180
Acceleration bounds 131
APPROACH 234, 245, 264
Approach vector 234, 246
Assembly 48
   stability model 136
Associated entity group 16
AUTOPASS 7

Bach 179
Balance 1
Barraquand 162, 169
Base coordinate frame 237, 240, 255
Basic shape 23
Bill of materials 9, 65
Bitmap 168–9, 176
Boissonnat 179
Boundary representation 11, 42
Bourtjault 43

CAM-I 12
CAD 8–9
   data in boundary form 131
   techniques 131
   specification approach 8
   system 8–9
Cartesian
   constraint direction 243
   control 208
   coordinates 210–11
   path control 248, 253
   space 207–9
Cell 159
Cell decomposition 8, 11, 159, 162, 169
   approaches 132
CMOVE 109
Co-face level 25

Co-level sub _feature 37
Collision repelling method 162–3
Colson 234
Command
   joint servo motion 233
   robot level 68, 233–4, 244–5, 264
   task level 69
Computation 1
Computational management 233
Configuration control and interlock
   commands 7
Configuration space
   advantages 156
   approach 68
   Cartesian 157
   creating 162, 163–4
   defined 154
   disadvantages 156
   free 169, 171, 172, 175
   obstacle 155, 168, 175
   representation 142
   slice 161, 164
Configuration space obstacle
   computed 155, 170
   defined 155
Configuration states 8, 135
Configurations
   machine 8
   of all objects 8
Constraint
   directions 237–8, 245
   graph 43
Control points 170
Convex
   cones 272
   hull 163, 168
   set 271
CSG 42

Database management 233
Degeneracy 237, 240

Degree of freedom 54, 153, 154, 234, 237, 242, 245, 255, 270, 281
   intersection of 56
Denavit Hartenberg parameters 151, 164
DEPART 234
Description
   geometric 8, 131–2
   kinematic 8, 131
   of manipulator characteristics 8, 131
   physical 8, 131, 134
   of the robot characteristics 135
   task 131
Design
   for assembly 278
   with features 223
   procedure 41
   with spatial relationships 41
Designing for stability 148
Dexterity 1
Differential
   displacement 241–2, 258
   rotation 235–7
   translation 236–7, 244, 259
Digital encoders 97
Direction
   assembly 266, 269
   disassembly 269
   extreme 271, 272
   optimal 269, 275
   roll 234
   tool approach 267, 268
Disassembly direction 66
DOF 217–19, 223, 227, 234, 254–5
   see also Degree of freedom
Duffy 265

Edge 11, 23
   concave 23
   convex 23
   cutting 16
   neutral 23
Editing and control of programs 5
End-effector 151, 173, 177, 207, 234, 237, 239–42, 259
   see also Hand
Entity
   extraction of servoing 124
   force 125
   torque 125

Envelope 22
   degenerate 23
   of a polyhedral object 126
   rule sets 28
Equations of motion 141–4
Error detection 9
Euler formula 14
Exclude-face-set (EFS) 27
Explicit feature representation 18
External feature bounding edge (EFBE) 24
External feature bounding loop (EFBL) 24
Extreme points 271

Face 11
   control 269
   mating 269
Face-based graph 22
Face bounded loop (SBL) 24
De Fazio 43
Feasible approach directions 48, 65
Feature 12, 13, 41, 48, 233–4
   bounding loop (FBL) 24
   decomposition 30, 37
   design with 11
   edge–face incidence matrix 33
   external 23
   extraction 22, 41
   face adjacency matrix 33
   face set (FSS) 23
   form 12
   internal 23
   neighboring faces set (FNF) 25
   reasoning 22
   recognition 22, 31
FFIM 13
FIND 107
Find space problem 152
Find path problem 152–3
Fits 48
Free design 11
Freeways 160
Function specifications 63

Gattrell 180
Generalized
   centroid 149–50
   cones 160
Generic feature primitives 34

Genus 14, 15
Geometric
  constraints 235
  data 178
  reasoning 11
Geometry 1
Glavina 158
Graph
  AND/OR 277, 287
  chain-type 37
  non-chain-type 37
  theory 278
GRASP 233–4, 245
Grasp
  analytic measures for 180
  approach direction for 188–9
  basic set of 182
  choice of 180
  feasible 83
  optimum configuration of 179
  encompass 184
  evaluation 190–1
  force 190–1
  lateral 184
  location 84
  manipulability in 183, 196–8
  parameters 180
  point 188
  precision 184
  radial rotatibility in 183, 198
  soft contact analysis for 192–3
  stability 178–9, 180, 183, 189
  torquability 183, 198
Grasp planner 178–9, 289
Grasp planning 9, 178–206
  design and implementation of 195
Grasping
  concepts 181
  connectivity in 180
  feature reasoning for 184, 200–1
  force closure in 180
  form closure in 180
  geometric constraints in 184–5, 201–2
  local accessibility in 186–7
  mutual visibility in 188
  parallelity and 184–5
  parallel jaw 179–80, 181
  reflective symmetry in 201
  rotational symmetry in 201
  world spatial relationship in 181

Gravity
  center of 96
Group theory 44

Hamiltonian 143
Hand 157, 173–4, 175
  control of the 6
  see also End-effector
  four-fingered Utah/MIT dextrous 179, 181
  multifingered robot 181
Handey 181
Hervé 44
Holonomic generalized active forces 141
Holonomic generalized inertia forces 141
Holonomic system 141
Homomorphic 14

IDEAS 205
IGES 11, 205, 233, 283
Implicit feature representation 18
Inertia ellipsoid 149–50
Intensity maps 98
Internal feature bounding loop (IFBL) 24
Interpreter
  method 4
  program 4
Inverse kinematic
  analysis 238, 263
  results 257
  solutions 234, 245, 254–5
IPIM 13

Jacobian 235, 237, 241–4, 254–5, 258, 262, 264–5
  positional 244, 255, 260, 264
Jameson 180
Joint
  coordinate frame 133
  knot 213
  limits 131, 135

Kane's Method 140
Khatib, O. 161
Kinematic
  implementation 244–5
  structure 233–4, 253–4, 263–4
Kinematics of linkages 134
Koditschek 161

Index

LAMA 7
Language
  machine 4
    explicit robot programming 3
  robot level 3
  task-level 7, 10, 71
  VAL 5
Laugier 180
LINEAR 234, 245, 246
Linear combination 271
Linear Variable Differential
    Transformers (LVDTs) 97
Link parameters 238, 255–6
List
  specific location 107–8
  servoing 109
Liu 180
LMOVE 109
Local minima 161, 162, 170, 174
Location determination 126
Locations definition 5
Loop 11
  empty 15
  solid 15
Lozano-Pérez, T. 66, 160
Lyons 181, 182

MacCullough 194
Machine world 8
2-Manifold 14
Manipulator program synthesis 8–9
Mating
  faces 66, 233
  frames 50, 60
Matrix
  homogeneous 236, 249
  homogeneous transformation 236, 246
  de Mello, Homem 43
Memory storage 248
Möbius bend 14
Model
  geometric 135
  kinematic 135
  liaison 276
  state 82
Monotonicity 268
Motion
  asteroid avoidance 154
  commands 6

  compliant 151, 154, 290
  connected bodies 153–4
  docking 151
  fine 151, 289
  gross 151
  guarded 290
  planning 9
  revolute planar 226
  route 151
MOVE 109, 233, 245, 263–4
  see also CMOVE; LMOVE
Mover's problem
  classical 152
  generalized 152

Nagai 180
k-Neighborhood 170
Newtonian reference frame 140

Object
  configurations 8
  coordinate frame 119
  features 8
  interpreter 88
Obstacle
  avoidance 67
  grown 163, 175, 177
  intersection with 164
Octal chains 123
Octants 132
Octree 160
  concept 132
  decomposition 132
  encoding scheme 8, 132
Open object 14
Operations specifications 135
Optimal fixture selection 139
Ordered _feasible 83

PAM 195
  see also Product Assembly Modeler
Parametric world modeler 133
Parser 88
Path 151
Pattern matching 36
Paul 234, 236, 248
PDES 12
Perreira 234
Plane
  supporting 26

Planner 151, 176
  general robot level 72
  general sensor 107
  generic robot level 72
  generic sensor 110
  global 162
  local 162
  mid level 72
  specific sensor 112
  task level 72
Polyhedral representation 66
Popplestone 41
Potential field
  artificial 161–2, 169, 176
  for the hand 174
  numerical 172, 177
Potentiometer 97
ppd 28
PPP 250, 263
PRR 250, 263
Precedence constraints 276
Precomputation time 248
Primitive instance 11
Process
  plan 1
  planner 9
  planning 269
Processor
  general level 125
  generic level 121
  specific level 121
Product Assembly Modeler 44
Product modeler 9
Product specifications 41
Programming
  explicit robot 4
  machine 2
  machine task-level 7, 9
  off-line 4
  on-line 4
  robot 2
    automatic robot 68, 151, 233
  task level 3, 151
PROLOG 91, 263
ProMod 11
PUMA 5
  robot 133

Quad-tree 160
  concept 8, 132

Quartenion 244, 253

RALPH 68, 91, 97, 137, 151, 181, 195, 233, 265
  domain of 182
RAPT 7, 44
Reasoning 1–2
  capability 1
  machine 2
  paradigm 8
Resolution 176
Response 1
Retraction 159, 170
  parameter (ρ) 170, 173, 177
Rimon 161
Robot
  independent plans 233
  industrial 234, 253
  revolute 233–4, 237
  SCARA 233, 264
  spherical 233, 263–4
Robot joint
  categories 165
  prismatic 163
  revolute 163
ROT 234, 253
Rotation
  on axis 165–6
  not on axis 166–7
RPP 249, 255–60, 264
RR 247
RRP 247
RRR 248, 255, 263–4

Saddle point 162
Sanderson 43
Search
  fine 107–8
  global 107–8
  location 107–8
  sphere 108
Sense 1
  cutaneous 94
  kinesthetic 94
Sensor
  characteristics 131
  contact 98, 105, 114, 121
  coordinate frame 119
  external 97
  force 95, 98, 103, 105

fusion 93–4
infrared 95, 97
internal 97, 106
matrix 94
noncontact 98, 104, 114, 120–1, 123
plan 103
planning 103
proximity 95
range 113
sets 106
tactile 94, 98, 103, 105
touch 94, 98
ultrasonic ranging 95
virtual 100
Sensor-tuned representation 100
Sensory inputs 178
Servoing
functions 102
information 125
Shannon 152
Sheinmann, Victor 5
Simons, Bruce 5
Singularity 237, 246, 255, 263–5
SO clauses 136
Spanning vector
algorithm 273, 274
background 271
definition 267
representation 275
Spatial enumeration 8
Spatial occupancy enumerations 11, 132
Spatial relationships 1, 9, 41, 48, 80, 131–2, 270
against 47, 50, 64
aligned 52, 65
parallel-offset 52, 64
parax-offset 52, 64
incline-offset 52, 64
include-angle 53, 65
symbolic 8–9
Spatial relationships engine 132
Stability reasoner 137
STEP 12
Stereovision 96
Stress patch 125
Subassembly 136–7, 139, 148, 150
Subassembly
stability 136–8
Sugimoto 265

Surface
mating 83
multicurved 11
non-orientable 14
parting 268
C-Surface 290
Sweep representation 11
Symbolic spatial relationships 135–6

Task planning 8, 67
Task requirement 182, 195–200
aspects of 182
basic attributes of 182
description of 182–3
format for 199–200
Task specification 8–9, 131, 135
Taylor 248, 253
Thomas 265
Timoshenko 194
Tokens
geometric 123
of information 121
Tolerance 63
Torque 176
Torras 44
Trajectory 176
bang-bang 229
cosine 229
cubic 214
exponential 229, 231
polynomial 231
sine on ramp 229
sinusoidal 231
TRANS 233, 245, 263
Transformation matrix 45, 164, 174
Translation
along axis 167
not along axis 167–8
Transmission ratios 148

UNGRASP 234, 245
Ultrasonic transducer, see Sensor, ultrasonic ranging

VAL 5
see also Victor's Assembly Language
VALII 5–7
Vertex 11
Victor's Assembly Language 5
Visibility graph 157

Vision 1, 8, 95
　information 131
　machine 96
　system 8, 131
VLSI circuit 153
Voronoi diagram 158–9, 173

Warren 162
Wavefronts 170
Wenhan 254–5
Whitney 43

Wolter 43, 180, 191, 192
Working envelope 245
Workspace 131, 149
　skeleton 170
World frame 80
World knowledge 8
World modeling 8, 131–2, 135
World objects 131–2

Yoshikawa 180
Yuchen 254–5